Knowledge Management in Modern Organizations

Murray E. Jennex, San Diego University, USA

IDEA GROUP PUBLISHING
Hershey • London • Melbourne • Singapore

Acquisition Editor:	Michelle Potter
Senior Managing Editor:	Jennifer Neidig
Managing Editor:	Sara Reed
Assistant Managing Editor:	Sharon Berger
Development Editor:	Kristin Roth
Copy Editor:	Sue Vanderhook
Typesetter:	Jamie Snavely
Cover Design:	Lisa Tosheff
Printed at:	Integrated Book Technology

Published in the United States of America by
Idea Group Publishing (an imprint of Idea Group Inc.)
701 E. Chocolate Avenue
Hershey PA 17033
Tel: 717-533-8845
Fax: 717-533-8661
E-mail: cust@idea-group.com
Web site: http://www.idea-group.com

and in the United Kingdom by
Idea Group Publishing (an imprint of Idea Group Inc.)
3 Henrietta Street
Covent Garden
London WC2E 8LU
Tel: 44 20 7240 0856
Fax: 44 20 7379 3313
Web site: http://www.eurospan.co.uk

Library of Congress Cataloging-in-Publication Data

Knowledge management in modern organizations / Murray Jennex, editor.
p. cm.
Summary: "This book discusses the philosophical foundations of knowledge management, serving as a viable resource for academicians, practitioners, researchers, and students. It also depicts a global perspective, as the authors come from Asia, Europe, the Middle East, and the United States, and adds value to any course focused on KM in organizations"--Provided by publisher.
Includes bibliographical references and index.
ISBN 1-59904-261-4 (hardcover) -- ISBN 1-59904-262-2 (softcover) -- ISBN 1-59904-263-0 (ebook)
1. Knowledge management. I. Jennex, Murray E., 1956-
HD30.2.K63725 2007
658.4'038--dc22
2006031353

British Cataloguing in Publication Data
A Cataloguing in Publication record for this book is available from the British Library.

Knowledge Management in Modern Organizations

Table of Contents

Preface

Knowledge management (KM) has been growing in importance and popularity as a research topic and business initiative since the mid-1990s. This is sufficient time for KM to grow into a discipline complete with its own journals. This book presents 20 chapters that discuss the theory and implementation of KM. The chapters come from articles published in Volume I of the *International Journal of Knowledge Management* (2005) and have been updated to reflect the current state of KM.

The purpose of this book is to document the state and key issues of KM in 2006. It is targeted to academics, practitioners, researchers, and students. Academics will get particular value from the foundational chapters in this book that discuss the philosophical foundations of knowledge and KM. Additionally, the first four chapters establish the foundation of KM as a discipline. This is done to lend legitimacy to research in KM and to help academics establish courses and degree programs that focus on KM. Practitioners will get special insight and value from the case studies and chapters on KM impacts on organizations and measurement of KM, as they are focused on successfully implementing KM in business organizations. Researchers and students will benefit from all chapters.

How is this book intended to be used? First it can be used as a reader in KM courses. It probably shouldn't be used as a sole textbook for a general KM course, but it would add value to any course focused on KM in organizations. Additionally, it is a good book for those wanting to keep current in KM or to begin a course of study or research in KM. Finally, it is good for business professionals just wanting to know how KM could help them run businesses and organizations more effectively.

The strength of *Knowledge Management in Modern Organizations* is in the diversity and strength of the chapter authors. The book has a global perspective as chapters come from Asia, Europe, the Middle East, and the U.S. The authors also reflect a blending of experiences; several are longtime KM researchers, while a few are just beginning their careers. Combining perspective and experience levels means that I am able to give the readers a truly global view of KM, something that I think is needed and will be appreciated.

Knowledge Management in Modern Organizations is organized into five sections. Section I, "What is Knowledge Management?" presents four chapters that define KM and establish KM as an academic discipline.

Chapter I, *What is Knowledge Management?*, summarizes the various definitions of knowledge and knowledge management into working definitions that are consistent with those used throughout the book.

Chapter II, *Knowledge Management as a Discipline*, uses Kuhn's (1996) definition of a discipline to show that knowledge management is a young discipline.

Chapter III, *A Birds-Eye View of Knowledge Management: Creating a Disciplined Whole from Many Interdisciplinary Parts*, uses the experience of creating the encyclopedia of Knowledge Management to illustrate the interdisciplinary nature of knowledge management and to show that the knowledge management community comes from all over the world and from a variety of backgrounds.

Chapter IV, *Knowledge Management Research: Are We Seeing the Whole Picture?*, investigates the knowledge management literature in order to identify areas of research focus and accepted research methodologies.

Section II, "Organizational Impacts of Knowledge Management," presents five chapters that discuss how KM impacts performance, competitiveness, trust, and communities of practice in organizations.

Chapter V, *Linking Knowledge to Competitiveness: Knowledge Chain Evidence and Extensions*, uses the knowledge chain model and a survey of knowledge management executives to establish linkages between knowledge management activities and organizational performance.

Chapter VI, *A Multi-Level Performance Framework for Knowledge Management*, uses case studies to construct a model indicating how knowledge management affects organizational performance.

Chapter VII, *The Influence of Organizational Trust on the Use of KM Systems and on the Success of KM Initiatives*, uses an empirical study of 97 organizations to explore the impact of organizational trust on the knowledge representation strategies of codification and personalization.

Chapter VIII, *Knowledge Management's Impact on Organizational Performance*, uses a Delphi study and a questionnaire to determine success indicators for knowledge management in a variety of organizations.

Chapter IX, *Factors that Contribute to the Success of Knowledge Management Communities of Practice*, uses a case study to identify critical success factors for a community of practice.

Section III, "Measuring Knowledge Management," presents four chapters that discuss KM metrics, success factors, success models, and adoption.

Chapter X, *Evaluation of Knowledge Management: A Review and Agenda for Future Research*, summarizes the literature on knowledge management metrics and provides a direction for future research.

Chapter XI, *Knowledge Management Success Factors and Models*, summarizes the literature on knowledge management critical success factors and applies the resulting generic knowledge management critical success factors to the evaluation of proposed knowledge management success and/or effectiveness models.

Chapter XII, *Knowledge Management Success: Empirical Assessment of a Theoretical Model*, uses survey results to validate portions of the Jennex and Olfman knowledge management success model (this model is based on the DeLone and McLean [2003] information systems success model).

Chapter XIII, *Knowledge Management Information Technology User Acceptance: Assessing the Applicability of the Technology Acceptance Model*, uses a survey to investigate the applicability of the technology acceptance model (TAM) to knowledge management.

Section IV, "Knowledge in Organizations," presents four chapters that define and describe knowledge, how knowledge is discovered, and how knowledge is transferred.

Chapter XIV, *The Role of Context and Its Explication for Fostering Knowledge Transparency in Modern Organizations*, uses case studies to investigate the value of context to the reuse of knowledge.

Chapter XV, *Toward the Multidimensional Conceptualization of Knowledge*, summarizes the literature to show that current classifications of knowledge are limited and proposes a four-dimensional model of knowledge and a view of knowledge as existing in a continuum along these dimensions.

Chapter XVI, *Eliciting Tacit Knowledge Using the Critical Decision Interview Method*, summarizes a variety of knowledge elicitation techniques and then uses a case study to propose and discuss the use of interview techniques in the identification of tacit knowledge.

Chapter XVII, *Knowledge Acquisition and Transfer in Developing Countries: The Experience of the Egyptian Software Industry*, uses quantitative and qualitative research to investigate factors hindering the acquisition and transfer of knowledge in firms.

Section V, "Experience with Knowledge Management," presents three chapters that discuss actual implementation of KM in an airline, a manufacturing/export firm, and a research laboratory.

Chapter XVIII, *Adopting Knowledge-Centred Principles in Innovation Pursuits: The Case of Singapore Airlines*, shows how knowledge management was used to foster innovation and to improve the competitive position of Singapore Airlines.

Chapter XIX, *Knowledge Management Gap: Determined Initiatives, Unsuccessful Results*, investigates the failure of a knowledge management initiative in a Hong Kong firm and identifies four lessons learned that are related to management support and organizational culture.

Chapter XX, *The Lifecycle of a Knowledge Management System for Organizational Learning: A Case Study*, discusses the construction and use of a knowledge system used to enhance organizational learning by helping to pass on organizational culture to new members at the Jet Propulsion Laboratory.

I hope you enjoy *Knowledge Management in Modern Organizations*. It has been a labor of love and is something I'm proud of. I think we have prepared a book that many will find valuable and enlightening.

Murray E. Jennex

August 2006

Acknowledgments

I'd like to thank the authors for their efforts; they are appreciated, especially when it is considered that book chapters carry little weight in decisions on promotions and tenure. I'd also like to thank the support I received from the publishing staff at Idea Group Inc.; they were helpful and encouraging. Finally, I'd like to thank my family for putting up with the mess around the house (20 chapters spread around the living room isn't a pretty sight) and the inattention I paid to them as I finished the book. I know they appreciate my efforts, and I truly appreciate their love and support.

Section I

What is Knowledge Management?

Chapter I

What is Knowledge Management?

Murray E. Jennex, San Diego State University, USA

Abstract

This chapter defines knowledge and knowledge management (KM) and establishes its roots. KM is not a brand new topic; organizational learning and organizational memory are related topics that have been fields of research for many years. This chapter relates these concepts to a relational model that shows that the three topics are related and influence organizational effectiveness. Additionally, this chapter explains that KM has become a research area due to a confluence of trends that have made KM necessary and technically useful.

Introduction

Knowledge management (KM) is a hot topic in many business communities. Although, the term knowledge management might suggest a rather simple definition, there are plenty of opinions on what exactly it is and how it should be used, if used at all. However, because of the ever-increasing pace of business development, the task of effective and competitive management of organizations becomes essential, and KM, if understood and implemented properly, may be a useful tool for business transformation as well as the key to competitive advantage. In this first chapter I would like to introduce the basic definitions and concepts of KM.

I thought this would be an easy chapter to write, as it seemed that we all knew what we were talking about when discussing KM. However, I became aware of the need to establish a definition of KM through the publication of an expert opinion in the Business Intelligence Journal. The editors asked three experts (this author included) about integrating KM and data warehouses. When the issue was released, I was surprised that the three experts all had different opinions on what KM was (Corral, Griffin, & Jennex, 2005). One expert described the purpose of KM as disseminating knowledge quickly and KM systems as essentially document management systems. The other expert considered KM as the process of handling unstructured knowledge. The final view (mine) combined technical and organizational initiatives to manage structured and unstructured knowledge in order to help the organization improve its effectiveness through improved retention and reuse of knowledge. Three experts expressed different definitions of KM and illustrated the need to define KM so that there is a common ground for discussion. The first step, though, is to define what is meant by the term *knowledge*, since this is the central theme in KM.

Knowledge

Davenport and Prusak (1998) view knowledge as an evolving mix of framed experience, values, contextual information, and expert insight that provides a framework for evaluating and incorporating new experiences and information. They found that in organizations, knowledge often becomes embedded in documents or repositories and in organizational routines, processes, practices, and norms. They also say that in order for knowledge to have value, it must include the human additions of context, experience, and interpretation. Nonaka (1994) expands this view by stating that knowledge is about meaning in the sense that it is context-specific. This implies that users of knowledge must understand and have experience with the context, or surrounding conditions and influences, in which the knowledge is generated and used in order for it to have meaning to them. This also implies that in order for a knowledge repository to be useful, it also must store the context in which the knowledge was generated. That knowledge is context-specific and argues against the idea that knowledge can be applied universally; however, it does not argue against the concept of organizational knowledge. This chapter considers organizational knowledge to be an integral component of what organizational members remember and use, meaning that knowledge is actionable.

Various knowledge taxonomies exist. Alavi and Leidner (2001) and Jennex and Croasdell (2005) found that the most commonly used taxonomy is Polyani's (1964, 1967) and Nonaka's (1994) dimensions of tacit and explicit knowledge. This book uses this taxonomy for knowledge. Tacit knowledge is that which is understood within a knower's mind. It consists of cognitive and technical components. Cognitive components are the mental models used by the knower that cannot be expressed directly by data or knowledge representations and also is known as unstructured knowledge. Technical components are concrete concepts that can be expressed readily and also is known as structured knowledge. Explicit knowledge also consists of these technical components that can be expressed directly by knowledge representations. Knowledge transfer in an organization occurs when members of an organization pass tacit and explicit knowledge to each other. Information technology (IT) assists knowledge transfer by providing knowledge repositories and methods for capturing and retrieving knowledge. The extent of the dimension of the knowledge being captured limits the effectiveness of IT in assisting KM. IT works best with knowledge that is primarily in the explicit dimension. Knowledge that is primarily in the tacit dimension requires that more context be captured with the knowledge in which context is the information used to explain what the knowledge means and how it is used. Nonaka and Takeuchi (1995) propose four modes of knowledge transfer and creation (known as the SECI model):

- Socialization is the process of sharing experiences and thereby creating tacit knowledge such as mental models and technical skills. Tacit knowledge can be obtained without using language through observation, imitation, and practice.
- Externalization is the process of articulating tacit knowledge in the form of explicit concepts, taking the shapes of metaphors, analogies, concepts, hypotheses, or models.
- Combination is the process of systemizing concepts into a knowledge system by combining different bodies of explicit knowledge. Explicit knowledge is transferred through media such as documents, meetings, and e-mail and/or phone conversations. Categorization of this knowledge can lead to the generation of new knowledge.
- Internalization is the process of converting explicit knowledge into tacit knowledge and is closely related to learning by doing.

These four modes or processes show that the transfer of knowledge is dependent upon the transfer of a common understanding from the knower to the user of the knowledge. Common understanding consists of the context (the story behind the knowledge, the conditions and situations that make the knowledge understandable) and the experience (those activities that produce mental models of how the knowledge should be used) expressed in a culturally understood framework.

Knowledge Management

Jennex (2005) defined KM as the practice of selectively applying knowledge from previous experiences of decision making to current and future decision-making activities with the express purpose of improving the organization's effectiveness. Also, Jennex (2005) viewed a KM system as that system created to facilitate the capture, storage, retrieval, and reuse of knowledge. This perception of KM and KM systems is that they holistically combine organizational and technical solutions to achieve the goals of knowledge retention and reuse in order ultimately to improve organizational and individual decision making. This is a Churchman (1979) view of KM that allows KM systems to take whatever form necessary to accomplish these goals. For some organizations, this may mean that the KM system essentially is a document management system. However, as a community, we don't want KM perceived essentially as a document management technology. Also, in some organizations, KM may be used only to manage unstructured knowledge. This may meet those organizations' needs, but again, the KM community is not comfortable accepting that KM handles only unstructured knowledge. Another key definition of KM includes Holsapple and Joshi (2004), who consider KM as an entity's systematic and deliberate efforts to expand, cultivate, and apply available knowledge in ways that add value to the entity in the sense of positive results in accomplishing its objectives or fulfilling its purpose. The entity's scope may be individual, organizational, transorganizational, national, and so forth. Finally, Alavi and Leidner (2001) in their seminal work concluded that KM involves distinct but interdependent processes of knowledge creation, knowledge storage and retrieval, knowledge transfer, and knowledge application.

It is important to note that none of these definitions of KM is purely technical in nature. They all include information system (IS) support, but they also include organizational considerations, and all include an impact on organizational productivity and effectiveness. This is important, because many experts consider KM to be a form of IS. However, while the IS component is important, in order for KM to be effective as a change or transformation tool, it must include more; it requires management support and an organizational culture.

A better understanding of KM is obtained by incorporating the concepts of organizational memory (OM) and organizational learning (OL). Jennex and Olfman (2002) found that the three areas are related and have an impact on organizational effectiveness. Organizational effectiveness is how well the organization does those activities critical to producing what the organization sells. OL is the process the organization uses to learn how to do these activities better. OL results when users utilize knowledge. That OL may not always have a positive effect is examined by the monitoring of organizational effectiveness. Effectiveness can improve, get worse, or remain the same. How effectiveness changes influences the feedback provided to the organization using the knowledge. KM and OM are the processes used to identify and capture critical knowledge. Knowledge workers and their organizations "do" KM; they identify key knowledge artifacts for retention and establish processes for capturing it. OM is what IT support organizations do; they provide the infrastructure and support for storing, searching, and retrieving knowledge artifacts. Figure 1 illustrates these relationships, and the following sections expand on these concepts.

Figure 1. The KM/OM/OL model (Jennex & Olfman, 2002)

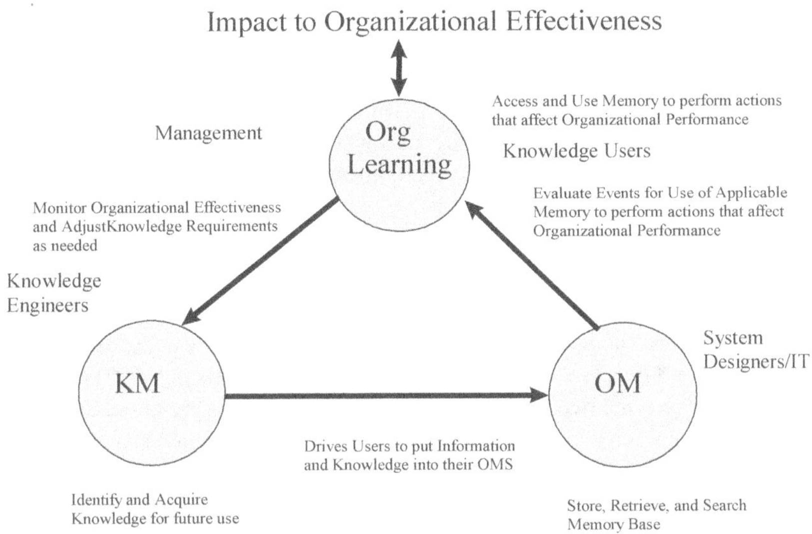

Organizational Learning

Organizational learning (OL) has been defined as a quantifiable improvement in activities, increased available knowledge for decision making, or sustainable competitive advantage (Cavaleri, 1994; Dodgson, 1993; Easterby-Smith, 1997; Miller, 1996). Malhotra (1998) defines Organizational Learning as the process of detection and correction of errors. In this view, organizations learn through individuals acting as agents for them. Individual learning activities are seen as being facilitated or inhibited by an ecological system of factors that may be called an organizational learning system. Learning in this perspective is based on Kolb's (1984) model of experiential learning in which individuals learn by doing.

Huber, Davenport, and King (1998) believe that an organization learns if, through its processing of information, its potential behaviors are changed. Huysman, Fischer, and Heng (1994) as well as Walsh and Ungson (1991) believe organizational learning has OM as a component. In this view, OL is the process by which experience is used to modify current and future actions. Huber (1991) considers four constructs as integrally linked to OL: knowledge acquisition, information distribution, information interpretation, and organizational memory. In this case, OM is the repository of knowledge and information acquired by the organization. Organizational learning uses OM as its knowledge base.

A different perspective on OL from Sandoe et al. (1998) is that organizations do not learn; rather, only individuals learn. During work, people gain experience, observe, and reflect in making sense of what they are doing. As they analyze these experiences into general abstractions, their perceptions on how work should be done changes. As these individuals influence their co-workers, the "organization" learns, and the process gradually is changed. Learning in this perspective also is based on Kolb's (1984) model of experiential learning.

Organizational Memory

Huber, Davenport, and King (1998) summarize organizational memory (OM) as the set of repositories of information and knowledge that the organization has acquired and retains. Stein and Zwass (1995) define OM as the means by which knowledge from the past is brought to bear on present activities, resulting in higher or lower levels of organizational effectiveness. Walsh and Ungson (1991) define OM as stored information from an organization's history that can be brought to bear on present decisions.

OM can be viewed as abstract or concrete. It is comprised of unstructured concepts and information that exist in the organization's culture and the minds of its members, and can be partially represented by concrete/physical memory aids such as databases. It also is comprised of structured concepts and information that can be represented exactly by computerized records and files. Sandoe and Olfman (1992) and Morrison (1997) describe these two forms of OM as having two functions: representation and interpretation. Representation presents just the facts (or knowledge or expertise) for a given context or situation. Interpretation promotes adaptation and learning by providing frames of reference, procedures, guidelines, or a means by which to synthesize past information for application to new situations. Comparing to the definition of knowledge, it is obvious that knowledge and OM are related through experience and learning. We consider knowledge to be a subset of OM and the processes of KM a subset of OM processes.

Knowledge Management Summary

To summarize knowledge is something that is actionable; that is, it is something that users can retrieve and apply to organizational activities. KM is the process used to make knowledge actionable to members of the organization. It involves capturing, storing, retrieving, and using knowledge. KM involves the creation of a KM system. The KM system includes an environment that promotes organizational learning. In this respect, KM can be an organizational change or transformation tool, as it can help management to create a learning organizational culture. Also, KM is a tool for improving organizational effectiveness, as it promotes knowledge reuse to improve decision making. Incorporating the concepts of OM, OL, and the previous discussion, this chapter proposes this working definition of KM:

> *KM is the practice of selectively applying knowledge from previous experiences of decision making to current and future decision-making activities with the express purpose of improving the organization's effectiveness.* (Jennex, 2005)

This definition, while perhaps not perfect, does represent the idea that when organizations implement KM, they are doing so with the expectation that organizational effectiveness will be improved and that organizational members will do this by acting on and using knowledge. The remaining chapters in this book may use differently worded definitions of KM, but all agree with this discussion and the spirit of this working definition.

This chapter concludes with a discussion on why KM has become an important organizational endeavor.

Why Knowledge Management?

This book is written with the assumption that KM is something that modern organizations need to do. It is expected that readers will read with the assumption that they need to implement KM in their organizations or that KM is something important to research. However, KM is a relatively new discipline (as discussed in the next couple of chapters), and it is worthwhile to conclude this chapter with a short discussion on why KM has become this important of an endeavor.

Jennex (2006) summarizes a workshop discussion at the 2006 Hawaii International Conference on System Sciences (HICSS) on the origins of KM and why KM came into prominence in the late 1990s. The following paragraphs summarize Jennex (2006).

Many consider the popularization of the term by Davenport and Prusak (1998), with the publication of "Working Knowledge," as the beginning of KM. However, there was a confluence of trends in the mid-1990s that actually caused KM to become an important topic.

The first trend was the fallout from the business process reengineering (BPR) fad of the early 1990s. This business process led many organizations to restructure their businesses by changing processes and reducing staff. These staff reductions led to many organizations discovering that they had lost key knowledge and were no longer as nimble and innovative as they needed to be. This loss of knowledge provided a driver for identifying and managing knowledge that was internal to the organization.

The second trend was technology. The mid to late 1990s saw accelerated advancement in the capability of the personal computer. Processing speed and memory grew rapidly (and continues to grow); also, the Internet became a practical reality, providing connectivity to many organizations and individuals. Additionally, the Year 2000 (Y2K) date problem was a driver for many organizations to replace older computers with current models as well as a driver to integrate and modernize organizational infrastructures. It was this trend that enabled KM to develop by providing organizations with technological capabilities for organizing, storing, searching, retrieving, and manipulating large amounts of structured and unstructured knowledge.

The third *trend* was the explosion in content, information, and knowledge caused by the rapid growth of the Internet and corporate intranets, data warehouses, and databases. This explosion in the amount of data, information, and knowledge and its easy availability to decision makers of all ranks led to a loss of information and knowledge control by managers. This loss of control served as a driver to managerial awareness that there was a problem that needed to be solved.

The fourth trend was organizational issues of maintaining business value and reducing risk of litigation and overall liability by managing human and intellectual capital better. A shift in many American businesses to a service-based economy led to an increasing focus on an organization's skills and capabilities as well as sustaining these capabilities by retaining high-value employees. This served as a driver to managerial awareness that employee knowledge was a valuable organizational asset.

The confluence of these trends led to the behaviors observed by Davenport and Prusak (1998) and to the emergence of knowledge management as a necessary discipline. Why do we care? One of the major concerns is the sustainability of KM. We don't want to waste our time and energy on a fad; we want to make sure that what we are doing is sustainable. Some of this concern is self-serving; we want to ensure that our peers recognize the importance of our research so that our research will be accepted and published and we can obtain promotion and tenure. Also, we want to ensure that we have a steady supply of students wanting to study our discipline and that these students have careers waiting for them in industry. Finally, we want to make a difference; we want to enable organizations to continue to prosper and better utilize their resources, as ultimately, this leads to a better society.

References

Alavi, M., & Leidner, D. E. (2001). Review: Knowledge management and knowledge management systems: Conceptual foundations and research issues. *MIS Quarterly, 25*(1), 107-136.

Cavaleri, S. (1994). Soft systems thinking: A pre-condition for organizational learning. *Human Systems Management, 13*(4), 259-267.

Corral, K., Griffin, J., & Jennex, M. E. (2005). Expert's perspective: The potential of knowledge management in data warehousing. *Business Intelligence Journal, 10*(1), 36-40.

Churchman, C. W. (1979). *The systems approach.* New York: Dell Publishing.

Davenport, T. H., & Prusak, L. (1998). *Working knowledge.* Boston: Harvard Business School Press.

Dodgson, M. (1993). Organizational learning: A review of some literatures. *Organization Studies, 14*(3), 375-394.

Easterby-Smith, M. (1997). Disciplines of organizational learning: Contributions and critiques. *Human Relations, 50*(9), 1085-1113.

Holsapple, C. W., & Joshi, K. (2004). A formal knowledge management ontology: Conduct, activities, resources, and influences. *Journal of the American Society for Information Science and Technology, 55*(7), 593-612.

Huber, G. P. (1991). Organizational learning: The contributing processes and the literatures. *Organization Science*, 2, 88-115.

Huber, G. P., Davenport, T. H., & King, D. (1998). Some perspectives on organizational memory [unpublished working paper]. In *Proceedings of the 31st Annual Hawaii International Conference on System Sciences,* Hawaii.

Huysman, M. H., Fischer, S. J., & Heng, M. S. H. (1994). An organizational learning perspective on information systems planning. *Journal of Strategic Information Systems, 3*(3), 165-177.

Jennex, M. E. (2005). What is KM? *International Journal of Knowledge Management, 1*(4), i-iv.

Jennex, M. E. (2006). Establishing the foundations of the knowledge management discipline. *International Journal of Knowledge Management, 2*(3), i-iv.

Jennex, M.E., & Croashell, D. (2005). Knoweledge Management: are we a discipline? *International Journal of Knowdge Management, 1*(1), pp. i-v.

Jennex, M. E., & Olfman, L. (2002). Organizational memory/knowledge effects on productivity, a longitudinal study. In *Proceedings of the 35th Hawaii International Conference on System Sciences.*

Kolb, D. (1984). *Experimental learning.* Englewood Cliffs, NJ: Prentice-Hall.

Malhotra, Y. (1998). *Knowledge management for the new world of business.* Retreived October 4, 2006 www.brint.com/km/whatis.htm.

Miller, D. (1996). A preliminary typology of organizational learning: Synthesizing the literature. *Journal of Management, 22*(3), 485-505.

Morrison, J. (1997). Organizational memory information systems: Characteristics and development strategies. In *Proceedings of the 30th Annual Hawaii International Conference on System Sciences.*

Nonaka, I. (1994). A dynamic theory of organizational knowledge creation. *Organization Science, 5*(1), 14-37.

Nonaka, I., & Takeuchi, H. (1995). *The knowledge-creating company—How Japanese companies create the dynamics of innovation.* Oxford: Oxford University Press.

Polanyi, M. (1964). *Personal knowledge: Toward a post-critical philosophy.* New York: Harper Torchbooks.

Polanyi, M. (1967). *The tacit dimension.* London: Routledge and Keoan Paul.

Sandoe, K., Croasdell, D. T., Courtney, J., Paradice, D., Brooks, J., & Olfman, L., (1998). Additional perspectives on organizational memory [unpublished working paper]. In *Proceedings of the 31st Annual Hawaii International Conference on System Sciences,* Hawaii.

Sandoe, K., & Olfman, L. (1992). Anticipating the Mnemonic Shift: Organizational remembering and forgetting in 2001. In *Proceedings of the Thirteenth International Conference on Information Systems.* ACM Press.

Stein, E. W., & Zwass, V. (1995). Actualizing organizational memory with information systems. *Information Systems Research, 6*(2), 85-117.

Walsh, J. P., & Ungson, G. R. (1991). Organizational memory. *Academy of Management Review, 16*(1), 57-91.

Ward, J., & Aurum, A. (2004). Knowledge management in software engineering—Describing the process. In *Proceedings of the @004 Australian Software Engineering Conference (ASWEC '04).*

Chapter II

Knowledge Management as a Discipline

Murray E. Jennex, San Diego State University, USA

David Croasdell, University of Nevada, USA

Abstract

This chapter presents arguments that show that knowledge management (KM) is a discipline. Kuhn's (1996) criteria for being a discipline are used as a framework for providing information showing KM to be a discipline. It was found that KM has interesting research questions, journals specific to KM, a body of accepted knowledge, professional societies, its own jargon and ontology, and its own degree programs. It also is concluded that KM is a young and growing discipline.

Introduction

Is knowledge management (KM) a discipline? Why do we care? Members of the KM community want recognition as a discipline and not as just a subset of the information systems community or as a subset of the organizational behavior community. While KM clearly relies on information systems, it is also a fusion of many disciplines and borrows from many more. The KM community believes that KM is a discipline in its own right, although admittedly, a discipline that heavily overlaps the IS and organizational behavior disciplines. Why we care was discussed in Chapter I but bears repeating; being recognized as a discipline gets us dedicated degree programs, continuing streams of students, and recognition of our research. This is important, as we need this recognition and resources in order to better serve the business community and to help organizations improve effectiveness.

Spiegler (2000) suggests that KM is just a new name for an old IS idea. While the idea may be old and does originate within IS, we believe that KM has emerged as a discipline based on new technologies, methodologies, and theories proposed and used by the KM community. To support this assertion, we refer to Kuhn (1996), who lists several criteria that define a discipline:

- Formation of specialized journals,
- Foundation of professional societies (or specialized interest groups [SIGs] within societies),
- Claim to a special place in academia (and academia's curriculum),
- An accepted body of knowledge for group members to build upon, eliminating having to build their field anew with each paper, and
- Promulgation of scholarly articles intended for and addressed only to professional colleagues, those whose knowledge of a shared paradigm can be assumed and who prove to be the only ones able to read the papers addressed to them (i.e., a specialized ontology).

We believe that KM meets these criteria, as discussed in the following paragraphs.

Formation of Specialized Journals

Established IS journals such as *MIS Quarterly*, *International Journal of Project Management*, *International Journal of Distance Education Technologies*, *International Journal of Management Science*, and the *Journal of Global Information Technology Management* have had special issues dedicated to KM over the last several years. However, journals dedicated to KM have been started. A search of the *Index of Information System Journals* has found several academic journals dedicated to general KM research (http://lamp.infosys.deakin.edu.au/journals/index.php). These include:

- *Electronic Journal of Knowledge Management*
- *International Journal of Knowledge Management*
- *Journal of Information and Knowledge Management*
- *Journal of Knowledge Management*
- *Knowledge Management Research and Practice*

Additionally, specialty journals exist within KM. These journals focus on topics such as knowledge representation and/or discovery, specialized KM technologies, and specialized knowledge uses. In many cases these journals cross discipline boundaries and incorporate multiple disciplines. These journals include:

- *Data and Knowledge Engineering*
- *Data Mining and Knowledge Discovery*
- *IEEE Transactions on Knowledge and Data Engineering*
- *Indilinga African Journal of Indigenous Knowledge Systems*
- *International Journal of Knowledge and Learning*
- *Interdisciplinary Journal of Information, Knowledge, and Management*
- *Interdisciplinary Journal of Knowledge and Learning Objects*
- *International Journal of Knowledge, Culture and Change Management*
- *International Journal of Knowledge-Based and Intelligent Engineering Systems*
- *International Journal of Software Engineering and Knowledge Engineering*
- *Knowledge and Information Systems*
- *Knowledge and Process Management: The Journal of Corporate Transformation*
- *Knowledge Engineering Review*

Given the extent of this list, we conclude that we meet the first criteria.

Foundation of Professional Societies

Several special interest groups, societies, and communities of practice have sprung up to address KM in both applied and theoretical settings. Exemplars of such groups include the Information and Knowledge Management Society, the Knowledge Board, the American Society for Information Science and Technology Special Interest Group on Knowledge Management, SIGs within specific domains (e.g., SIGKDD and SIGDSS within AIS, SIGKM within NIH, etc.), and communities of practice consisting of researchers grouped around major conferences. Additionally, professional certification of KM practitioners is available and claims more than 2,500 KM practitioners trained (eKnowledgeCenter, 2006).

The establishment and growth of these groups address the second criteria regarding professional societies.

Academic and practitioner research communities of practice have been established at several conferences, some focused solely on KM, others more general but with strong KM tracks or mini-tracks. Attendance at these conferences and review of their proceedings clearly show KM research communities. Examples of these conferences include tracks or mini-tracks at major conferences such as the Hawaii International Conference on System Sciences, HICSS; the International Conference on Information Systems, ICIS; the Americas, European, and Pacific and Asia Conferences of the Association of Information Systems, AMCIS, ECIS, PACIS; and the Information Resource Management Association (IRMA) conference. Additionally, conferences dedicated to KM exist, including the International Conference on Information and Knowledge Management (CIKM), the European Conference on Knowledge Management (ECKM), the International Conference on Practical Aspects of Knowledge Management (PAKM), the European Conference on Organizational Knowledge, Learning, and Capabilities (OKLC), and the Australian Conference on Knowledge Management and Intelligent Decision Support (ACKMIDS). Curiously though, there appears to be little shared awareness among these communities, as evidenced by the relatively few papers referenced or built upon from other conferences. It is expected that this will change as the KM discipline matures.

Given the existence of professional KM communities, we meet the criteria of special interest groups, but we do need to do a better job of building the KM community.

Academic Curricula

A Web search or perusal of college catalogues yields a list of many universities offering courses in KM. A cursory scan of dissertation abstracts over the last five years show a number of works that focus on KM and related topics. The growth of KM courses and graduate theses legitimizes the claim that KM has found a place in academia, thus satisfying the third requirement. Additionally, graduate degree programs in KM have been implemented in many universities in many countries. Examples of these programs are found at the California State University at Northridge, the University of Oklahoma, George Mason University, Tilborg University, McMaster University, the University of South Australia, and Cranfield University.

The existence of specialized degree programs and courses indicates acceptance of KM as an academic topic worthy of inclusion in the curricula and also indicates that KM is a discipline.

Body of Knowledge

Publishing high-quality research supports the establishment and growth of the KM body of knowledge and the establishmenta common ontology. Building the body of knowledge and common ontology needs to be a priority for the KM community. We reviewed the citations from 204 KM-related papers presented at HICSS from 1998 through 2006 to identify seminal works. HICSS papers were chosen for their availability and widespread coverage of the KM field. The HICSS review presents the viewpoint that KM includes the topics of organizational memory, OM, and organizational learning (OL). We accept this view and categorized the papers as KM, OM, or OL. The 204 papers yielded more than 4,000 citations, of which approximately 500 focused on the following key areas:

- Cognitive theories of knowledge management and organizational memory,
- Basic theories associated with knowledge creation, knowledge management, and organizational memory,
- Use of semantic networks, topic maps, the Internet, digital documents, XML, taxonomies, ontologies, and other technologies in order to implement knowledge management systems,
- Organizational culture impacts on knowledge management,
- Design of information and communication systems that facilitate knowledge transfer and sharing,
- Enablers and inhibitors of knowledge sharing and knowledge transfer behaviors,
- Knowledge transfer and sharing behaviors within emergent organizational forms such as virtual communities,
- Knowledge reuse in organizations,
- Organizational and economic incentive structures for knowledge sharing and use,
- Knowledge acquisition and transfer processes,
- Case studies of knowledge management and organizational memory systems,
- Metrics and effectiveness of knowledge management and organizational memory systems,
- Knowledge management in small and medium enterprises,
- Methodologies and processes for developing knowledge management systems,
- Global issues in knowledge management and organizational memory,
- Knowledge management strategy,
- Issues related to the capture, storage, search, retrieval, and use of knowledge and organizational memory,
- Knowledge management training issues, and
- Organizational learning.

Table 1 presents the top nine cited KM manuscripts from 173 of the 204 papers determined to be predominately KM-focused. "Cited" refers to the number of times the citation is listed in the references; "Years" reflects the number of years the citation appeared in at least one paper. The most heavily cited work appears in 54 of 173 papers, while the second, third, and fourth most citations appear in 57, 45, and 43 papers, respectively (citation frequencies of 36%, 33%, 26%, and 25%). Many authors cited either Nonaka and Takeuchi (1995) or Nonaka (1994) or other Nonaka-based works, giving more power to the ideas proposed by Nonaka (approximately 70% of all papers cited some work with Nonaka as an author).

Additionally, The works of Polanyi were cited in various papers and, when combined, were found in 46 papers, or a 27% citation frequency. We propose that the top four citations are seminal works and that terms such as *tacit* and *explicit* knowledge are common ontology. However, there is little consensus on other items of interest. Table 1 indicates an existing but weak body of knowledge and little common ontology.

Table 2 presents the top four OM manuscripts cited in the 32 papers determined to be predominately OM-focused. The table headings are the same as Table 1. Note that the top two

Table 1. Top KM citations

References	Cited	Years
Nonaka, I., & Takeuchi, H. (1995). *The knowledge-creating company: How Japanese companies create the dynamics of innovation.* Oxford: Oxford University Press.	63	9
Davenport, T. H., & Prusak, L. (1998). *Working knowledge: How organizations manage what they know.* Boston: Harvard Business School Press.	57	9
Nonaka, I. (1994). A dynamic theory of organizational knowledge creation. *Organization Science, 5*(1), 14-37.	45	8
Alavi, M., & Leidner, D. (2001). Review: Knowledge management and knowledge management systems: Conceptual foundations and research issues. *MIS Quarterly, 25*(1), 107-136.	43	5
Polanyi, M. (1967). *The tacit dimension.* London: Routledge and Kegan Paul.	32	9
Hansen, M. T., Nohria, N., & Tierney, T. (1999, March-April). What's your strategy for managing knowledge? *Harvard Business Review*, 106-116.	20	7
O'Dell, C., & Grayson Jr., C. J. (1998). *If only we knew what we know.* New York: The Free Press.	18	6
Davenport, T. H., DeLong, D. W., & Beers, M. C. (1998). Successful knowledge management projects. *Sloan Management Review, 39*(2), 43-57.	17	6
Polanyi, M. (1964). *Personal knowledge: Towards a post-critical philosophy.* New York: Harper Torchbooks.	14	6

Table 2. Top OM citations

Citation	Cited	Years
Stein, E. W., & Zwass, V. (1995). Actualizing organizational memory with information systems. *Information Systems Research, 6*, 85-117.	25	8
Walsh, J. P., & Ungson, G. R. (1991). Organizational memory. *Academy of Management Journal, 16*, 57-91.	19	7
Morrison, J., & Weiser, M. (1996). A research framework for empirical studies in organizational memory. In *Proceedings of the 29th Annual Hawaii International Conference on System Sciences. IEEE Computer Society Press* (pp. 178-187).	10	6
Tuomi, I. (1999). Data is more than knowledge: Implications of the reversed knowledge hierarchy for knowledge management and organizational memory. In *Proceedings of the 32nd Hawaii International Conference on System Sciences.* IEEE Computer Society Press.	10	6

Table 3. Top OL citations

Citation	Cited	Years
Brown J. S., & Duguid, P. (1991). Organizational learning and communities of practice: Towards a unified view of working, learning and organization. *Organization Science, 2*(1), 40-57.	27	8
Senge, P. M. (1990). *The fifth discipline—The art and practice of the learning organization.* Random House.	23	9
Huber, G. P. (1991). Organizational learning: The contributing processes and the literatures. *Organization Science, 2*(1), 88-115.	21	9
Argyris, C., & Schon, D. A. (1978). *Organizational learning: A theory of action perspective.* Reading, MA: Addison-Wesley.	21	8
Wenger, E. (1998). *Communities of practice: Learning, meaning, and identity.* Cambridge University Press.	17	5

citations appear in 25 and 19 papers, respectively, while the remaining citations are in 10 papers and are conference proceedings, mostly from HICSS. There is agreement that two papers (78% and 59%) are OM seminal works and provide a consistent definition of OM as bringing knowledge from past decisions to bear on current decision making.

Table 3 presents the top five OL manuscripts cited in the 40 predominately OL-focused papers. These citations have citation frequencies of 68%, 58%, 53%, 53%, and 43%. Additionally, six papers cited other works from Argyris and Schon, indicating that 68% of the OL papers cited their work. All are considered OL seminal works, including Wenger (1998), as it is a citation released during the first year of the study and, given time, should have the same citation frequency as the other citations.

The citation counts show that there are only a few models and theories and little common ontology. However, the body of literature is growing rapidly. More than 500 KM, OM, and OL citations were identified. Many of the cited manuscripts have been published since 2001, and many were refereed conference proceedings. Converting these citations into journal articles takes time, but papers originally appearing at HICSS (and other conferences) are showing up in peer-reviewed journals on a regular basis.

In order to address this shortcoming, efforts are being made to formalize the KM ontology and body of knowledge. A workshop was held at the 2006 HICSS conference dedicated to identifying the KM core ontology and body of knowledge and will result in a series of published articles. Additionally, the *Encyclopedia of Knowledge Management*, David Schwartz, editor, was published by Idea Group Publishing in 2006. This book will be discussed in the next chapter, but it is representative of efforts to document the KM ontology and body of knowledge.

While the existence of seminal works and accepted theories exist, we conclude that we need to do more in order to establish our body of knowledge and ontology.

Promulgation of Scholarly Articles

The last criterion is the promulgation of scholarly articles intended for the KM community. This criterion is met by the listed academic journals. These journals publish articles that only members of the KM community can fully understand and use. This criterion is considered met.

Conclusion

The question was asked, is KM a discipline? It is concluded that we are a young discipline. The criteria listed by Kuhn (1996) defining a discipline are met. This book and the listed journals, particularly the *International Journal of Knowledge Management* (from which this book is drawn), is dedicated to ensuring that KM grows as a discipline and will support this by helping to build the KM research community, body of knowledge, and ontology.

References

EknowledgeCenter. (2006). *KM professional certification.* Retrieved March 15, 2006, from http://www.eknowledgecenter.com

Kuhn, T. S. (1996). *The structure of scientific revolutions* (3rd ed.). University of Chicago Press.

Chapter III

A Birds-Eye View of Knowledge Management: Creating a Disciplined Whole from Many Interdisciplinary Parts

David G. Schwartz, Bar-Ilan University, Israel

Abstract

Knowledge management is a fragmented field, whether of necessity or of design. In this chapter, we present and discuss data that maps out a number of the characteristics of the field. We then discuss trends that indicate how knowledge management is evolving into a discipline in its own right and present some thoughts on what the dominant characteristics of that discipline need to be.

Introduction

Defining a new discipline is no easy task, and establishing one is yet a harder task. Yet from the past decade of research, either directly called or indirectly related to knowledge management (KM), emerges a discipline. However, KM remains a fragmented field with multiple, often conflicting terminologies and goals. In this chapter, we present and discuss data that map out a number of the characteristics of the field. We then discuss trends that indicate how knowledge management is evolving into a discipline in its own right and present some thoughts on what the dominant characteristics of that discipline need to be.

One may be tempted to learn from the example of information systems as a discipline. But after more than 40 years of Information Systems research, there still remains great divergence and diversity in how accurately to define this important field. Banville and Landry (1989), Backhouse, Liebenau, and Land (1991), Vessey, Ramesh, and Glass (2002), Adam and Fitzgerald (2000), Baskerville and Myers (2002), and Avison (2003) are but six of the many attempts to reach a broadly accepted definition. Fortunately, the lack of acceptance of any such definition has in no way hampered the development of the field. On the contrary, some, such as Frank (1998), question whether a common profile for Information Systems research is even desirable.

This same sort of qualification process might be applied to the endeavor of knowledge management, and we could ask what constitutes the field of KM, what common profile can be ascribed to KM researchers, and if we, in fact, can consider knowledge management as a discipline in its own right. Jennex and Croasdell (2005) have called for a determination that knowledge management be considered a discipline. As they discuss, meeting Kuhn's (1996) criteria for the establishment of a discipline may be a necessary step; it is clearly not sufficient. The actual nature, characteristics, behaviors, and interaction of those researchers identifying themselves as KM researchers ultimately will determine whether we emerge a discipline or not. The analysis presented in this chapter moves us a step forward in that direction by taking a broad analytical view of KM research underway from both departmental and geographic standpoints.

In this chapter, we will focus to a large extent on the publication process of one of the most comprehensive knowledge management works of recent years, the *Encyclopedia of Knowledge Management* (Schwartz, 2006). We will present some of the findings from the editorial process first reported in Schwartz (2005) and draw insights regarding the global knowledge management community. We will begin with a number of findings based on the initial response to a call for papers for the *Encyclopedia of Knowledge Management* that was issued in October 2003 (Schwartz, 2003). We will present some descriptive statistics that form what in essence is a profile of the self-described knowledge management community. We will then present an overview of the actual contents of the encyclopedia and analyze it from a high-level perspective in an attempt to map out the field itself. By presenting a layered view of knowledge management in which different streams of research are categorized, we believe a clearer picture of the state of KM emerges. Not only that, but by following a layered systematic approach, we can draw implications for both the study of KM and the adoption of KM in modern organizations.

Background and Motivation

Managing the editorial process to create the *Encyclopedia of Knowledge Management* meant creating an overall map of research being conducted that impacts KM both directly and indirectly. It required reaching out to practitioners and academics in a wide range of disciplines in order to elicit their views on what makes knowledge management the pursuit that it is. And it meant attempting to organize that knowledge in a meaningful way so that it can be delivered to and made use of by KM researchers and practitioners in the future. In essence, the same acquire-organize-distribute model (Schwartz, Divitini, & Brasethvik, 2000) that can be used to manage the knowledge of a single enterprise was modified and applied to a multi-organizational and multi-party knowledge management editorial task.

In an attempt to provide as broad of coverage as possible for KM, the call for papers, including a detailed list of topics and subtopics (Figure 1) was prepared in consultation with the international editorial advisory board (http://www.idea-group.com/reference/details.asp?ID=4464&v=editorialBoard). It was through the interactions of the EAB that the CFP metamorphosed from what was originally a very IT-centric worldview to the knowledge and organization-centric view of its final form. Further modifications (shown in italics) were the result of feedback from potential contributors subsequent to the release of the CFP.

Soliciting Contributions

Proposals for contributions to the EKM were solicited through five main channels:

1. The ISWORLD mailing list,
2. The DBWORLD mailing list,
3. The knowledge acquisition/modeling/management (KAW) mailing list,
4. The publisher's (IGI) master mailing list, and
5. The editorial advisory board; each member of the editorial advisory board was asked to distribute the CFP through his or her personal mailing list of relevant researchers.

Departmental Affiliation

One place to start understanding the directions being taken in knowledge management research is the departmental affiliation of those authors working in an area that they identify as relevant to knowledge management.

Authors affiliated with 29 distinct disciplines found it relevant to contribute article proposals. Table 1 shows the main departmental affiliation of proposal authors from the preliminary round of submissions to the *Encyclopedia of Knowledge Management*. Where an author indicated multiple affiliations, the first affiliation listed was used.

Figure 1. Detailed major topics from the Encyclopedia of Knowledge Management *CFP, each of which divides into multiple subtopic entries*

Theoretical Aspects of Knowledge Management
Defining and Understanding Knowledge
Types of Knowledge
Philosophical underpinnings
Ontologies of Knowledge Management
Historical Underpinnings
Organizations and the Inquiring Organization
The People Perspective
Knowledge Management Models

Processes of Knowledge Management
Knowledge Creation
Knowledge Discovery
Knowledge Acquisition
Knowledge Classification
Knowledge Verification and Validation
Knowledge Codification
Knowledge Calibration
Modeling Knowledge
Knowledge Integration
Knowledge Sharing
Knowledge Dissemination
Knowledge Maintenance

Organizational and Social Aspects of Knowledge Management
Knowledge Transfer
Corporate Culture
Motivation
Organizational Memory
Organizational Learning
Cross-border knowledge
Innovation Processes
Social Capital
Social Network Analysis
Community-based knowledge
Organizational Structure

Managerial Aspects of Knowledge Management
KM Strategies and Practices
KM Systems Analysis and Design
KM Systems Management and Lifecycle
Human Resource Management

Legal Aspects of Knowledge Management
Intellectual Property/Capital
Privacy Issues
Digital Rights Management
Liability and the Reliance upon KM Systems
Ethics

Technological Aspects of Knowledge Management
Knowledge Representation
Knowledge Organization and Indexing
Meta-knowledge and Metadata
Storage and Retrieval
Presentation and Application Integration
Artificial Intelligence in KM
Computational Experimentation
Data Mining in KM
Other specific technologies impacting KM

Application-specific Knowledge Management Issues
Biomedical Knowledge Management
Commercial and Financial KM
Industrial Knowledge Management
Military Knowledge Management
Mobile Knowledge Management
Safety-Critical Systems
Customer Knowledge Management
Mathematical Knowledge Management
KM in Counter-terrorism
Higher Education
Workflow Systems
Engineering Design
Legal Knowledge Management
Social Welfare Organizations
Franchise KM
Software Maintenance Knowledge

Noteworthy Knowledge Management Systems and Initiatives

The top four affiliations show an overwhelming concentration in the fields in which knowledge management has been addressed actively over the past decade.

These top four affiliations reflect what may be expected from most of the knowledge management community. Of greater interest perhaps is the participation in KM research in what can be termed *nontraditional* KM affiliations.

Table 1. Departmental affiliation of responding authors

Rank	**Main Departmental Affiliation**	Count	%
1	Information Systems	111	44.58%
2	Computer Science	39	15.66%
3	Information and Library Science	15	6.02%
4	Management	12	4.82%
5	Communications	6	2.41%
6	Economics	6	2.41%
7	Marketing	6	2.41%
8	Cognitive Science	5	2.01%
9	Management Science	5	2.01%
10	Philosophy	5	2.01%
11	Engineering Management	4	1.61%
12	Social Psychology	4	1.61%
13	Information Management	3	1.20%
14	Organizational Science	3	1.20%
15	Sociology	3	1.20%
16	Education	2	0.80%
17	Engineering	2	0.80%
18	Finance	2	0.80%
19	Human Resource Management	2	0.80%
20	Innovation Studies	2	0.80%
21	Mathematics	2	0.80%
22	Media Management	2	0.80%
23	Technology Management	2	0.80%
24	Banking	1	0.40%
25	Business Administration	1	0.40%
26	Cultural Studies	1	0.40%
27	Real Estate	1	0.40%
28	Science and Technology	1	0.40%
29	Statistics	1	0.40%
		249	100%

A second point of interest from Table 1 is the wide range of departmental participation, lending strength to the interdisciplinary nature of KM and providing an indication as to what types of courses a form program knowledge management studies might need to include.

Also of note is the complete lack of any departmental affiliation specific to knowledge management. While a number of authors were associated with KM Research Labs or facilities, these were clearly research-oriented initiatives and not teaching initiatives or programs.

Geographic Distribution

Another area of interest is that of geographic distribution. Here, we see concentrations of KM research by country and geographic region.

Table 3 presents the total number of authors by country in which they work (i.e., main university/employer affiliation).

Table 2. Division of respondents into traditional and nontraditional IS/management fields

Traditional Information and Management-Related Fields		Nontraditional Fields	
Information Systems	44.6%	Economics	2.4%
Computer Science	15.7%	Marketing	2.4%
Information Science	6.0%	Cognitive Science	2.0%
Management	4.8%	Philosophy	2.0%
Communications	2.4%	Social Psychology	1.6%
Management Science	2.0%	Sociology	1.2%
Engineering Management	1.6%	Education	0.8%
Information Management	1.2%	Engineering	0.8%
Organizational Science	1.2%	Finance	0.8%
Human Resource Management	0.8%	Innovation Studies	0.8%
Media Management	0.8%	Mathematics	0.8%
Technology Management	0.8%	Banking	0.4%
Business Administration	0.4%	Cultural Studies	0.4%
		Real Estate	0.4%
		Science and Technology	0.4%
		Statistics	0.4%
Total	82.3%	Total	17.7%

Table 3. National affiliation of responding authors

Rank	Author Affiliation by Country	Count	Percent
1	United States	76	30.52%
2	England	26	10.44%
3	Italy	17	6.83%
4	Germany	16	6.43%
5	Netherlands	16	6.43%
6	Israel	15	6.02%
7	Australia	13	5.22%
8	France	13	5.22%
9	Ireland	10	4.02%
10	Spain	10	4.02%
11	Canada	9	3.61%
12	Brazil	5	2.01%
13	Singapore	4	1.61%
14	Switzerland	4	1.61%
15	Denmark	2	0.80%
16	Hong Kong	2	0.80%
17	India	2	0.80%
18	Norway	2	0.80%
19	South Korea	2	0.80%
20	Austria	1	0.40%
21	Greece	1	0.40%
22	Japan	1	0.40%
23	Macau	1	0.40%
24	South Africa	1	0.40%
		249	100.00%

Table 4. Regional affiliation of responding authors

Geographic by Region	Count	Percentage
EMEA	98	39%
North America	85	34%
UK	36	14%
Asia Pacific	25	10%
South America	5	2%

The Resulting Volume

The actual table of contents, by topic, of the encyclopedia is presented in the "Appendix" section. The six logical topics form a structured framework for the research and study of knowledge management. Within each main section, specific subtopics are addressed, and within each subtopic, there appear a number of articles addressing different aspects and perspectives. This result was achieved after two rounds of review and consultation with the editorial advisory board.

Discussion

The contents, as described previously, necessitated a new way to think about how all these topics and subtopics interrelate. And to that end, we developed the diagram shown in Figure 2. The first five sections of the *Encyclopedia of Knowledge Management* are the result of what can be characterized as a *layered approach* to the discipline of knowledge management.

The holistic view of knowledge management and its foundations can be used as a guide for research as well as study. The central core of philosophies (the middle) must inform our choice of practical knowledge management processes (the first ring). These processes must be implemented and adapted in order to address organizational, social, and managerial needs (the second ring). Finally, the implementation of KM process to meet our organizational needs must be supported by and implemented through a set of relevant information technologies (the outer ring).

The primary processes that make up knowledge management in practice ideally should derive from the core theories. Figure 2 illustrates a number of philosophers whose theories of knowledge, economics, and business form the core of knowledge management. Understanding these philosophies is fundamental to creating a lasting discipline. Without grounding our processes in their theoretical soil, we run the very real risk of simply cobbling together processes on an opportunistic basis. In a disciplined manner, we must turn to our theoretical core in determining the essential processes of KM. In cases in which experience begets a process that has yet to be identified with a core theory, one must not belittle the need to eventually discover that grounding. At the end of the day, this is what will help to distinguish fad from enduring science.

The layer of knowledge management processes presents one view of the different stages, activities, and cycles that comprise knowledge management. Up to 30 different processes can be identified, depending on the model of KM chosen. We have based our diagram on those processes identified and discussed in Schwartz et al. (2000). Processes need to be pragmatic in terms of our ability to implement them, comprehensive so that we can achieve an end-to-end solutions, replicable and generalizable so they can be applied across a wide range of organizations. Processes are:

- Technology-independent,
- Application-independent,
- Founded on theory, and
- Generalizable.

That is not to say that these processes should be devoid of organizational context. On the contrary, it is the function of the third layer, organizational, social, and managerial (OSM) considerations, to mold, combine, and innovate using the KM processes in order to meet their well-defined theory-driven goals. OSM elements are:

- Technology-independent,
- Application-specific,
- Founded on practice, and
- Organization-specific.

Encasing all is the outer ring—the enabling technologies that so often seem to be driving KM rather than facilitating it. Figure 2, of course, is representative rather than exhaustive. Additional technologies and new applications of existing technologies will continue to expand this layer. The technologies are:

- Possibly interchangeable,
- Application-independent,
- Founded on practice, and
- Organization-independent.

Moving from Layer to Layer

One benefit from taking this holistic view of the field is that it enables a novel KM implementation strategy. It allows us to focus at an operative level on the issues we need to address in practice; for example, addressing the organizational, social, and managerial elements of the third ring. Then, we can move either up or down to determine (a) which KM processes are

Figure 2. Layer upon layer of knowledge management

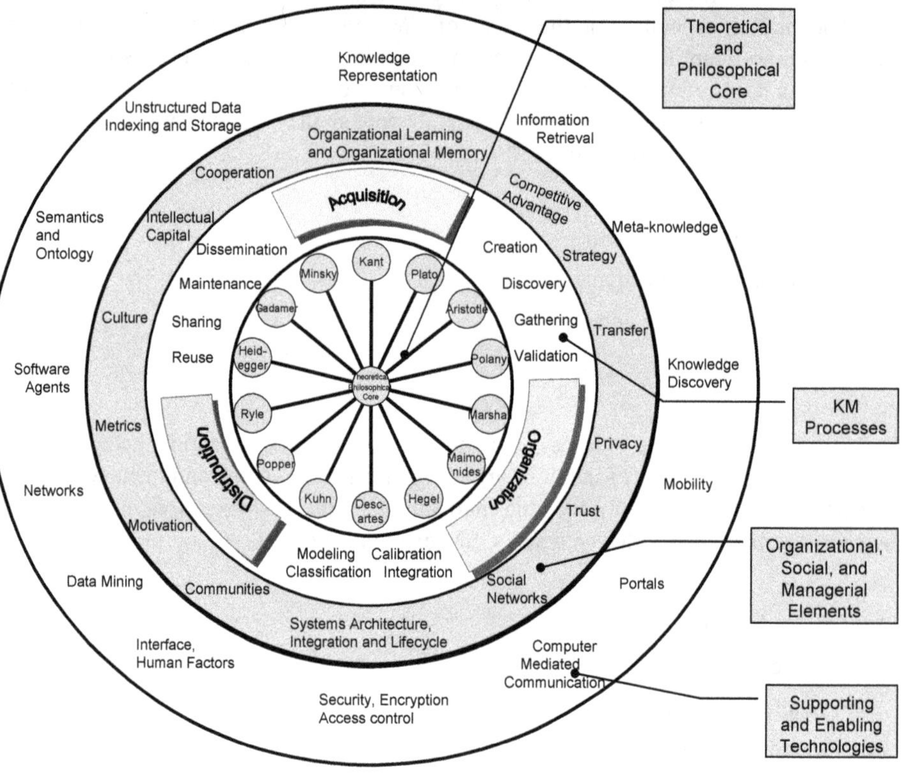

needed to support the OSM elements selected and (b) which technologies can be leveraged to implement the required processes.

Another benefit to this layered view is helping to focus the study of knowledge management in an academic environment. Programs of study first need to provide specialization in each of the layers and, therefore, must include a philosophical component; an introduction to process and process-oriented thinking and planning; OSM theory and practice courses; and finally, enough of a foundation in the wide range of technologies to allow for intelligent evaluation and adoption. Second, a KM program of study needs to provide tools and understanding in order to enable students to interrelate the layers and to follow how technology, OSM element, and process complete each other.

In summary, the advantages of this approach are as follows:

- Each layer can be isolated and studied on its own.
- We can focus on needs and requirements of each layer independently.
- Enables one to consider how each layer is supported by the others.

- Creates a flexible conceptualization of KM that is anchored in research but supportable in practice.
- Once a layer focus is determined, we can look at connections with other layers.

Conclusion

There is no question that knowledge management has extended its reach into a staggering number of areas of study. While the fields of computer science, library science, sociology, psychology, business strategy, and the like will remain disciplines in their own rights, there clearly are benefits to be gained by starting to view knowledge management as a discipline separate from the others.

We believe that the layered view of knowledge management can be of great help in establishing and advancing the discipline.

- The central core of philosophies must inform our choice of practical knowledge management processes. Primary processes that make up KM in practice ideally should derive from the core theories. Without grounding our processes in theory, we risk cobbling together processes on an opportunistic basis.
- These processes must be implemented and adapted to address managerial, social and organizational needs. The layer of processes we have chosen presents just one view of the different stages, activities, and cycles that comprise knowledge management. Processes need to be:
  - Pragmatic in terms of our ability to implement them.
  - Comprehensive so we can achieve end-to-end solutions.
  - Replicable.
  - Generalizable to be applied across a wide range of organizations.
- The third layer, organizational, social and managerial considerations, uses an organizational context to mold, combine, and innovate using KM processes.
- Finally, the implementation of the KM process to meet our organizational needs must be supported by and implemented through a set of relevant technologies. The outer layer is the enabling technologies that so often seem to be driving KM rather than facilitating it.

We have taken some of the first steps to enable this to happen by mapping out the distinct elements that comprise knowledge management, from each of its contributing disciplines. We then have shown how these elements are grouped into logical and interrelated layers. The structured interrelation between the layers not only serves to create a conceptual framework for research in KM but also serves as a guide for developing programs of study and as a basis to develop novel implementation strategies for KM. Developing KM strategies and plans by moving among layers and taking the necessary elements for a given KM situation that each layer has to offer will better equip the modern organization to deal with what can be an overwhelming field.

References

Adam, F., & Fitzgerald, B. (2000). The status of the IS field: Historical perspective and practical orientation. *Information Research, 5*(4). Retrieved from http://informationr.net/ir/5-4/paper81.html

Avison, D. E. (2003). Is IS an intellectual subject? (Response to opinion piece.) *European Journal of Information Systems, 12*(3), 229-230.

Backhouse, J., Liebenau, J., & Land, F. (1991). On the discipline of information systems. *Journal of Information Systems, 1*(1), 19-27.

Banville, C., & Landry, M. (1989). Can the field of MIS be disciplined? *Communications of the ACM, 32*(1), 48-60.

Baskerville, R. L., & Myers, M. D. (2002). Information systems as a reference discipline. *MIS Quarterly, 26*(1), 1-14.

Frank, U. (1998). *Reflections on the core of the information systems discipline* [report no. 14]. Institute fur Wirtschaftsinformatik, University of Koblenz-Landau.

Jennex, M. E., & Croasdell, D. (2005). Is knowledge management a discipline? *International Journal of Knowledge Management, 1*(1), i-v.

Kuhn, T.S. (1996). *The structure of scientific revolutions* (3rd ed.). Chicago: University of Chicago Press.

Schwartz, D. G. (2003). Call for papers. *The Encyclopedia of Knowledge Management.* Retrieved October 1, 2003, from http://faculty.biu.ac.il/~dgs/ekm/EKM-CFP.pdf

Schwartz, D. G. (2005). The emerging discipline of knowledge management. *International Journal of Knowledge Management, 1*(2), 1-11.

Schwartz, D. G. (2006). *Encyclopedia of knowledge management.* Hershey, PA: Idea Group Reference.

Schwartz, D. G., Divitini, M., & Brasethvik, T. (2000). *Internet-based organizational memory and knowledge management.* Hershey, PA: Idea Group Publishing.

Vessey, I., Ramesh, V., & Glass, R. L. (2002). Research in information systems: An empirical study of diversity in the discipline and its journals. *Journal of Management Information Systems, 19*(2), 129-174.

Appendix

Table of Contents, Encyclopedia of Knowledge Management

Chapter IV

Knowledge Management Research:
Are We Seeing the Whole Picture?

Todd Peachey, Auburn University, USA

Dianne Hall, Auburn University, USA

Casey Cegielski, Auburn University, USA

Abstract

Although knowledge management (KM) is maturing as a research topic, there is no agreement on what constructs constitute its foundation. Because the topic has received increasing attention in academic journals, it is important for researchers to be aware of the research streams associated with KM. Accordingly, this chapter reviews the knowledge management literature published in top-tier journals from 2000 to 2005. These articles then are classified by knowledge management construct and by research methodology. The results indicate that the majority of knowledge management research has examined the construct of knowledge transfer. Formal theory/frameworks and field studies using primary data are the dominant methodologies. Trends of published KM research, gaps, and inconsistencies in the examined literature and triangulation of applied research methodologies are discussed.

Introduction

Managing knowledge never has been more important than in today's competitive environment (Desouza, 2003). The quantity of published research from the mid-1990s though 2005 shows increased effort to better understand knowledge management (KM). Despite this increased attention, several issues remain. There is dissention by both academia and practitioners over the true definition of knowledge management. There are also questions about the relevant constructs that comprise knowledge management and where our collective research has taken us in our efforts to discover those constructs. This chapter describes current research in knowledge management with regard to which constructs are most researched and how these constructs are studied.

Research in knowledge management has increased dramatically in recent years. From 1990 to 1995, a search of the ABI/Inform database using the keywords *knowledge management* returned 43 articles. From 1995 to 2000, the number of articles increased to more than 700, and from 2000 to 2005, the number of articles increased yet again to well over 2,000. This research is published to varying degrees in a wide variety of disciplines, including management; hospitality; economics; healthcare; and, naturally, information systems. If we examine the sample of published research in this study with consideration for the total number of articles published, we see that approximately 7% of this research is published in what are considered by many to be the leading IS journals. This is interesting since some of what are considered the leading IS journals are cross-discipline journals, such as *Decision Sciences and Management Science*. From this, we can infer that not only does research in knowledge management often originate outside the information systems discipline, but also knowledge management research appears in a wide variety of journals with differing readerships and theoretical foundations.

We propose that understanding the future direction of research in knowledge management requires that we first know what constructs in knowledge management have received the most attention from researchers and where there currently are gaps in the published literature. Given the quantity of current literature, there is an adequate sample size to determine the coverage of our collective research efforts.

Second, it is critical for researchers to understand the strengths and weaknesses of research methodologies used in KM research, the subsequent implications of chosen methodologies, and the results of those choices. Each research strategy has inherent strengths and weaknesses. However, when a topic is triangulated properly by different research strategies, it becomes more robust and refined. Triangulation occurs when researchers use different methods to study the same topic in order to reduce the inherent weaknesses in a specific technique. McGrath (1982) stated that it is impossible to do an unflawed study; however, by triangulating strategies to ensure full spectrum coverage, these weaknesses can be reduced significantly.

Knowledge Management Contructs

A coherent review emerges only from a coherent conceptual structuring of the topic itself (Bem, 1995). There is a number of frameworks available for classifying knowledge management research; one literature review identified 26 different frameworks from both practitioners and academics (Rubenstein-Montano et al., 2001). Some frameworks address specific concepts within the overall discipline of knowledge management. For example, Griffith, Sawyer, and Neale (2003) developed a framework to better facilitate understanding of knowledge transfer among groups and teams. Holsapple and Joshi (2001) proposed a framework to better understand an organization's knowledge resource hierarchy. These frameworks are significant contributions to the literature in their focused area. However, because a goal of this study is to identify trends prevalent to the overall concept of knowledge management, the authors determined that the Alavi and Leidner (2001) and Davenport and Prusak (1998) frameworks are aligned most appropriately with our goal. Both of these frameworks are parsimonious in their structure and relevant to academics and practitioners alike. Additionally, these are two of the most cited knowledge management works (Jennex & Croasdell, 2005). The Alavi and Leidner (2001) framework separates knowledge management research into four constructs: creation, storage and retrieval, transfer, and application. Alavi and Leidner (2001) provide extensive definitions of the four areas of knowledge management that were used in this study to categorize research articles into construct categories. In addition to the articles reviewed for this study, source articles described in the Alavi and Leidner (2001) framework also were reviewed for consistency of categorization. Some of those source articles are described next in relation to specific definitions of the constructs used by Alavi and Leidner.

The Alavi and Leidner (2001) framework is formed around four constructs in the organizational knowledge management process. They state that "organizations as knowledge systems consist of four sets of data: (1) creation (also referred to as construction), (2) storage/retrieval, (3) transfer, and (4) application. The four constructs of this model are essential to effective organizational knowledge management" (Alavi & Leidner, 2001, p. 115).

Davenport and Prusak's (1998) research suggests another set of constructs: knowledge generation, knowledge codification and coordination, knowledge transfer, and knowledge roles and skills. They also provide extensive definitions and examples of the different constructs. Knowledge generation, knowledge codification and coordination, and knowledge transfer are the key processes of knowledge management; these processes are critical for an organization's successful management of knowledge (Davenport & Prusak, 1998). Knowledge roles and skills, along with technology, are enablers of knowledge management (Davenport & Prusak, 1998). This is particularly notable as a construct because, of the 26 frameworks identified by Rubenstein-Montano et al. (2001); none identified the supporting roles and skills necessary to make a knowledge management initiative successful. We believe, however, that we cannot reasonably separate the people or the technology from the overall process of knowledge management; the Davenport and Prusak (1998) framework was chosen as the second reference framework for this specific reason. Based on the definitions provided by the authors of both frameworks, we were able to develop a five-construct model for categorization. Figure 1 outlines the combined framework.

Figure 1. Five construct categorization framework

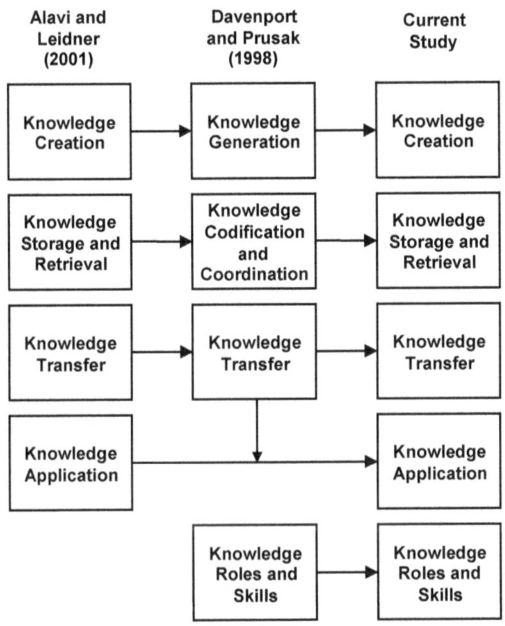

In the following section, the constructs are defined. These definitions are used for the classification of the papers that are the subject of this research. The definitions were developed from the material in the two reference frameworks. When noted, the original citation was reviewed to ensure that the authors of this research correctly grasped the intent of the frameworks' authors.

Creation

Alavi and Leidner (2001) present organizational knowledge creation as "involving a continual interplay between the tacit and explicit dimensions of knowledge and a growing spiral flow as knowledge moves through individual, group, and organizational levels" (p. 166). One dimension of the knowledge creation process can be drawn from the distinction between tacit and explicit knowledge. Explicit knowledge can be transmitted by formal, semantic means, while tacit knowledge has a personal quality that makes it difficult to formalize and communicate (Nonaka, 1994).

Davenport and Prusak (1998) also examine knowledge creation, focusing on the conscious and intentional generation of knowledge in the organizational context. They posited that knowledge generation can take many forms and specifically identified five of these forms. Knowledge can be acquired by an organization as well as developed from within it, or knowledge can be rented in the form of research grants and consulting contracts. R&D departments and other specialized units that focus on knowledge generation are dedicated resources within the organization. Fusion intentionally introduces conflict and complexity into the process to develop synergies. An organization's ability to adapt is critical to its long-term survival; employees are adaptive resources that can acquire new knowledge and skills. Employees also may form informal or formal learning networks and groups focused on knowledge generation.

Storage and Retrieval

Storing, organizing, and retrieving organizational knowledge from organizational memory (Stein & Zwass, 1995; Walsh & Ungson, 1991) is a process that is critical to the organization's ability to learn and make informed decisions. Organizational memory is the means by which knowledge from the past is brought to bear on present activities, thus resulting in higher or lower levels of organizational effectiveness, depending on application (Stein & Zwass, 1995). Data mining and learning tools are examples of supporting technologies of knowledge storage and retrieval; collaborative systems is another that has gained popularity. These systems enable organizations to create organizational memory in the form of both structured and unstructured information and to share this information across time and space (Walsh & Ungson, 1991).

Codification is the act of arranging organizational knowledge into a form that makes it accessible to those who need it; it is categorized and often digitized into a form that is organized, explicit, portable, and easy to understand (Davenport & Prusak, 1998). Mapping knowledge is also important to the codification process. The primary purpose and benefit of a knowledge map is to organize and classify knowledge within the organization (Davenport & Prusak, 1998).

Transfer

The ability to transfer knowledge is critical for an organization and occurs at many levels. Knowledge can be transferred among individuals, between individuals and explicit sources, between individuals and groups, among groups, and between groups and the organization (Alavi & Leidner, 2001). Knowledge transfer among organizations is also possible, particularly in areas such as enterprise resource planning and supply chain management.

Knowledge transfer may be affected by many things within an organization, such as the inclination of the knowledge holder to divulge it, the recipient's ability or desire to receive it, or perceived value (Gupta & Govindarajan, 2000). However, whether managed in a formal process or not, knowledge transfer happens routinely in organizations, such as in informal exchanges between organizational members. Perhaps because of its inherently complex nature, knowledge transfer is the first construct that the two frameworks view in a marginally different form. While Alavi and Leidner (2001) separate transfer and application, Davenport and Prusak (1998) posit that knowledge transfer is incomplete until the receiver has both internalized it and used it. Keeping transfer and application (use) separate, however, allows a slightly more precise classification of the knowledge transfer concept.

Application

Knowledge application is a source of competitive advantage and is based on the application of the knowledge rather than simple possession of the knowledge according to Grant (1996), who defined four mechanisms for integrating specialized knowledge: (a) rules and directives, (b) sequencing, (c) routines, and (d) group problem solving and decision making. Grant (1996) defines rules and directives as those things in an organization that govern the interactions among individuals. Sequencing is the structuring of a process so that each specialist's knowledge occurs independently (Grant, 1996). Routines are simple chains of interactions and complex interactions among individuals that are not governed by rules and directives. Grant (1996) states that group problem solving and decision making is based on maximizing the use of rules and routines in order to reduce communication and knowledge transfer and only use group problem solving in the most extreme, important, and unusual tasks. Davenport and Prusak (1998) state that improving an organization's capabilities is the goal of knowledge transfer and that it is of no value if it does not lead to some change in behavior. They close their thoughts on knowledge transfer with the succinct statement that "knowing is not doing" (Davenport & Prusak, 1998, p. 102).

Roles and Skills

If knowledge management is to thrive, organizations must create a set of roles and skills to do the work of capturing, distributing, and using knowledge (Davenport & Prusak, 1998). For instance, many organizations have created positions specifically to manage the information process (e.g., chief information officer) and, more recently, to manage the knowledge process (e.g., chief knowledge officer). While these leadership roles are important, many others in the organization play vital roles. For instance, a manager who routinely makes decisions about processes based on information embedded in the organization's memory is performing a task of a knowledge worker. Individuals in specialized roles such as innovation and creation (e.g., product development) are also knowledge workers. Each of these depends on knowledge management processes to keep organizational memory accurate and timely, to maintain expert directories, and to engage in acquisition from outside the organization. Davenport and Prusak (1998) define four levels of knowledge management roles: (a) line workers who must manage knowledge within their own jobs, (b) knowledge management workers, (c) knowledge project managers, and (d) senior knowledge executives, along with differing skills required of each position.

Research Strategies

Triangulation within and between projects allows researchers to draw more bias-free conclusions from their work. Researchers can determine the triangulation of a topic within a discipline by examining the research strategies used and by understanding their inherent strengths and weaknesses (Scandura & Williams, 2000; Palvia, Mao, Salam, & Soliman,

Table 1. Quadrants

Quadrant I	Quadrant II	Quadrant III	Quadrant IV
Field Experiment	Experimental Simulation	Judgment Task	Formal Theory/ Frameworks*
Field Study: Primary Data	Laboratory Experiment	Sample Survey	Computer Simulation
Field Study: Secondary Data*			

*Note: * Modified from the original based on Scandura and Williams (2000)*

2003). Use of more than one research strategy or approach indicates, at least to a degree, the existence of triangulation (Scandura & Williams, 2000; Palvia et al., 2003). Other researchers also have called for increased triangulation in research in general (Kaplan & Duchon, 1988; Mingers, 2001).

The eight research strategies and four quadrants described by McGrath (1982) are listed in Table 1. Eight types of research are readily distinguishable. Each methodology is closely related to others and has its own strengths and weaknesses. McGrath (1982) grouped these strategies into four quadrants based on each strategy's similarity to others. The groupings are listed in Table 1 and then are discussed in some detail.

The quadrants exist in a circular, spiraling model. As research in a topic moves from one quadrant to another, new findings may require new theoretical development. The research stream would not return to the original starting point (unless, of course, it is unequivocally refuted) but, instead, would move back to a stage of theoretical development, grounded on but expanding the previous research in some manner. The research stream is logically directional and systematically circular (McGrath, 1982). For example, once a theory is posed in Quadrant IV, researchers usually will go into the field to examine the phenomena in a realistic context. As these findings are analyzed in context of the theory, the next step is to improve the precision of the research stream using Quadrant II methodologies. Once an adequate precision is obtained, then researchers can use the methodologies in Quadrant III to examine the generalizability of their work. When the first circle is complete, researchers move back to Quadrant IV to review the original theories and to adjust them appropriately as justified by their findings.

Quadrant I

Field studies and field experiments are closely related in that they take place in settings that are real for the participants (McGrath, 1982). The primary difference is that field studies are generally unobtrusive to the participants. Data collection in a field study may be collected wither through unobtrusive observation, as described by McGrath (1982), or through the use of secondary data such as information from a census or a public source such as the *Wall Street Journal.*

A study similar to this one, based on management research, further differentiates field studies by their data collection method (Scandura & Williams, 2000). Based on that study, we have expanded field studies to differentiate between primary data collection, such as would be done in a case study, to include secondary data collection, which more often is used in studies in which data are not directly collected by the researcher. In field experiments, the researcher will control some aspect of the study to increase precision. The realism of context is at its maximum in these types of research (McGrath, 1982).

Quadrant II

Laboratory experiments and experimental simulations are similar in that they involve artificial settings that are not real for the participants (McGrath, 1982). They differ in their level of realism; in a laboratory experiment, the setting and behavior under study is completely artificial, while in an experimental simulation, some aspects of realism of context are retained. In this quadrant, the researcher tries to maintain some realism of content while giving up realism in context. The major benefit of research in this quadrant is the precision of measurement of behavior (McGrath, 1982).

Quadrant III

Two strategies, sample surveys and judgment tasks, balance context and generalizability. In a sample survey, the researcher studies behaviors by neutralizing context (McGrath, 1982). In contrast, in a judgment task, the researcher tries to nullify the context of all extraneous conditions. These two techniques are far more rigorous in their sampling. The sample survey strives to obtain effective sampling of populations in order to increase generalizability, whereas in a judgment task, only a few "judges" are used under the assumption they are generic judges. Research in this quadrant is focused on generalizability (McGrath, 1982).

Quadrant IV

Two types of research strategies are not empirical in that no behavior occurs (McGrath, 1982). However, while formal theory and computer simulations are low in realism of context, they increase population generalizability. Formal theory articles pose a view of the universal vs. the particular, while computer simulations are an attempt to model a concrete system (McGrath, 1982). Following Scandura and Williams (2000), we have combined formal theory and literature reviews. In addition, as explained later in this chapter, we also have included frameworks in this category.

Table 2. Leading information systems journals

Peffers and Ya (2003)	Mylonopoulos and Theoharakis (2001)
MIS Quarterly	*MIS Quarterly*
Information Systems Research	*Information Systems Research*
Journal of Management Information Systems	*Journal of Management Information Systems*
Decision Support Systems	*Decision Support Systems*
Information and Management	*Information and Management*
Information Systems Journal	
Journal of the AIS	
*International Journal of Electronic Commerce***	
Information Systems	
European Journal of Information Systems	
	*Communications of the ACM**
	IEEE Transactions
	*Harvard Business Review**
	Decision Sciences
	Management Science

*Note: *Not used in this study because of the practitioner angle represented by the journal; **included in the study but contained no KM articles and, therefore, is not reflected in the findings*

Methodology

Major contributions in any domain are more likely to be represented in leading journals of that field (Webster & Watson, 2002). knowledge management research has been published in a wide variety of journals in many disciplines; however, our focus is specifically that increasing amounts of research in a specific topic that is accepted into leading IS journals is generally a signal of a topic's increasing maturity and acceptance within mainstream academia. In order to determine which journals are generally considered to be leaders, we used two recently published articles that rank IS journals to determine the list of publications to search for the relevant literature. The first is a recent article by Peffers and Ya (2003), and the second is an article by Mylonopoulos and Theoharakis (2001). Using two ranking schemes allows us to increase the probability that the journals chosen for this study research do, in fact, represent the leading IS journals.

The Peffers and Ya (2003) article in the *Journal of Information Technology Theory and Application* (JITTA) identified 10 top-rated journals: *Information Systems Research*; *MIS Quarterly*; *Journal of Management Information Systems*; *European Journal of Information Systems*; *Decision Support Systems, Information and Management*; *Information Systems Journal*; *Journal of the Association for Information Systems*; *International Journal of Electronic Commerce*; and *Information Systems*. In this chapter, Peffers and Ya (2003) used several ranking schemes. For our purposes, we chose the ranking scheme that used the average weighted perceived value rating of journals as outlets for information systems research. In this scheme, Peffers and Ya (2003) (see Table 6) use only rankings for journals that were rated by at least 10% of their respondents. This avoids the problem of artificially inflating a journal's ranking by a small number of researchers (Peffers & Ya, 2003).

Five of the journals in this study are also included in the article by Mylonopoulos and Theoharakis (2001): *Information Systems Research, MIS Quarterly, Journal of MIS, Information and Management*, and *Decision Support Systems*. Rounding out their top 10 are the *Communications of the ACM, Management Science, IEEE Transactions, Harvard Business Review*, and *Decision Sciences*.

The journals chosen to represent IS also cross other disciplines and represent practitioners. For instance, *Decision Sciences* and *Management Science* are leading journals in the Productions and Operations Management disciplines as well as IS. Journals outside the IS discipline are beyond the scope of the current investigation. Additionally, given the nature of the second part of this study, we did not include practitioner journals such as *Communications of the ACM* and *Harvard Business Review*, given their tendency not to include information about the methodology of their work. Table 2 summarizes the journals reflected in the aforementioned articles and indicates those not chosen for this study.

The current study examines a five-year window. The ABI/Inform database was used to search the journals listed in Table 2 from January 1, 2000 through December 31, 2005. Each selected journal was searched, using the advanced search option that allowed the authors to restrict the search dates and publication source; however, there were no applicable articles that appeared in the *International Journal of E-commerce*, and therefore, this journal will not appear in the following discussion.

After applying date and publication constraints, the authors further restricted the search to the keyword knowledge management. This keyword was chosen because we were specifically searching for articles in which the authors explicitly stated they were researching knowledge management. We found that this term allowed us to find articles specific to individual constructs (i.e., knowledge transfer) as well as articles that may or may not represent constructs but are published under the general knowledge management umbrella. We also found many articles whose authors specifically used Knowledge Management as a keyword and who were not in the mainstream of knowledge management research. For example, one article focused on the effects of animation on information seeking on the World Wide Web (Zhang, 2000). Had we chosen only to search on specific construct terms, these more general articles may have been missed. The process outlined previously resulted in a sample of 158 articles.

The identified articles were reviewed to determine which framework construct or constructs were represented in the research, and each was coded according to the number of constructs represented in the article, with a maximum sum of one. For example, an article describing research in knowledge creation was coded as a one. We expected, and found, that many articles examined more than one construct. When this was the case, the value was divided by the number of constructs addressed in the article. For example, if an article contained research in both knowledge creation and transfer, each construct was coded with 0.5, giving the total for the article a sum of 1. In order for a construct to receive credit from a specific article, the study had to examine all coded constructs thoroughly. Several articles briefly discussed one of the other functions without including it in the research or conceptual development. This coding process was conducted individually by our research team, based on the collective understanding of the construct definitions and the example articles from our two frameworks. Cross-checking was conducted in order to identify inconsistencies in coding; although few were found, each was addressed, and a consensus was reached as to the correct coding scheme.

Results

The number of articles in the publications ranged from a high of 33 to a low of three. Knowledge transfer was the most frequently researched topic, with knowledge storage and retrieval being the topic secondmost frequently researched. The constructs of creation, application, and roles/skills were represented about equally across the publications.

Decision Support Systems and *Management Science* published the most articles on knowledge management, with 33 and 30 articles, respectively. However, each of these journals published a special knowledge management issue. *Decision Support Systems* published nine articles in its knowledge management issue in May 2001; *Management Science* published 13 articles in its Special Issue on Managing Knowledge in Organizations: Creating, Retaining, and Transferring Knowledge in April 2003; and *MIS Quarterly* published 13 articles in two special issues in early 2005. Removing these special issue articles indicates that these journals have published about the same number of articles on knowledge management as the next highest journals, *European Journal of Information Systems* and *Journal of Management Information Systems*.

We also investigated the result of removing the special issues from the aforementioned numbers. We removed a total of 35 articles: nine from the May 2001 issue of *Decision Support Systems*, 13 from the April 2003 issue of *Management Science*, seven from the March 2005 issue of *MISQ*, and six from the June 2005 issue of *MISQ*. In doing so, we found that the percentage of Storage and Retrieval articles increased by approximately 3%, and Transfer decreased by slightly less than 3%. The other constructs changed by less than 1%. This suggests that the constructs in the special issues were covered similarly to those in the remaining publications when the special issues are removed. Thus, the articles from the special issues are included in the following analysis.

Table 3 shows the complete distribution of the relevant articles published in the journals reviewed. The Number of Articles column contains the total number of articles published in the specific journal over the period in our study. The number in each of the other five columns (creation, storage and retrieval, transfer, application, and roles and skills) reflects how many of the articles were coded as reflecting a specific knowledge management construct. Fractions may occur, depending on the number of articles included that represent more than one construct. The second number in the Totals row shows the percentage of coverage of each construct. Research in Transfer was covered at twice the rate of any of the other constructs. Storage and Retrieval and Application were covered at a similar rate, while Creation was covered at a slightly lower rate. Roles and Skills received the least coverage in this sample.

Table 3. Leading IS journals and knowledge management articles

Journal	# of Articles	Creation	Storage/ Retrieval	Transfer	Application	Roles/ Skills
Decision Sciences	9	1	0	4.5	3.5	0
Decision Support Systems *	33	1.25	12.25	7.75	10.75	1
European Journal of Information Systems	11	1.75	0.75	3.75	4.75	0
Information and Management	16	2.75	2.08	5.08	1.83	4.26
IEEE Transactions	6	1	3	1.5	0.5	0
Information Systems	3	0	3	0	0	0
Information Systems Journal	4	0	2	2	0	0
Information Systems Research	7	1	0.5	4.5	1	0
Journal of AIS	4	0	0	1	3	0
Journal of MIS	16	2.95	5.45	4.95	1.45	1.20
*MIS Quarterly**	20	3.75	.75	8.75	4.25	2.5
*Management Science**	29	4.33	1.83	18.83	1.5	2.5
Totals	158	19.78 12.68%	31.61 20.26%	62.61 40.13%	32.53 20.85%	11.46 7.35%

*Note: * This journal published at least one special issue on knowledge management*

Table 4 outlines the total use of different methodologies in this sample. When examining research on KM from this sample, we find that a majority of the research is conducted using Quadrant I methodologies, which focus on realism of context. Field studies using primary data were the most frequently used methodology (33.33%) in this quadrant and across the entire spectrum. The second highest level of activity was in Quadrant IV. Within Quadrant IV, formal theory and frameworks were the dominant methodology, comprising 28.40% of the sample. Formal Studies and Frameworks were also the secondmost often used methodology across the spectrum. Quadrants II and III, which maximize generalizability and realism, show a very similar level of activity. Given the circular nature of research, it is likely that research in Quadrants II and III will trail behind the other quadrants, except in all but the most mature disciplines.

Table 5 outlines the research strategies applied in different KM constructs. Research in Creation and Storage and Retrieval shows a similar effort in Quadrants I and IV. Research on the constructs of Transfer, Application, and Roles and Skills appears to be dominated by a desire for realism evidenced by the high percentage of work in Quadrant I. However, note

Table 4. Methodologies per research quadrant

Methodology	Overall	Quadrant
Field Experiment	1.23%	
Field Study: Primary Data	33.33%	I
Field Study: Secondary Data *	11.11%	45.67%
Experimental Simulation	1.85%	II
Laboratory Experiment	7.41%	9.26%
Judgment Task	2.47%	III
Sample Survey	11.73%	14.20%
Formal Theory/Frameworks *	28.40%	IV
Computer Simulation	2.47%	30.87%

*Note: * Modified from the original based on Scandura and Williams (2000)*

Table 5. Methodologies per KM construct

Methodology \ Construct	Creation	Storage/ Retrieval	Transfer	Application	Roles/ Skills
Field Experiment	5.06%	3.16%			
Field Study: Primary Data	14.31%	29.52%	37.26%	38.43%	52.44%
Field Study: Secondary Data	20.22%	8.95%	14.90%	5.63%	
Quadrant I	**39.59%**	**41.63%**	**52.16%**	**44.05%**	**52.44%**
Experimental Simulation		1.58%	2.40%	3.07%	
Laboratory Experiment		4.75%	9.58%	12.30%	4.36%
Quadrant II	**0.00%**	**8.67%**	**11.98%**	**15.37%**	**4.36%**
Judgment Task		6.33%		3.07%	8.73%
Sample Survey	20.22%	1.58%	13.5%	14.60%	10.9%
Quadrant III	**20.22%**	**7.91%**	**13.58%**	**17.68%**	**19.63%**
Formal Theory/ Frameworks	30.08%	44.13%	22.28%	16.75%	23.56%
Computer Simulation	10.11%			6.15%	
Quadrant IV	**40.19%**	**44.13%**	**22.28%**	**22.90%**	**23.56%**

Table 6. Methodology by level of analysis

Methodology \ Level	Individual	Group	Organization
Field Experiment	2.00%	3.45%	0.00%
Field Study: Primary Data	44.00%	27.59%	31.58%
Field Study: Secondary Data	12.00%	13.79%	10.53%
Subtotal Quadrant I	**58.00%**	**44.83%**	**42.11%**
Experimental Simulation	2.00%	3.45%	1.32%
Laboratory Experiment	8.00%	27.59%	3.95%
Subtotal Quadrant II	**10.00%**	**31.04%**	**5.27%**
Judgment Task	2.00%		3.95%
Sample Survey	16.00%	3.45%	13.16%
Subtotal Quadrant III	**18.00%**	**3.45%**	**17.11%**
Formal Theory/ Literature Reviews	10.00%	20.69%	36.84%
Computer Simulation	4.00%		2.63%
Subtotal Quadrant IV	**16.00%**	**20.69%**	**39.47%**

that there were very few articles published that investigated the Roles and Skills construct. There is somewhat of a shortage of research in all constructs in Quadrant II to maximize precision of the measures.

Table 6 shows the research strategy measured against the level of the analysis. The percentages under each level are for research in that category only. In the Individual unit of analysis, the quantity of research using Quadrant I is much larger than the other quadrants. This phenomenon is also apparent in the Group level of analysis, although to a lesser degree. The balance in the Group level also suffers from a very small percentage of research using Quadrant III methodologies, which maximize generalizability. In contrast, at the Organization level of analysis, the gap is in Quadrant II, in which the percentage is much smaller than the all the other quadrants. However, in the Organization level of analysis, there is a much better balance between Quadrants I and IV.

Table 7 shows the research levels of the dependent construct vs. KM constructs. As might be expected, there is a large quantity of research at the organization level for Creation. Research in Storage and Retrieval appears to focus on the two extreme levels of the unit of analysis: the individual and the organization. Research in Transfer is the only construct that seems to have a reasonable balance of effort among the three levels of the unit of analysis. Research in Application shares a distribution similar to Creation in which most of the effort has been at the Organization level of the unit of analysis. Finally, a majority of research in Roles and Skills has been at the organization level of the unit of analysis.

Table 7. Level of analysis by construct

Construct / Level	Creation	Storage/ Retrieval	Transfer	Application	Roles/ Skills
Individual	22.12%	35.11%	40.06%	21.65%	0%
Group	11.06%	14.47%	26.97%	18.03%	34.11%
Organization	66.81%	50.42%	32.98%	60.32%	65.89%

Discussion

This research raised almost as many questions as it answered. Why is knowledge transfer the most heavily researched construct within knowledge management? Is knowledge transfer more interesting to researchers than the other constructs? Is it easier to operationalize and examine? One explanation may be that information technology supports the knowledge transfer construct, possibly more than other constructs; thus, many IS researchers choose to investigate the medium rather than the process of transfer. However, as noted earlier from both Alavi and Leidner (2001) and Davenport and Prusak (1998), knowledge transfer is more than just the channel, medium, or technology. In fact, it is estimated that much, if not most, knowledge transfer happens during informal communication (e.g., the water-cooler theory) and is separate from technology. Further, research suggests that tacit to explicit knowledge conversion is difficult, if not impossible, despite advances in research in communication technology (Nonaka & Takeuchi, 1995). Channel and media richness has been researched, as has collaborative systems, but less research has been done in the realm of learning behaviors. Given the results of this study, it appears that IS researchers are looking at knowledge management through a technical lens and, thus, that research is published more frequently in IS-oriented journals.

Another question that this study begets is whether the balance of the literature still would lean toward knowledge transfer if other journals that publish knowledge management articles but were not included in the scope of this research were included in the study. To answer that question, we examined articles in top management journals as ranked by Johnson and Podsakoff (1994). The journals identified by this ranking were *Administrative Science Quarterly, Research in Organizational Behavior, Academy of Management Review, Academy of Management Journal, Journal of Applied Psychology, Strategic Management Journal, Organizational Behavior and Human Decision Processes, Industrial and Labor Relations Review*, and *American Sociological Review*. Using the same constraints and keyword search as those used for IS journals, we found approximately 20 articles in these 9 journals that referenced knowledge management. Of those articles, approximately 75% addressed the issue of knowledge transfer. The May 2003 issue of *Organization Behavior and Human Decision Processes* was composed almost exclusively of articles that addressed knowledge transfer in one form or another; however, if we remove this issue from the sample, knowledge transfer is the topic of almost half of the remaining articles. Apparently, knowledge transfer also is the most commonly researched construct from a management viewpoint, albeit in a less technical view than within the IS community.

We believe that transfer as well as storage and retrieval remain topics of interest in the IS community because of their obvious tie with information technology. It is telling that academic journals publish articles on knowledge roles/skills least often, whereas practitioner journals published that construct as often as both creation and application (Peachey, Hall, & Cegielski, 2005). While both creation and application also have direct ties to information technology (e.g., data mining, discovery, expert systems), roles and skills are associated less directly. Nonetheless, information systems are constructed to support people; skills necessary to succeed with new technology and roles designed to shepherd the process are necessary and should be included in research done by the IS community.

When examining research methodologies used in the KM articles (see Table 4) in this sample, one may surmise that there has been an adequate amount of formal theory and frameworks published, and now, researchers are taking these propositions to the field to maximize realism of context. Quadrants II and III maximize precision and generalizability, respectively. It seems somewhat intuitive that as the discipline of KM moves through the spiral research process, Quadrants II and III would lag behind the efforts in Quadrants I and IV. This is not an entirely surprising result; when compared to research on policy, organizational theory, organizational behavior, and human resources, the distribution of research per quadrant is very similar (see Table 2) (Scandura & Williams, 2000). The similarity of the distributions of KM to policy and organizational theory is almost exact and only slightly different from the distributions of organizational behavior and human resources. A relatively new field would go through a phase in which theories from Quadrant IV are first tested in the field. As more field studies are completed, then researchers may move to improve the generalizability and precision of research within the discipline.

Table 5 shows the percentage of each methodology compared within each KM construct. Given the small number of articles examining the construct of Creation, it appears that the research is fairly well distributed across the quadrants, with the exception of Quadrant II, which maximizes precision of measurement. The constructs of Creation and Storage and Retrieval are both well represented in Quadrants I and IV. This could be somewhat problematic in that research in these two constructs currently is lacking in both precision and generalizability relative to the distributions of the other constructs.

Generally, the results of comparing the methodologies against the level of the unit of analysis (see Table 6) seem appropriate given the circular nature of research, although two areas are lacking. As when comparing the constructs, there are distinct weaknesses in both generalizability and precision, partly because of a relative shortage of work in Quadrants II and III. These quadrants showed a significantly lower level of activity than would be expected. Overall, this analysis shows a relatively good distribution of the research effort when triangulating on the unit of analysis.

When analyzing the KM constructs against the level of the unit of analysis (see Table 7), there are two significant shortages. Both creation and application suffer from a lack of work at both the individual and group levels. Organizational Storage and Retrieval of knowledge has been a dynamic topic for many years. The level of activity is fairly well-balanced between the individual and the organization but somewhat smaller at the group level. This actually seems relatively intuitive that the work in Storage and Retrieval at the group level is slightly lower than at the other two levels. One possible cause for this is that groups create and transfer knowledge but store it in individuals or at the organizational level, given that most groups are not ongoing entities and that the individual or the organization wants to retain the knowledge after the group project ends. Knowledge transfer is almost perfectly balanced across the three levels of the unit of analysis. One reason for this is a large quantity of research in Group Support Systems and similar topics that focus on transfer of knowledge among group members. This work balances the deficiencies apparent in both Creation and Storage and Retrieval. The Roles and Skills construct suffers from an exceedingly small sample size, and thus, we cannot draw too many conclusions from the numbers. One might wonder why there was no work at the individual level in this construct, given the recent attention to the role of the CKO. The most probable explanation for this lack is the practitioner focus of the work on the individual roles. Since *Harvard Business Review* and *Communications of the ACM* were omitted from this study, this result is not entirely surprising. There has been work published on the role of the CKO, but it has been published predominantly in practitioner journals that were outside the scope of this work.

Specific data on the samples used in these articles were collected for field studies, sample surveys, and laboratory experiments. Many of the field studies presented a mix of data gathering. Often, the study started with interviews and the richness of data they provide and then followed up with surveys to a larger population but still in the limits of the organization within the study. A survey of a specific population within an organization suffers from a somewhat lower generalizability than a pure sample survey. Combined with another data-gathering methodology within a study, however, it is usually quite beneficial to the research. A field study can take place in a small or large organization, with each study providing some rich data in a realistic context, although with a certain lack of generalizability. One concern from the sample survey is that the mean sample size is barely large enough to use advanced statistical techniques such as Structural Equation Modeling.

There are two other interesting trends in this data set. The first is the large number of frameworks published examining some part of KM. This study included 17 frameworks to examine different parts of KM. The majority (13) of these frameworks was developed as formal theory. The other four were developed through field studies, two with primary data and two with secondary data. From this, we can conclude that KM is either a very diverse topic that needs a wide variety of ways to examine and categorize its phenomena or that there is a lack of agreement on the key principles underlying KM.

Implications

This research has several interesting implications for the KM community, but three findings seem to be of particular interest. First, knowledge management research seems dominated by its focus on knowledge transfer. Second, the methodological triangulation is similar to that of more mature disciplines. Finally, while published research has uncovered many valuable relationships for both academics and practitioners, there is much left to uncover.

Knowledge transfer is the subject of almost twice the amount of research in the other constructs, as outlined by Alavi and Leidner (2001) and Davenport and Prusak (1998). Storage and Retrieval and Application show a similar level of activity, which is interesting in that a previous study showed application covered at a lower rate (Peachey et al., 2005). Reviewing the data showed a high level of research in Application in 2005, which was not included in that study. The majority of this activity was in *Decision Support Systems* and *MISQ* that published two special issues on KM in 2005. This research benefits practitioners by examining a facet of knowledge management that is related more directly to the bottom line than previous research in other constructs.

As a research topic, knowledge management is perhaps more mature than some researchers would believe. A comparison of the results of this study with Scandura and William's (2000) study of management topics show that the triangulation is similar. However, there is still a long way to go, given a topic that bridges technology, process, and people to the degree evident in knowledge management research. There are dramatic weaknesses in both generalizability and precision in this sample. As this research stream progresses, it is important to address these issues.

Practitioners have provided much valuable support to this research stream, given its applied nature. However, given the weaknesses in Quadrant III, which maximize context and generalizability, academic researchers need more access to the practitioner community. Without participation from a broad range of organizations, we are limited in the generalizability and precision of our research. Perhaps the research community must be more forthcoming in its dissemination of findings to practitioners. There is evidence that organizations could benefit from practitioner-oriented studies, particularly in the construct of Roles and Skills. There has been a trend toward more Application-oriented studies; this should continue, as it also provides a more direct benefit to organizations.

Conclusion

This study found that current research has investigated the construct of knowledge transfer more frequently than the other constructs in knowledge management, as developed from the Alavi and Leidner (2001) and Davenport and Prusak (1998) frameworks. While this currently is not a significant problem, it could be in the future. In order for knowledge management to continue to mature within the IS research discipline, we must present well-rounded yet diverse research to the rest of the IS community. If one of the constructs within knowledge management is researched extensively while the others are less developed, the topic as a whole will suffer from an imbalance. As a community of researchers, we know that knowledge management is more than just knowledge transfer. In order for practitioners to deploy effective knowledge management systems, the other constructs must be more fully developed. The Davenport and Prusak (1998) construct of Knowledge Roles and Skills adds a critical area to the topic of knowledge management. In order for knowledge management to be truly effective, few could argue against the need for appropriate supporting skills. Researchers should investigate more fully the other constructs that form the foundation of knowledge management.

This research raised a number of interesting questions about the research effort that is apparent in published works. As a community of researchers, we know that knowledge management is more than regeneration or integration of other more mature topics such as expert systems or decision support systems. Additional research in the other core constructs, while framed under the topic of knowledge management, will serve to enhance the understanding of the breadth of knowledge management and ensure that it remains a dynamic research topic in the future while simultaneously providing value to practitioners and organizations.

A list of the articles reviewed for classification may be obtained from the authors.

References

Alavi, M., & Leidner, D. E. (2001). Review: Knowledge management and knowledge management systems: Conceptual foundations and research issues. *MIS Quarterly, 25*(1), 107-136.

Bem, D. (1995). Writing a review article for psychological bulletin. *Psychological Bulletin, 118*(2), 172-177.

Davenport, T. H., & Prusak, L. (1998). *Working knowledge.* Boston: Harvard Business School Press.

Desouza, K. C. (2003). Strategic contributions of game rooms to knowledge management: Some preliminary insights. *Information & Management, 41*(4), 63-75.

Grant, R.M. (1996). Towards a knowledge based theory of the firm. *Strategic Management Journal, 17*, 109-122.

Griffith, T. L. Sawyer, J. E., & Neale, M. A. (2003). Virtualness and knowledge in teams: Managing the love triangle of organizations, individuals, and information technology. *MIS Quarterly, 27*(2), 265-288.

Gupta, A. K., & Govindarajan, V. (2000). Knowledge flows within multinational corporations. *Strategic Management Journal, 21*, 473-496.

Holsapple, C. W., & Joshi, K. D. (2001). Organizational knowledge resources. *Decision Support Systems, 31*(1), 39-55.

Jennex, M., & Croasdell, D. (2005). Editorial preface: Is knowledge management a discipline? *International Journal of Knowledge Management, 1*(1), i-v.

Johnson, J. L., & Podsakoff, P. M. (1994). Journal influence in the field of management: An analysis using Salancik's index in a dependency network. *Academy of Management Journal, 37*(5), 1392-1407.

Kaplan, B., & Duchon, D. (1988). Combining qualitative and quantitative methods in information systems research: A case study. *MIS Quarterly, 12*(4), 571-586.

McGrath, J. E. (1982). *Dilemmatics: The study of research choices and dilemmas.* Newbury Park, CA: Sage.

Mingers, J. (2001). Combining IS research methods: Towards a pluralist methodology. *Information Systems Research, 12*(3), 240-259.

Mylonopoulos, N. A., & Theoharakis, V. (2001). On-site: Global perceptions of IS journals. *Communications of the ACM, 44*(9), 29-33.

Nonaka, I. (1994). A dynamic theory of organizational knowledge creation. *Organization Science, 5*(1), 14-37.

Nonaka, I., & Takeuchi, H. (1995). *The knowledge creating company*. New York: Oxford University Press.

Palvia, P., Mao, E., Salam, A. F., & Soliman, K. S. (2003). Management information systems research: What's there in a methodology? *Communications of the Association for Information Systems*, *11*, 289-309.

Peachey, T., Hall, D., & Cegielski, C. (2005). Knowledge management and the leading information systems journals: An analysis of trends and gaps in published research. *International Journal of Knowledge Management*, *1*(3), 55-69.

Peffers, K., & Ya, T. (2003). Identifying and evaluating the universe of outlets for information systems research: Ranking the journals. *Journal of Information Technology Theory and Application*, *5*(1), 63-84.

Rubenstein-Montano, B., Liebowitz, J., Buchwalter, J., McCaw, D., Newman, B., Rebeck, K., et al. (2001). A systems thinking framework for knowledge management. *Decision Support Systems, 31*(1), 5-16.

Scandura, T. A., & Williams, E. A. (2000). Research methodology in management: Current practices, trends, and implications for future research. *Academy of Management Journal*, *43*(6), 1248-1264.

Stein, E. W., & Zwass, V. (1995). Actualizing organizational memory with information systems. *Information Systems Research*, *6*(2), 85-117.

Walsh, J. P., & Ungson, G. R. (1991). Organizational memory. *Academy of Management Review*, *16*(1), 57-91.

Webster, J., & Watson, R. T. (2002). Analyzing the past to prepare for the future: Writing a literature review. *MIS Quarterly*, *26*(2), xiii-xxiii.

Zhang, P. (2000). The Effects of animation on information seeking performance on the Web: Securing attention or interfering with primary tasks. *Journal of the Association for Information Systems*, *1*(1), 1-28.

Section II

Organizational Impacts of Knowledge Management

Chapter V

Linking Knowledge to Competitiveness:

Knowledge Chain Evidence and Extensions

Clyde Holsapple, University of Kentucky, USA

Kiku Jones, University of Tulsa, USA

Meenu Singh, Murray State University, USA

Abstract

Knowledge management (KM) initiatives are undertaken in order to improve organizational performance. The goal of such improvement is to make an organization more competitive in delivering value to its customers, employers, and stakeholders. However, without a plan that links KM activities to organizational performance, the time, effort, and money devoted to a KM initiative may yield little benefit. Thus, understanding this linkage is crucial to competitiveness of knowledge-based organizations. This chapter uses the knowledge chain model as the theoretical base for an empirical study of the linkage between KM activities and approaches to competitiveness. It finds that every one of the nine knowledge chain activities can be performed in ways that improve organizational competitiveness in any of four ways: enhanced productivity, agility, innovation, and reputation. Aside from offering empirical support for the knowledge chain model, the primary finding of this research is that each knowledge chain activity deserves to be considered as a possible means for implementing each of these four approaches to improving organization performance.

Introduction

Knowledge management (KM) rapidly is becoming an integral function for many organizations as they realize that competitiveness hinges on effective management of their knowledge resources (April, 2002; Carneiro, 2000; Grover & Davenport, 2001; Holsapple & Whinston, 1987). In an economy in which the only certainty is uncertainty, one source of lasting competitive advantage is knowledge and its manipulation (Nonaka, 1991). Researchers in the field of sustainable competitive advantage have discovered that knowledge, which includes what the organization knows, how it uses what it knows, and how fast it can know something new, is the only thing that offers an organization a competitive edge (Adams & Lamont, 2003; Prusak, 1996; Sharkie, 2003). Critical to our understanding of the value of knowledge management is its link to organizational performance outcomes (Kalling, 2003).

Contentions about competitiveness achieved through knowledge management (KM) are found to be in harmony with results of empirical studies conducted by organizations such as Delphi (1997), Hughes Space and Communications Company, and Ford Motor Company (Ward & Le, 1996). User success stories via knowledge management abound in the "KM in Practice" section of the *KMWorld* periodical. On the other hand, unmanaged organizational knowledge led to failures such as those concerned with risk management at Barings Bank, Kidder Peabody, and Metallgesellschaft (Marshall, Prusak, & Shpilberg, 1996), and insufficient knowledge management activity was a contributing factor to the 9/11 disaster (Spender, 2003).

All businesses involve creation, dissemination, renewal, and application of knowledge toward organizational sustenance and survival in the face of increasingly discontinuous environmental change (Malhotra, 1998, 2003). However, except for an assortment of success stories supporting the macro-level contentions that KM initiatives are undertaken for the purpose of achieving better organizational efficiency and effectiveness, there has been little investigation of the connections between KM and competitiveness. There is a dearth of formal research in this area (Grover & Davenport, 2001). Therefore, in this regard, an interesting question is, What are the specific KM activities that can be contributors for achieving superior organizational performance? One answer to this question is furnished by the knowledge chain model, which identifies and characterizes nine key KM activities Holsapple & Singh, 2000a). This model asserts that these nine key KM activities deserve careful attention in an organization's quest to leverage its knowledge into a competitive advantage.

The knowledge chain (KC) model is grounded in a descriptive KM ontology that was developed collaboratively via a Delphi study involving an international panel of KM experts (Holsapple & Joshi, 2002a, 2004; Joshi, 1998). It is somewhat analogous to Porter's value chain model. The KC model posits nine distinct, generic activities that an organization performs in the course of managing its knowledge resources. It contends that these are focal points for achieving competitiveness through knowledge management in the sense that an organization can perform one or more of them better than competitors can in order to achieve a competitive edge. Anecdotal evidence has been reported that illustrates the direct role of each of the nine KC activities in adding value to an organization and in increasing its competitiveness (Holsapple & Singh, 2000b).

This chapter takes a step beyond the anecdotal evidence, further substantiating the knowledge chain model via an empirical investigation of the relationship between each KM activity in the KC model and the organizational performance achieved through four approaches to competitiveness: productivity, agility, innovation, and reputation (collectively referred to as the PAIR approaches). The study uses the perceptions of chief knowledge officers (CKOs) and other leaders of KM initiatives. The results provide evidence that every KM activity in the KC model can provide enhanced organizational performance via each of the four competitive approaches. In addition, this chapter provides a description of a recent extension to the knowledge chain model that uncovers more than 60 distinct types of activities found within the nine primary and secondary activities of the knowledge chain model.

The primary implication of these results is that the knowledge chain model furnishes an empirically supported structure for KM research and practice. Researchers can use it to frame their designs of investigations into KM activity and KM outcomes. Practitioners can use the KC Model to systematically assess their own practices with respect to each of the nine activities; to identify where there are deficiencies or opportunities in their organizations' conduct of KM; and to better understand KM linkages to organizational performance in directions of improved productivity, agility, innovation, or reputation. Educators can use it to structure KM coverage.

The rest of the chapter is organized into five sections. The first section reviews relevant background research literature, including an overview of the knowledge chain model. The second section discusses the research methodology employed in this study, and the third section presents the results and findings. The fourth section discusses the extended knowledge chain model. Finally, the concluding section points out strengths and limitations of this study, along with directions for future research.

Background

In the knowledge economy, the value of knowledge as input and output is growing, knowledge is a key ingredient of what is bought and sold (both explicitly and implicitly), knowledge resources are rising in importance relative to traditionally recognized resources, and new technologies and techniques for managing knowledge resources are emerging (Stewart, 1998). Knowledge management aims to ensure that the right knowledge is available in the right representation to the right processors (humans or machines) at the right time for the right cost. Execution of KM activities undertaken in pursuit of this objective involves a panorama of knowledge flows and processing within a knowledge-based organization. In many cases, the manipulation activities and the flows that connect them can be performed, enabled, or facilitated electronically.

We contend that one key to more fully exploiting the competitive potential of knowledge management is a model that identifies value-adding KM activities. Practitioners could use the model to structure their consideration and evaluation of KM initiatives. Researchers could use the model to structure their exploration of connections between KM and competitiveness. Educators and students could use the model to help structure coverage of KM activities and their impacts. These motivations, coupled with the absence of such a model in the literature, lead us to investigate further the knowledge chain model.

The Knowledge Chain Model

The knowledge chain model is based on a KM ontology developed via a Delphi study involving an international panel of prominent KM practitioners and academicians (Holsapple & Joshi, 2002a; Joshi, 1998). This ontology identifies five major knowledge manipulation activities (acquisition, selection, generation, assimilation, emission) that occur in various patterns within KM episodes. (In this chapter, we use the terms *assimilation* and *emission* as synonymous substitutes for Joshi's original terminology *internalization* and *externalization* in order to avoid possible confusion with Nonaka's terms.) The ontology also identifies four major managerial influences (leadership, coordination, control, measurement) on the conduct of knowledge management. Respectively, these form the five primary and four secondary KM activities in the knowledge chain model (Holsapple & Singh, 2000a).

As Figure 1 suggests, these activities lead to four organizational performance implications: productivity, agility, innovation, and reputation. Referred to as the PAIR approaches to competitiveness, these four are discussed in greater detail later in this section. Prior anecdotal analysis indicates that the model's set of nine interrelated activities appears to be common across diverse organizations (Holsapple & Singh, 2001). The KC model contends that these are the major KM activities with which a chief knowledge officer needs to be concerned. KM skills of an organization's participants need to be cultivated, harnessed, and organized in the performance of these activities. KM technologies need to be identified and adopted in support of these activities. The KC model theorizes that the specific ways in which the

Figure 1. Primary and secondary knowledge management activities in the knowledge chain

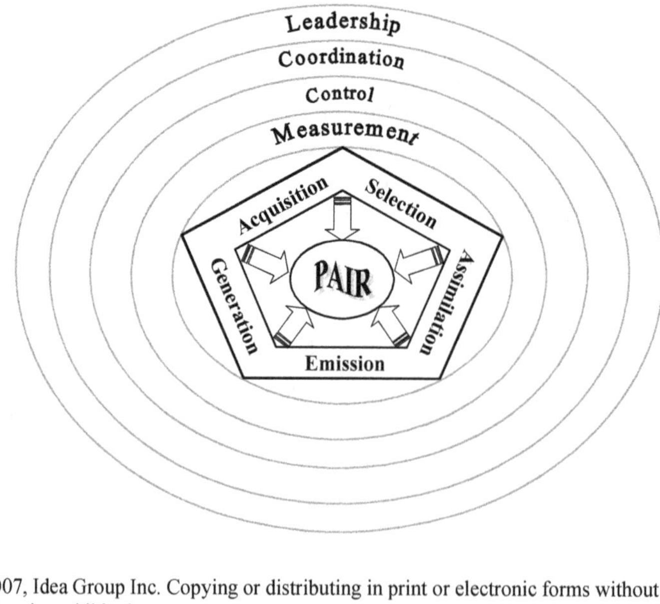

nine KM activities are implemented in using an organization's knowledge resources can lead to enhanced competitiveness through the PAIR approaches.

The primary activities that an organization's knowledge processors perform in manipulating knowledge resources are summarized in Table 1. One or more processors operate when a particular instance of one of these knowledge manipulation activities occurs in an organization. Some processors are human, and others may be computer-based. Many processors may be capable of performing a given type of primary activity. On the other hand, a given processor may perform multiple types of primary activities. Moreover, each primary activity involves subactivities (Holsapple & Joshi, 2002b).

An organization may possess the best knowledge resources and the best knowledge manipulation skills, but they are of no use until they are applied effectively during the conduct of KM (Holsapple & Joshi, 2000; Joshi, 1998). The Delphi study identified three major kinds of forces that conspire to influence how the conduct of KM ultimately unfolds in an organization: managerial influences, resource influences, and environmental influences (Joshi, 1998). Because the managerial influences denote meta activities that impact or determine the deployment of resources and patterns of manipulation activities, they are included as secondary activities in the knowledge chain model. Table 2 summarizes these secondary KM activities that support and guide the performance of primary KM activities. There is also interaction among the secondary activities; one may support or guide the performance of another.

Table 1. Primary activities in the knowledge chain (Adapted from Holsapple & Singh, 2000b)

Knowledge Acquisition	Acquiring knowledge from external sources and making it suitable for subsequent use.
Knowledge Selection	Selecting needed knowledge from internal sources and making it suitable for subsequent use.
Knowledge Generation	Producing knowledge by either discovery or derivation from existing knowledge.
Knowledge Assimilation	Altering the state of an organization's knowledge resources by distributing and storing acquired, selected, or generated knowledge.
Knowledge Emission	Embedding knowledge into organizational outputs for release into the environment.

Table 2. Secondary activities in the knowledge chain (Adapted from Holsapple & Singh, 2000b)

Knowledge Leadership	Establishing conditions that enable and facilitate fruitful conduct of KM.
Knowledge Coordination	Managing dependencies among KM activities to ensure that proper processes and resources are brought to bear adequately at appropriate times.
Knowledge Control	Ensuring that needed knowledge processors and resources are available in sufficient quality and quality, subject to security requirements.
Knowledge Measurement	Assessing values of knowledge resources, knowledge processors, and their deployment.

Anecdotal Support for the Knowledge Chain

As management scholars have pointed out, espoused theory tells us little about real behavior; we need to study theory in practice (i.e., view the actions that reflect managerial conduct) (Leonard-Barton, 1995). In this spirit, anecdotal evidence has been collected from the literature that illustrates each KC activity's role in adding value to an organization in order to increase its competitiveness (Holsapple & Singh, 2001). Some of the evidence involves uses of technology to better perform a KM activity; some are not technology-based.

Competitiveness due to KM practices can manifest itself in such ways as increasing profits and bolstering an organization's reputation, employees' creativity, productivity, efficiency, flexibility, and innovation (Ward & Le, 1996). Analysis of the anecdotal evidence gathered from the literature has revealed that KC activities can yield four important approaches to high performance: productivity, agility, innovation, and reputation (Holsapple & Singh, 2001). Therefore, we examine the competitive role of each KM activity in the knowledge chain in terms of one or more of these PAIR approaches.

Four Specific Approaches to Competitiveness

Productivity. Productivity is the rate at which goods and services are produced per unit cost. Although it commonly is defined in terms of labor, it also can be seen as the value people contribute to business processes (Delio, 2000). Productivity is a major concern for any organization, because it deeply influences the well being of the organization and its members (Turban, McLean, & Wetherbe, 1999). At the root of organizations' need to be more competitive, to perform better, is their need for improved productivity (Markland, Vickery, & Davis, 1995).

Companies always have looked for ways to improve productivity by reducing the time and costs of product development (Fruin, 1997), and KM is increasingly seen as a potential contributor to productivity (Wiig & Jooste, 2003). Cisco Systems, for example, uses knowledge management techniques effectively to improve productivity (Sherman, 2000). The company has opened Internet and intranet portals for its customers, suppliers, and employees. By doing so, it has expanded access to its intellectual capital, increased operational efficiencies, and reduced its costs. Cisco's innovative knowledge practices to improve productivity foster a fiercely competitive, highly profitable organization.

Agility. In an economy that is becoming increasingly driven by new paradigm shifts, it is no longer possible for an organization to seek out a static position with its product and process structures. The organization routinely must be repositioned in the competitive niches that it occupies, adjusting its organizational structures, managerial practices, product and service offerings, business processes, personnel, technologies, and marketing strategies to a dynamic marketplace. Due to the increasing dynamism of the global market, the competitive advantage provided by agility has emerged as an important priority. The ability to react rapidly to demand variability is becoming so critical in today's environment that it outshines all other competitive weapons (Fliedner & Vokurka, 1997).

Rick Dove, Director of Strategic Analysis at the Agile Manufacturing Enterprise Forum, defines agility as "the ability to thrive in an environment of unpredictable and constant change" (Stewart & Pinholster, 1994). To compete effectively in many developed and emerging global markets, organizations must develop the capacity to react quickly and successfully to change—they must become agile (McGaughey, 1999). An agile enterprise must be appropriately organized to thrive on change and uncertainty (Devor, Graves, & Mills, 1997). An agile organization has the ability to respond quickly to competitive threats and market opportunities. As any part of the firm's environment changes, it must be sufficiently flexible to reorganize its human systems and technical systems not just to adapt to change but also to take advantage of change.

Being agile means being proficient at change and allows an organization to do anything it wants to do whenever it wants to do it (Dove, 1994). This proficiency is, at least in part, a function of an organization's conduct of knowledge management (Dove, 2003). Agility challenges the strategic priorities of the 1980s through the ability to respond rapidly to any change in market demand, whether it is a change in product characteristics, customer orders, or internal company conditions (Fliedner & Vokurka, 1997). It is interesting to note that the difference between agility and flexibility is whether or not the change in market demand has been predicted. Flexibility refers to the capability of changing rapidly from one task to another when changing conditions are defined ahead of time; agility provides the firm with the ability to respond quickly to unanticipated marketplace changes (Fliedner & Vokurka, 1997).

Innovation. Some organizations constantly create new and better products, services, and processes, delighting their customers and jolting their competitors. By breaking the rules of the game and thinking of new ways to compete, an organization strategically can redefine its business and catch its bigger competitors offguard; and the trick is not to play the game better than the competition but to develop and play an altogether different game (Markides, 1997). Businesses face many strategic challenges in the 21st century, such as accelerating rates of change, increasing competition, rapidly advancing technology, a more diverse workforce, and a change from an industrial to a knowledge-based economy. These conditions have led researchers and practitioners alike to call for more creativity in management practices, products, services, and production processes (Lengnick-Hall, 1992). Creativity produces innovation, but the organization's culture also must foster creativity and then turn it into the innovation that leads to competitive advantage (Higgins, 1995).

Innovation is the means by which organizations exploit change as an opportunity for a different business or a different service, and it is capable of being learned and practiced (Drucker, 1986). The innovation process is defined as bringing ideas to market (Amidon, 1997). Now and in the future, more than at anytime in history, the secret to competitive advantage is innovation (Higgins, 1995; Trudel, 1998). For any industry, competitive advantage achieved through innovation often proves to be so profound that entire markets are disrupted and destabilized. Innovation can lead to quantum leaps in organizational performance.

Knowledge can spur and drive innovation (Adams & Lamont, 2003; Amidon & Mahdjoubi, 2003; Smith, 1998). Wayne Tomes of Delphi Consulting Group says, "The single differentiator that is likely to last is innovation, and the raw material of innovation is knowledge" (Evans & Wurster, 1997). Carneiro (2000) points out the importance of knowledge development and the role of knowledge management in order to assure competitiveness. Nonaka and Takeuchi (1995) claim that the use of knowledge is the primary reason Japanese companies foster creativity and innovation for competitive advantage. By triggering insights and new approaches and by leveraging experiences and hard-earned lessons, knowledge management is all about the pursuit of that most valuable of capabilities in today's frenetic business world: innovation. In a top-line finding of an Ernst & Young survey (1997), executives see innovation as the greatest payoff from knowledge management, even though KM efforts have so far concentrated on achieving productivity gains.

Reputation. Jeffrey Brown from Opinion Research Corporation states, "Think of your reputation as a reservoir of goodwill. You can only go to the well so often before it dries up. Protect your reputation whenever you can" (Garone, 1998, p.19). *Webster's Dictionary* defines r*eputation* as "a place in public esteem or regard." Reputation derives from the interpretation by a public of a particular set of knowledge cues emanating from an organization; thus, an organization's conduct of KM activity impacts its reputation.

Microsoft, Intel, Sony, Dell, and Lucent have the best reputations among companies in the digital economy (TechWeb News, 1999). Respondents rated these companies on emotional appeal, quality of products, durability, friendly service, speed of service, financial performance, vision and leadership, workplace environment, and social responsibility. Few business owners would disagree that one of their most valuable assets is the reputation of their business, because a stellar reputation builds a competitive advantage (Kartalia, 2000). For knowledge-based organizations, competitive advantage and profits are generated through the successful management of intangible assets such as reputation (Sveiby, 1997).

For many organizations, favorable corporate reputation has become the key advantage that sets their organizations apart from the competition and motivates stakeholder decisions (Perrin, 2000). Reputation goes beyond logo, name, and packaging design. It emanates from the fabric of an organization: corporate values and practices, corporate brand, product brand, financial and nonfinancial performance, and stakeholder relationships. Two elements paramount to corporate reputation are the corporate brand and the product brand (Garone, 1998). Differentiating a brand without losing its purpose and reason is tricky, but it is necessary for survival. Companies that adopt efficient differentiation of their brands, products, and services often gain competitive advantage over their rivals (MacMillan & McGrath, 1997).

Managing reputation is becoming increasingly important for three reasons, according to Charles Fombrun, professor of management at New York University's Stern School of Business (Garone, 1998):

- Increasing commodization, which restricts a company's ability to differentiate its goods.
- Increasing globalization, which furthers commodization across borders with overlapping goods.
- Increasing information, which overloads people with hard data, leaving them hungry for the soft, intangible aspects.

For these reasons, reputation is becoming central in the language of competition and strategy. Fombrun also asserts that a strong reputation can help an organization to:

- Attract resources (e.g., new employees, customers, and investors),
- Improve employee loyalty and morale,
- Secure customer retention and loyalty,
- Increase sales and income, and
- Develop a competitive advantage.

Research Study

Anecdotal evidence can be a very useful part of the methodology by which we come to understand new uncharted areas. Although such evidence can be a strong indication of KM initiatives undertaken for the purpose of achieving better organizational performance, it can be complemented usefully by a survey that studies perceptions of KM leaders toward the connection between each KM activity and organizational performance. Accordingly, we conducted a study to ascertain answers to the following questions with regard to each of the nine knowledge chain activities: Does the activity contribute to a competitive advantage by:

1. Improving productivity (e.g., lower cost, greater speed)?
2. Enhancing reputation (e.g., better quality, dependability, brand differentiation)?
3. Enhancing organizational agility (e.g., more alertness, rapid response ability, greater flexibility and adaptability)?
4. Fostering innovation (e.g., inventing new products, services, processes)?

Essentially, we are exploring whether each knowledge chain activity exists as a lever for improving firm performance (i.e., competitiveness) and, if so, via which of the PAIR approaches. The exploration involves collecting and analyzing perceptions from experienced leaders of KM initiatives.

Instrument Construction

The methodology chosen is a field survey involving one instrument. The instrument begins with brief instructions to respondents. Next, a two-page quick reference guide is provided, including an overview of the knowledge chain model and brief descriptions of its activities. The main body of the instrument is structured in nine similar sections, one for each of the nine knowledge chain activities. At the beginning of each section, a description of the nature of its knowledge chain activity (and its subactivities, if applicable) is given. The questions about relationships of this activity to competitiveness follow. For each activity, the questions shown in the previous section are answered on a seven-point Likert-type scale. The evaluative characterizations for the scale responses are not at all (1), moderately (4), and extensively (7). Characterizations of the remaining scale levels (i.e., 2, 3, 5, 6) are not defined explicitly on the questionnaire. The survey concludes with general demographic questions. A cover letter and survey instrument were mailed to each of the potential respondents in the sample. To encourage response, each respondent was promised a report of the findings. Confidentiality of responses was guaranteed.

Pilot Testing

For the validation of the survey items, the survey instrument was pilot tested with two academics and two experienced practitioners in the KM area. All pilot-test candidates were sent a cover letter inviting them to participate in pilot testing, a model cover letter to be used during the survey, and the questionnaire. In the cover letter, each tester was asked to:

- Complete the initial version of the questionnaire and to critique it with regard to overall style, clarity, and organization. A structured comment sheet was enclosed for convenience.
- Provide us with their permission to identify them as "pretesters" in the cover letter to be sent during the actual survey.

The cover letter also promised to share results of this study with the pilot testers. All testers agreed to be identified in the cover letter used in the actual survey. Their comments and suggestions were used to refine, extend, and correct the questionnaire as needed.

Identification of Potential Respondents

In this study, we are interested solely in organizations with ongoing knowledge management initiatives. We set the qualification of a potential survey candidate to be someone in the leadership position for a KM initiative within an organization. Candidates should hold positions such as CKO (chief knowledge officer), CLO (chief learning officer), director of KM, knowledge officer, vice president of intellectual capital, knowledge leader, or positions that report to these. Based on these criteria, we searched a variety of sources to get a mailing list for leaders of knowledge management initiatives. In this regard, companies that publish directories of top business executives such as Leadership Directories, Inc., Applied Computer Research, Demand Research, D&B Million Dollar Database, CIO Communications, Inc., Corporate 1000, Inc., NorthernLight.Com, and so forth were contacted. Unfortunately, none of the contacted companies could provide a ready list of people experienced in leading KM initiatives. Moreover, efforts at obtaining such a list from trade publications in the KM field (e.g., *Knowledge Management, KMWorld*) also proved fruitless.

This unsuccessful search for a mailing list compelled us to take what might be called a brute-force strategy to identify potential candidates. We searched for and accumulated names of people reported to be in leadership positions in KM initiatives from such sources as trade publication articles, Web searches, keynote speakers, and participants from KM conferences. Care was taken to check their biographies and profiles to confirm their KM backgrounds. From among this tentative list of candidates, we selected only those for whom contact information could be found. One hundred and two potential respondents thereby were identified. The survey instrument was sent to these 102 candidates. To improve the response rate, reminders were issued a couple of weeks after the due date to respondents who failed to return their surveys on the designated date stated in the cover letter. After five months elapsed, 32 had responded, for a response rate of 31.4%.

Profile of the Research Sample

Demographic information gathered from the respondents indicates diversity in job titles, type of industry, size of organization, geographic region, and participants' background and experience. The distribution of respondents' job titles showed that 51% are CKO or CKO types (e.g., director of KM, knowledge officer, vice president of intellectual capital, knowledge leader). Consultants comprise 31%. Given the relative novelty of formal KM for many organizations, it is common for them to seek leadership for their initiatives from consultants. Each consultant who participated was asked to respond on the basis of his or her experience in the organization in which the most competitive advantage was realized from KM. In some cases, this turned out to be his or her own organization, and his or her KM initiatives used proprietary systems. Other titles make up the remaining 18%. An interesting array of titles is found in this category: IT architect, CEO, chairman, president, chief innovation officer, director of people strategy, among others. The bulk of respondents' KM initiatives occurred in North America (85%), with 9% in Europe, and 6% elsewhere.

Table 3 displays the variety of fields that describe respondents' backgrounds and training in approaching KM. Respondents report using a combination of fields for approaching KM in

Table 3. Fields describing respondents' background, training, and experience

Field	Frequency
Information Systems	50%
Others	38%
Communication	28%
Marketing	25%
Philosophy	22%
Human Resources	19%
Library Science	13%
Service Operations	9%
Sociology	9%
Accounting, Finance, Administration	9%

Table 4. Experience in KM practice

KM Experience (Years)	Frequency
Less Than 2	25%
2 - 4	22%
5 - 10	31%
Greater than 10	22%

their organizations. Half of the respondents approach KM from a background and training in information systems. In the *Others* category, diverse backgrounds and training, such as law, artificial intelligence, systems engineering, electronics engineering, political science, and systems science, are reported. Accordingly, about 66% of the participants report that they approach KM from a technical angle. Respondents' backgrounds include considerable doses of nontechnical fields as well, totaling in excess of 100%.

The demographics also are indicative of participants' interests, experience, and involvement in the KM field. Experiences of respondents in KM practice ranges from one to 40 years. The mean is seven years, and the median is five years. The distribution of experience levels is shown in Table 4.

Analysis Approach

For expository convenience, we define some qualifiers that are used in the ensuing data analysis and discussion. As mentioned previously, when eliciting respondent perceptions of the degree to which a knowledge chain activity was performed in a way that yielded competitive advantage, the survey instrument used a Likert scale of 1 through 7. The degrees of contribution to competitive advantage are represented by 1 being *not at all*, 4 being *moderately*, and 7 being *extensively*. For analyzing and explaining the empirical evidence, we aggregate the degrees of contribution of each KM activity into three categories, as shown in Table 5: Weak, Moderate, and Strong, with the latter two being referred to as substantial.

For each primary and secondary activity in Tables 1 and 2, we are seeking evidence that it deserves to be included in the knowledge chain model. In order to be included, it must exist as an activity that can be performed in ways that enhance an organization's performance sufficiently to yield a competitive advantage for the organization. Our analysis reports means and standard deviations for each of the knowledge chain activities on each of the PAIR dimensions, along with relative frequency distributions.

Table 5. Degree of contribution for each KM activity to competitiveness

	Degree of Contribution		
	Not Substantial	Substantial	
	Weak	Moderate	Strong
Likert Scale	1 to 3	4	5 to 7

A mean in the strong category tends to support the contention that the activity deserves inclusion in the KC model (i.e., on average, respondents recognize that the way this activity is performed in their KM initiatives makes it a strong contributor to competitiveness along the respective PAIR dimension). On the other hand, if an activity has means in the weak range for all four PAIR approaches to competitiveness, then we should be cautious about whether that activity deserves inclusion in the KC model (i.e., on average, respondents perceive that the way it is performed in their KM initiatives does not make it a strong contributor to competitiveness along any of the PAIR dimensions). However, the activity cannot be dismissed on this basis alone. A weak mean very well could conceal a situation in which a relatively small yet appreciable portion of respondents have experienced that the way the activity is performed has strong positive impacts on competitiveness, even though most respondents have not found or exercised ways to perform the activity that give them competitive edges. Such would tend to be the case in which there is a relatively large standard deviation and can be readily seen by examining the relative frequency distribution of responses.

In considering the relative frequency distribution for a given activity with respect to a particular PAIR dimension, there is the issue of how to interpret the observed pattern. Does it provide evidence that an appreciable portion of respondents have realized strongly positive performance on the PAIR dimension by virtue of the way in which that activity was implemented? If so, then existence of this activity in the knowledge chain is supported. The threshold for an appreciable portion is not definitive. It could be that a 5% or 10% threshold is more than sufficient in order to be confident about the existence of a phenomenon, as in the case of various rare diseases. However, because this study is dealing with perceptions (that can be faulty), we contend that a relatively high threshold should be set in order to be confident about the existence of cases in which the way an activity is performed has a strong positive impact on performance of a PAIR dimension. In the analysis of results, we adopt a fairly high 25% threshold for existence. However, failure to meet this threshold does not necessarily imply the absence of potentially strong connections between performing the activity and its competitive impact. It just may be rare. Indeed, it could reflect a major opportunity to be one of the few who has discovered ways to perform the activity so that a strongly positive competitive impact is realized.

Results and Discussion

Performance Implications for Primary Knowledge Management Activities

For each primary activity, Table 6 shows means and standard deviations for the degrees of contribution to competitiveness along each PAIR dimension (i.e., productivity, agility, innovation, reputation). Figure 2 graphically depicts these contributions in terms of the relative frequencies for the weak, moderate, and strong contribution categories. The percentages of "strong contribution" responses are shown in Table 7. Ranges of strong contribution percentages for the four PAIR approaches to competitiveness are as follows: Productivity (37.0%-58.6%), Reputation (36.0%-65.5%), Agility (31.0%-65.5%), and Innovation

(42.9%-67.9%). In all cases, the 25% threshold is surpassed. Each of the primary activities even passes a 50% threshold for at least one of the PAIR dimensions. We conclude that there can be no question about the existence and inclusion of the five primary activities in the knowledge chain model. This result is wholly consistent with previously cited anecdotal support for the model's primary activities.

Table 6. Means (standard deviations) of contributions made by primary activities along each pair dimension

Primary Activity	Contribution to Productivity	Contribution to Agility	Contribution to Innovation	Contribution to Reputation
Acquisition	4.47 (1.3)	4.17 (1.6)	4.67 (1.8)	4.80 (1.6)
Selection	4.33 (1.8)	4.33 (1.7)	4.33 (1.9)	4.47 (1.9)
Generation	4.72 (1.4)	4.90 (1.4)	4.57 (1.7)	4.86 (1.6)
Assimilation	3.85 (1.8)	3.83 (1.8)	4.15 (1.9)	3.72 (2.0)
Emission	3.89 (1.8)	3.54 (1.4)	3.50 (1.8)	4.14 (1.9)

Figure 2. Degree of contribution by each primary activity to PAIR

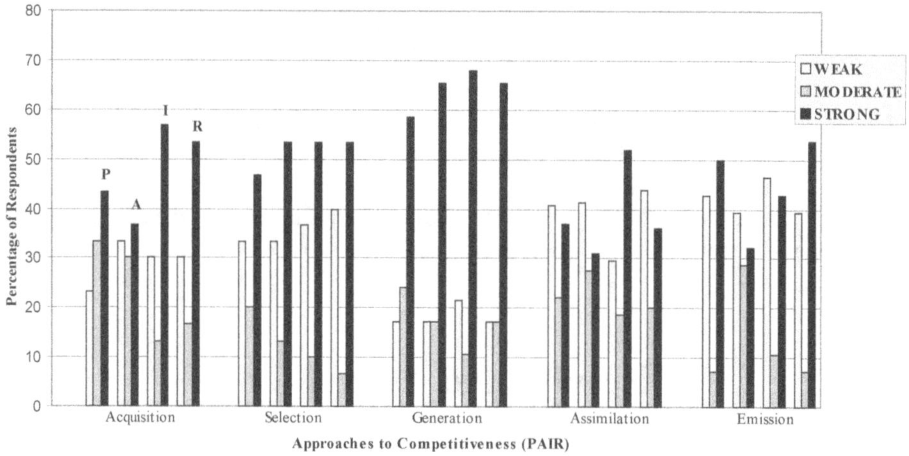

Table 7. Percentages of "strong contribution" responses for each primary activity

Primary Activity	Productivity	Agility	Innovation	Reputation
Acquisition	43.3	36.7	56.7	53.3
Selection	46.7	53.3	53.3	53.3
Generation	58.6	65.5	67.9	65.5
Assimilation	37.0	31.0	51.9	36.0
Emission	50.0	32.1	42.9	53.6

Looking at knowledge acquisition, the highest strong performance implication is for innovation. Acquisition brings fresh, stimulating ideas into the organization from the external environment, and the innovation process has been defined as "bringing ideas to market" (Amidon, 1997). For selection, there is little difference among the four approaches. Interestingly, Generation has the highest strong percentages for all the four approaches among all the primary activities in terms of current practices: productivity (58.6%), reputation (65.5%), agility (65.5%), and innovation (67.9%). The way in which knowledge generation contributes strongly to competitive advantage is not so much by being productive as by allowing the organization to be more agile and innovative. This, in turn, fosters reputation (Higgins, 1995).

Assimilation and acquisition operate in tandem in that both have the highest percentage for fostering innovation. This is understandable because, while acquisition acquires fresh new ideas from the outside, Assimilation exposes participants to existing in-house knowledge, thus rendering provocative and stimulating effects that consequently promote innovation. As for Emission, competitive advantage is achieved more through productivity and reputation than through agility and innovation. Knowledge emission results in projections. When an organization transfers an output (e.g., in the form of products, services, and knowledge artifacts), it is projecting. The process of effective projection adds value to an organization. The empirical evidence suggests that the added value comes more strongly in the forms of productivity and reputation and less so with agility and innovation. The justification may be that, in many cases, agility is something that happens prior to emission. Similarly, in many organizations, learning (i.e., innovation) happens prior to emission. Therefore, it is understandable that agility and innovation take a back seat to productivity and reputation. Even so, the evidence confirms that emission can lead to competitive advantage via agility and innovation.

If an organization is planning a competitive strategy that identifies agility, for instance, as a focal point for improving performance, then the question is how to implement this strategy. In knowledge management terms, Table 7 shows that knowledge generation is the activity most commonly performed in ways that strongly lead to competitive advantage through agility. ANOVA results, testing for equality of means across the five activities for the agility dimension, show that the differences among these means are statistically significant (p-value of .02). Although ways (i.e., methodologies, technologies) for performing knowledge generation are most widely associated with superior agility, devising ways to perform any of the other four primary activities should not be overlooked in strategic planning for agility. As Figure 2 illustrates, achieving better agility by particular practices for any of these other activities is not at uncommon.

Similar ANOVA tests for productivity, innovation, and reputation do not reveal statistically significant ($\alpha \leq 0.10$) differences among the means of the five activities, although the p-values for reputation and innovation are .11 and 0.12, respectively. This suggests that it would be prudent to consider all of the model's five primary knowledge management activities in designing competitive strategies that focus on a productivity, innovation, or reputation dimension of performance. For each of these activities, the consideration seeks to find methods and/or technologies that are sufficiently distinct from and superior to those of competitors in terms of positive effects on the selected PAIR dimension(s).

ANOVA also is used to check for statistically significant differences among the PAIR means for each of the five primary activities. For instance, is knowledge acquisition performed in

Table 8. Means (standard deviations) of contributions made bysecondary activities via each PAIR dimension

Primary Activity	Contribution to Productivity	Contribution to Agility	Contribution to Innovation	Contribution to Reputation
Leadership	4.39 (1.5)	4.32 (1.6)	4.43 (1.7)	4.68 (1.7)
Coordination	3.74 (1.6)	3.93 (1.7)	3.93 (1.8)	3.59 (1.8)
Control	3.52 (1.4)	3.52 (1.4)	3.48 (1.5)	3.69 (1.6)
Measurement	2.75 (1.7)	2.93 (1.8)	2.86 (1.9)	2.68 (1.7)

ways associated with major differences in the mean competitiveness realized for productivity vs. agility vs. innovation vs. reputation? The ANOVA answer is no; we do not observe statistically significant ($\alpha \leq 0.10$) difference in the PAIR means for knowledge acquisition. The same result holds for the four other primary activities of the knowledge chain model. This result suggests that, in general, focusing on any particular primary activity as a means toward improved competitiveness neither favors nor excludes any of the PAIR avenues toward that end. However, in the context of a particular firm's situation (i.e., resources and environing conditions), it is possible that practices (current or potential) for some primary activities may align more readily with one or another of the PAIR dimensions.

Performance Implications for Secondary Knowledge Management Activities

For each secondary activity, Table 8 shows means and standard deviations for the degrees of contribution to competitiveness along each PAIR dimension (i.e., productivity, agility, innovation, reputation). Figure 3 graphically depicts these contributions in terms of the rela-

Figure 3. Degree of contribution by each secondary activity to PAIR

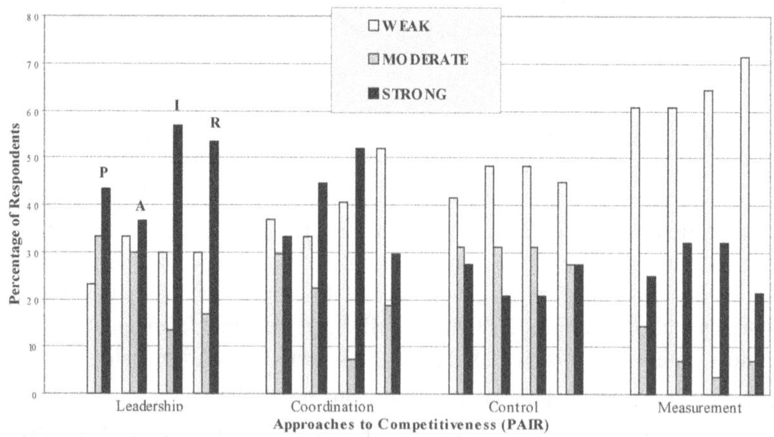

Table 9. Percentage of "strong contribution" responses for each secondary activity

Primary Activity	Productivity	Agility	Innovation	Reputation
Leadership	50.0	42.9	46.4	42.9
Coordination	33.3	44.4	51.9	29.6
Control	27.6	20.7	20.7	27.6
Measurement	25.0	32.1	32.1	21.4

tive frequencies for the weak, moderate, and strong contribution categories. The percentages of "strong contribution" responses are shown in Table 9. Ranges of percentages for strong contributions for the four approaches are as follows: productivity (25.0%-50.0%), reputation (21.4%-42.9%), agility (20.7%-44.4%), and innovation (20.7%-51.9%). Observe that each of the four secondary activities garners a "strong contribution" of more than 40% for at least one of the PAIR approaches, thereby surpassing the 25% threshold. In all cases, at least 20% of respondents recognized that the secondary activity is executed in ways that yield strong contributions to competitiveness for any of the PAIR approaches. We conclude that there can be no question about the existence and inclusion of the five secondary activities in the knowledge chain model. This result is wholly consistent with previously cited anecdotal support for the model's secondary activities.

Looking at leadership, the most frequent strong approach to achieving competitive advantage comes from improving productivity at 50%. This suggests that it is fairly common to find ways to accomplish knowledge leadership so that organizational knowledge resources and associated knowledge manipulation skills are deployed to lower costs and to reduce cycle times in respondents' organizations. CEOs come to realize that they have to manage their organization's intellectual assets the same way in which they manage physical assets. This means finding, understanding, and reusing best practices for bringing products to market, cutting cycle time, improving defect analysis, and boosting customer service better than their competitors. Anecdotes are found in literature in which good leadership improves employees' morale and helps to shape a coherent positive culture in organizations. This, in turn, could foster employees' creativity and innovation capabilities.

Strong contribution by knowledge leadership to enhanced performance via innovation is a close second at 46.4%. The other two approaches, reputation and agility, are not far behind in frequency of strong contribution at about 43%. An ANOVA test for differences in leadership's PAIR means shows no statistically significant difference ($\alpha \leq 0.10$) among them, generally suggesting that focusing on leadership practices as a way toward improved competitiveness neither favors nor excludes any of the PAIR avenues. However, in the context of a particular firm's situation (i.e., resources and environing conditions), it is possible that knowledge leadership practices (current or potential) may align more readily with one or another of the PAIR dimensions.

As for the knowledge coordination activity, the strongest performance implication is in the direction of innovation (at 51.9%). Coordination techniques suggested and used include linking incentives to desired KM behaviors and outcomes. Anecdotes found in the literature suggest that innovation could follow from activities such as establishing incentives for appropriate KM behaviors, determining appropriate communication channels for knowledge flows, and installing programs to encourage learning (Holsapple & Singh, 2001). Leadership and coordination operate in parallel; both foster innovation. This is understandable, because one creates and shapes the culture/environment conducive to innovation and the other manages dependencies and marshals sufficient skills for executing various patterns of primary activities. For coordination, reputation is the least frequent strong contributor at about 30%. Productivity is the second lowest at 33.3%. Although a minority of respondents do not think that the way in which coordination activity is carried out in their organizations contributes to competitive advantage via reputation or productivity, the 30%+ level is suggestive that these linkages are worthy of study and consideration. As with leadership, ANOVA testing shows no statistically significant difference ($\alpha \leq 0.10$) among coordination's PAIR means.

As far as knowledge control is concerned, strong contribution was reported by a range of about 20% to 28% of respondents for the four approaches. ANOVA testing shows no statistically significant difference ($\alpha \leq 0.10$) among control's PAIR means. Even though no single approach is prominent in its contribution, respondents report sufficient value for all PAIR approaches so that none should be ignored in KM planning and operation. Those who are realizing strong performance levels through a particular PAIR approach well may be a leader in best methods or best technologies for accomplishing knowledge control. Or they may have special situational aspects (e.g., resources, environment) that are conducive to implementing knowledge control activities in ways that are conducive to impacts on a particular PAIR dimension.

Innovation and agility, at 31% each, are the most frequently recognized strong beneficiaries of knowledge measurement methods and technologies. As with other secondary activities, however, ANOVA testing of measurement's PAIR means show no statistically significant ($\alpha \leq 0.10$) difference. Thus, in general, focusing on any particular secondary activity neither favors nor excludes any of the PAIR avenues as a means toward better organizational performance. However, in the context of a particular firm's situation, it is possible that practices (current or potential) for some secondary activities may align more readily with one or another of the PAIR dimensions.

If an organization has selected any one of the PAIR approaches for its competitive strategy, it might as well pay attention to what other organizations that have explicitly launched the KM initiatives are doing in this regard. This includes both competing and noncompeting organizations. Table 9 provides benchmarking data in terms of each secondary activity. In every PAIR case, ANOVA testing for equality of means across the four secondary activities shows that differences among these means are statistically significant ($\alpha \leq 0.01$). This implies that devising better ways to perform any of the four secondary activities should not be overlooked in strategic planning for competitive advantage along any of the PAIR dimensions. As with the knowledge chain model's primary activities, it is prudent to consider all of the model's four secondary activities in designing competitive strategies that focus on a productivity, innovation, or reputation dimension of performance. For each of these activities, the consideration seeks to find methods and/or technologies that are sufficiently distinct from and superior to those of competitors in terms of positive effects on the selected PAIR dimension(s).

Extending the Knowledge Chain

We now have solid evidence from leaders of KM initiatives that every KC activity can affect an organization's performance in terms of PAIR approaches to competitiveness. Building on this, a next logical step is to develop the knowledge chain in more detail by identifying what specific actions within the realm of the nine activity classes are candidates for affecting an organization's performance. Taking this step, we find that the nine KC activities of the original KC Model are actually activity classes comprised of 61 distinct types of KM activities, each of which is a candidate for improving an organization's competitiveness (Holsapple & Jones, 2004, 2005). Here, we briefly describe these distinct types, each of which can be tested for its connection to the PAIR model.

Primary KM Activities

Knowledge acquisition is found to encompass 10 distinct activities. Each of these 10 activities falls into one of two categories: direct or indirect acquisition (see Table 10). Activities in which the organization takes an active role in gaining knowledge residing in the external environment are categorized as direct acquisition. Activities such as participating in collaborative acquisition and obtaining/licensing data sets fall into this category. In contrast, activities through which knowledge is acquired as a result of or in conjunction with some other action are considered indirect acquisition. Examples of these activities include indirectly acquiring knowledge en masse and indirectly acquiring knowledge on an individual level.

Five distinct activities are found to comprise knowledge selection. Each of these activities is categorized either as action-oriented or archival-oriented (see Table 10). The action-oriented activities represent activities of capturing knowledge as it is being used. An example of this type of activity is participating in in-house training. Archival-oriented activities, on the other hand, represent activities in which knowledge is retrieved from some type of repository. An example of archival-oriented knowledge selection is retrieving knowledge from a technological repository.

Knowledge generation encompasses nine distinct activities. These activities are categorized into two classes: discovery or derivation (see Table 10). Discovery activities are those activities in which knowledge is generated in less structured ways. Examples include devising or developing strategies and generating knowledge through collaboration. In contrast, derivation activities utilize process knowledge and descriptive knowledge to produce new processes, rules, and/or descriptive knowledge. Examples of derivation activities include analytical derivation and inferential derivation. Two of the nine distinct knowledge generation activities fall somewhere between the subcategories of learning lessons and sensemaking, and making decisions. Both of these activities can be exploited in very structured or unstructured manners.

Four distinct activities are found within knowledge assimilation. These activities fall into two categories: formal or informal (see Table 10). Formal assimilation activities are those that follow a well-defined, institutionalized method for distributing and storing knowledge. Informal assimilation activities, on the other hand, are more ad hoc and opportunistic. Both of these categories can include publishing (i.e., an emphasis on archiving) and interaction (i.e.,

Table 10. Distinct types of primary KM activities (Adapted from Holsapple & Jones, 2004)

Primary KC Activity	Subcategory	Distinct Type
Acquisition	Direct	Obtaining/Licensing data sets
	Direct	Obtaining/Licensing patents, copyrights
	Direct	Using competitive intelligence, looking for windows of opportunities, obtaining trade secrets
	Direct	Soliciting knowledge from external sources
	Direct	Reviewing professional literature
	Direct	Monitoring technological advances
	Direct	Receiving external training
	Direct	Participating in collaborative acquisition
	Indirect	Indirectly acquiring knowledge en masse
	Indirect	Indirectly acquiring knowledge on an individual level
Selection	Action-Oriented	Participating in in-house training
	Action-Oriented	Seeking out people's know-how, know-what, and know-why
	Action-Oriented	Awareness of processes and events in the organization, looking for windows of opportunities, observing behaviors of participants in the organization
	Archival-Oriented	Recalling from a technological repository
	Archival-Oriented	Recalling from a nontechnological repository
Generation	Discovery	Devising or developing strategies
	Discovery	Developing products/processes
	Discovery	Mining
	Discovery	Creating
	Discovery	Generating through collaboration
	Discovery/Derivation	Learning lessons, sensemaking
	Discovery/Derivation	Making decisions
	Derivation	Analytical derivation
	Derivation	Inferential derivation
Assimilation	Formal	Formal internal publishing
	Formal	Formal internal interaction
	Informal	Informal internal publishing
	Informal	Informal internal interaction
Emission	Formal	Formal external publishing
	Formal	Formal external interaction
	Informal	Informal external publishing
	Informal	Informal external interaction

Table 11. Distinct types of secondary KM activities (Adapted from Holsapple & Jones, 2005)

Secondary KC Activity	Subcategory	Distinct Type
Leadership	Planning	Analyzing the business case
	Planning	Aligning KM with business strategies
	Planning	Establishing KM guidelines
	Executing	Creating a KM culture
	Executing	Delegating activities
	Executing	Sharing a leader's knowledge
Coordination	Structuring Efforts	Establishing communication patterns
	Structuring Efforts	Building infrastructure
	Structuring Efforts	Structuring knowledge work
	Structuring Efforts	Allocating knowledge workers
	Securing Efforts	Explaining KM to employees
	Securing Efforts	Establishing incentives and motivating employees
	Securing Efforts	Securing sponsorship
Control	KM Resource Control	Controlling financial resources available for KM
	KM Resource Control	Controlling KM processors
	KM Resource Control	Controlling quality
	KM Resource Control	Auditing knowledge
	Process Governance	Protecting/Providing access controls
	Process Governance	Using a risk management standard
	Process Governance	Managing/Monitoring KM
Measurement	Applying Measures	Measuring knowledge resources
	Applying Measures	Measuring KM abilities/skills
	Applying Measures	Measuring KM activities
	Applying Measures	Tracking stakeholder information
	Applying Measures	Valuing knowledge
	Applying Measures	Managing/Monitoring KM
	Applying Measures	Measuring effects of KM
	Determine/Develop	Determine/Develop quantitative measures
	Determine/Develop	Determine/Develop qualitative measures

an emphasis on channel usage) activities. Some examples include publishing a policy manual (i.e., formal internal publishing) and storytelling (i.e., informal internal interaction).

Knowledge emission mirrors knowledge assimilation in its number and type of categories: formal and informal (see Table 10). The meanings for these categories are the same as knowledge assimilation; however, the target for the knowledge is external rather than internal. Publishing and interaction activities are also present in knowledge emission. Some examples here include developing an advertisement (i.e., formal external publishing) and giving lectures (i.e., formal external interaction).

Secondary KM Activities

Knowledge leadership is found to encompass six distinct activities. Each activity falls into one of two categories: planning or executing (see Table 11). Activities performed before a KM implementation is done are considered planning activities. Examples of planning activities include analyzing the business case and establishing KM guidelines. Activities involved in carrying out the KM initiative are considered executing activities. These activities include creating a KM culture and delegating activities.

Seven distinct activities are found within knowledge coordination. These activities are further divided into two categories: structuring efforts or securing efforts (see Table 11). Structuring efforts entails dealing with the establishment of the structure by which KM will be implemented. Examples of this type of activity include establishing communication patterns and allocating knowledge workers. Activities dealing with ensuring that the proper incentives for executing the KM initiative are in place and that management and employees are motivated to participate are considered securing efforts. Examples include explaining KM to employees and securing sponsorship.

Seven distinct activities were found within knowledge control. Each of these activities can be placed into one of two categories: KM resource control or process governance. Activities dealing directly with KM resources are placed in the KM resource control category. Examples of these activities include controlling financial resources available for KM and auditing knowledge. Activities concerned with governing the KM initiative are placed in the process governance category. Some activities here include protecting/providing access controls and using a risk management standard.

Knowledge measurement is found to encompass nine distinct activities. These activities can be grouped further into two categories: determining/developing measures or applying measures (see Table 11). Activities in the determining/developing measures category are concerned with deciding what truly needs to be measured and how those measures are to be developed. Examples of this category include determining/developing qualitative measures and determining/developing qualitative measures. Once these activities have been completed, the next category of activities is used. The applying measures activities actually use the measures that were developed. Examples of these activities include valuing knowledge and measuring effects of KM.

Conclusion

A major contribution of this research study is that it provides evidence from leaders of KM initiatives that confirms the existence and inclusion of each of the nine activities in the knowledge chain model. Thus, it confirms earlier anecdotal findings (Holsapple & Singh, 2001). In a very practical sense, this study tells us that each of the nine knowledge chain activities is linked to organizational competitiveness. Each can be conceived and implemented in ways that strongly contribute to an organization's competitiveness in terms of PAIR approaches. Each deserves consideration in planning an organization's competitive strategy.

Rather than just saying that KM in general is essential for competitiveness, we show that each specific knowledge chain activity is a candidate for improving organizational performance via any of four approaches to competitiveness. Thus, the KC model provides structure to researchers and practitioners for considering what specific activities are candidates for competitiveness. It is up to the individual organization to determine which of these to focus on, given its present resource base and situational circumstances. It is also up to the individual organization to identify and institute the specific methods and technologies that will work best, given its resource and environment situation.

While the contributions of this study are important, there is a limitation. This study could not find a ready mailing list for leaders in organizations with ongoing knowledge management initiatives and proceeded with 102 candidates found in various ways. Even though the response rate was 31.4%, the study could benefit by having a larger pool of candidates and respondents. Thus, the present study is relatively exploratory.

A next logical step to this research is further study of the extension of the knowledge chain model that encompasses more than 60 distinct KM activities organized into the nine classes shown in Tables 1 and 2 (Holsapple & Jones, 2004, 2005). Future research needs to investigate each of the distinct activities identified in the extended knowledge chain model, looking for their connections to the PAIR. This will provide practitioners with guidance in determining the degrees to which specific types of activities within the overall knowledge chain classes of activities lead to the various PAIR competitiveness indicators.

The knowledge chain model is descriptive in nature. The intent is to identify KM activities that researchers and practitioners need to consider in managing knowledge in order to achieve competitiveness. Therefore, future research should focus on providing practitioners with such prescriptive guidance as identifying best methodologies and best technologies for contributing to competitiveness via the PAIR approaches. Future research could focus on the technological aspect of KM. Using the knowledge chain model and the extended knowledge chain model as the basis, the linkage between KM technology and competitive advantage could be investigated. This, perhaps, could help to pioneer new KM products and approaches that fuel the current knowledge economy.

The percentages shown in Tables 8 and 9 are, or course, susceptible to change. For instance, a low percentage well may increase in the future as new practices and technologies are deployed to implement that particular activity for achieving the corresponding PAIR benefit. Conversely, a high percentage may decline as best methods/technologies for implementing a KM activity become widespread. That is, the competitive advantage evaporates, but nevertheless, the activity will still remain a focal point for the PAIR benefit in order to avoid slipping into a position of competitive disadvantage.

Acknowledgments

This research has been supported in part by the Kentucky Initiative for Knowledge Management (established in 1988 at the University of Kentucky) and by a grant from the Kentucky Engineering and Science Foundation.

Referneces

Adams, G. L., & Lamont, B. T. (2003). Knowledge management systems and developing sustainable competitive advantage. *Journal of Knowledge Management*, *7*(2), 142-154.

Amidon, D. M. (1997). *Innovation strategy for the knowledge economy: The ken awakening*. Boston: Butterworth-Heinemann.

Amidon, D. M., & Mahdjoubi, D. (2003). An atlas for knowledge innovation: Migration from business planning to innovation strategy. In C. Holsapple (Ed.), *Handbook on knowledge management: Knowledge directions*. (pp. 331-352). Berlin: Springer.

April, K. D. (2002). Guidelines for developing a k-strategy. *Journal of Knowledge Management*, *6*(5), 445-456.

Carneiro, A. (2000). How does knowledge management influence innovation and competitiveness? *Journal of Knowledge Management*, *4*(2), 87-98.

Delio, M. (2000). Proving we're productive. *Knowledge Management*, *3*(7), 47-50.

Delphi. (1997). Not just another fad. *PC Week*, *14*(53), 73.

Devor, R., Graves, R., & Mills, J. J. (1997). Agile manufacturing research: Accomplishments and opportunities. *IIE Transactions*, *29*(10), 813-823.

Dove, R. (1994). *The meaning of life and the meaning of agile*. Retrieved from http://catalog.com/napmsv/agile.htm

Dove, R. (2003). Knowledge management and agility: Relationships and roles. In C. Holsapple (Ed.), *Handbook on knowledge management: Knowledge directions* (pp. 309-330). Berlin: Springer.

Drucker, P. (1986). *Innovation and entrepreneurship*. New York: HarperBusiness.

Ernst & Young. (1997). *Knowledge management*. Retrieved from http://www.businessinnovation.ey.com/research/electr/overview.html

Evans, P. B., & Wurster, T. S. (1997). Strategy and the new economics of information. *Harvard Business Review*, *75*(5), 71-82.

Fliedner, G., & Vokurka, R. (1997). Agility: The next competitive weapon. *APICS, Online Edition*, *7*(1), 56-59. Retrieved from http://www.apics.org/magazine/jan97/agility.htm

Fruin, M. (1997). *Knowledge works: Managing intellectual capital at Toshiba*. New York: Oxford University Press.

Garone, S.J. (1998). Managing reputations with image and brands. *The Conference Board Report* (Rep. No. 1212-98-CH).

Grover, V., & Davenport, T. (2001). General perspectives on knowledge management: Fostering a research agenda. *Journal of Management Information Systems, 18*(1), 5-21.

Higgins, J. M. (1995). Innovate or evaporate: Seven secrets of innovative corporations. *The Futurist, 29*(5), 42-48.

Holsapple, C. W., & Jones, K. (2004). Exploring primary activities of the knowledge chain. *Knowledge and Process* Management, *11*(3), 155-174.

Holsapple, C. W., & Jones, K. (2005). Exploring secondary activities of the knowledge chain. *Knowledge and Process Management, 12*(1), 3-31.

Holsapple, C. W., & Joshi, K. (2000). An investigation of factors that influence the management of knowledge in organizations. *Journal of Strategic Information Systems, 9*(2-3), 237-263.

Holsapple, C. W., & Joshi, K. (2002a). A collaborative approach to ontology design. *Communications of the ACM, 45*(2), 42-49.

Holsapple, C. W., & Joshi, K. (2002b). Knowledge management activities: Results of a Delphi study. *Information and Management, 39*(6), 477-490.

Holsapple, C. W., & Joshi, K. (2004). A formal knowledge management ontology: Conduct, activities, resources, and influences. *Journal of the American Society for Information Science and Technology, 55*(7), 593-612.

Holsapple, C. W., & Singh, M. (2000a). The knowledge chain model. In *Proceedings of the Third Annual Conference of the Southern Association for Information Systems,* Atlanta, Georgia.

Holsapple, C. W., & Singh, M. (2000b). The knowledge chain model: Activities for competitiveness. *International Journal of Expert Systems, 20*(1), 77-98.

Holsapple, C. W., & Whinston, A. (1987). Knowledge-based organizations. *The Information Society, 5*(2), 77-90.

Joshi, K. (1998). *An investigation of knowledge management characteristics: Synthesis, Delphi study, analysis* [doctoral dissertation]. University of Kentucky.

Kalling, T. (2003). Knowledge management and the occasional links with performance. *Journal of Knowledge Management, 7*(3), 67-81.

Kartalia, J. (2000). Managing your most valuable asset: The corporate reputation. *Entegra.* Retrieved from http://www.senet.com/articles_managing_assest.htm

Lengnick-Hall, C.A. (1992). Innovation and competitive advantage: What we know and what we need to learn. *Journal of Management, 18*(2), 399-429.

Leonard-Barton, D. (1995). *Wellsprings of knowledge: Building and sustaining the sources of innovation.* Boston: Harvard Business School Press.

MacMillan, I. C., & McGrath, R. G. (1997). Discovering new points of differentiation. *Harvard Business Review, 75*(4), 133-145.

Malhotra, Y. (1998). Knowledge management for the new world of Business, *@BRINT, Online Articles on Knowledge Management.* Retrieved from http://www.brint.com/km

Malhotra, Y. (2003). Why knowledge management systems fail: Enablers and constraints on knowledge management in human enterprises. In C. Holsapple (Ed.), *Handbook on knowledge management: Knowledge matters* (pp. 577-600). Berlin: Springer.

Markides, C. (1997). Strategic innovation. *Sloan Management Review, 38*(3), 9-23.

Markland, R., Vickery, S., & Davis, R. (1995). *Operations management: Concepts in manufacturing and services*. New York: West Publishing Company.

Marshall, C., Prusak, L., & Shpilberg, D. (1996). Financial risk and the need for superior knowledge. *California Management Review, 38*(3), 77-101.

McGaughey, R.E. (1999). Internet technology: Contributing to agility in the 21st century. *International Journal of Agile Manufacturing Systems, 1*(1), 7-13.

Nonaka, I. (1991). The knowledge creating company. *Harvard Business Review, 69*(6), 96-104.

Nonaka, I., & Takeuchi, H. (1995). *The knowledge-creating company: How Japanese companies foster creativity and innovation for competitive advantage*. New York: Oxford University Press.

Perrin, T. (2000). *Corporate image 2000: Strategies for managing your corporate and brand reputation*. Conference program sponsored by Towers Perrin and Reputation Management, Toronto, Canada.

Prusak, L. (1996). The knowledge advantage. *Strategy & Leadership, 24*(2), 6-8.

Sharkie, R. (2003). Knowledge creation and its place in the development of sustainable competitive advantage. *Journal of Knowledge Management, 7*(1), 20-31.

Sherman, L. (2000). A matter of connections. *Knowledge Management, 3*(7), 42-47.

Smith, L. R. (1998). Let's discuss. ... *Journal of Innovative Management, 4*(1), 23-27.

Spender, J. C. (2003). Knowledge fields: Some post-9/11 thoughts about the knowledge-based theory of the firm. In C. Holsapple (Ed.), *Handbook on knowledge management—Knowledge matters* (pp. 59-72). Berlin: Springer.

Stewart, S. L., & Pinholster, G. (1994). Systems integration needs of U.S. manufacturers, conference report. *Journal of National Institute of Standards Technology, 99*(5), 687.

Stewart, T. A. (1998, October 12). Knowledge, the appreciating commodity. *Fortune, 138*(7) 199-200.

Sveiby, K. (1997). *What is knowledge management?* Retrieved from http://www.sveiby.com.au/KnowledgeManagement.html

TechWeb News. (1999). *Microsoft tops reputation survey.* Retrieved from http://www.techweb.com/wire/story/TWB19991122S0004

Trudel, J.D. (1998). Trudel to form. *Electronic Design, 46*(17), 48j. Retrieved from http://www.elecdesign.com/Globals/PlanetEE/Content/1267.html#9

Turban, E., McLean, E., & Wetherbe, J. (1999). *Information technology and management* (2nd ed.). New York: John Wiley.

Ward, A., & Le, V. (1996). Lessons learned on the knowledge highways and byways. *Strategy & Leadership, 24*(2), 16-20.

Wiig, K., & Jooste, A. (2003). Exploiting knowledge for productivity gains. In C. Holsapple (Ed.), *Handbook on knowledge management: Knowledge directions* (pp. 289-308). Berlin: Springer.

Chapter VI

A Multi-Level Performance Framework for Knowledge Management

Anne Massey, Indiana University, USA

V. Ramesh, Indiana University, USA

Mitzi Montoya-Weiss, North Carolina State University, USA

Abstract

Knowledge management (KM) has gained increasing attention since the mid-1990s. A KM strategy involves consciously helping people share and put knowledge into action. However, before an organization can realize the promise of KM, a fundamental question needs to be asked: What performance goal(s) is the organization trying to achieve? In this chapter, we develop and offer a multi-level framework that provides a view of the performance environment surrounding organizational knowledge work. We illustrate the KM framework using two organizational case studies. Then, based on the KM framework and further insights drawn from our case studies, we offer a series of steps that may guide and assist organizations and practitioners as they undertake KM initiatives. We further demonstrate the applicability of these steps by examining KM initiatives within a global software development company. We conclude with a discussion of implications for organizational practice and directions for future research.

Introduction

Knowledge management (KM) is a topic that has gained increasing attention since the mid-1990s. Knowledge about customers, products, processes, and past successes and failures are assets that may produce long-term sustainable competitive advantage for organizations (Huber, 2001; Leonard & Sensiper, 1998; Stewart, 2001). KM proponents argue that these assets are as important as managing other organizational assets like labor and capital. A survey conducted by *Knowledge Management* magazine and the International Data Corporation suggests that KM is evolving from a discrete undertaking to a strategic component of business solutions (Dyer & McDonough, 2001).

A KM strategy entails consciously helping people share and put knowledge into action by creating access, context, infrastructure, and simultaneously shortening learning cycles (Alavi & Leidner, 2001; Davenport, DeLong, & Beers, 1998; Davenport & Prusak, 1998; O'Dell & Grayson, 1998). It takes place within a complex system of organizational structure and culture and often is enabled through information technology (IT) (Alavi, 2000; Alavi & Leidner, 2001). While technology drove the initial interest in KM, both academics and practitioners have begun to realize that effective KM initiatives and solutions require a broader understanding of knowledge work environment (Grover & Davenport, 2001; Holsapple & Joshi, 2002; Massey & Montoya-Weiss, 2002; Rubenstein-Montano et al., 2001). Specifically, before an organization can realize the promise of KM, a fundamental question needs to be asked: What performance goal(s) is the organization trying to achieve? Addressing this question will direct the organization to what knowledge should be managed and how it should be managed.

Improving customer service, shortening product development cycles, growing revenues, and improving profits commonly are cited as goals motivating KM initiatives. If the intent of a KM initiative is to enhance organizational performance, organizations first need to understand the performance environment surrounding and driving the underlying knowledge work. For example, improving customer service or shortening product development cycles requires that firms look to their processes, which may be reengineered to capitalize on or expand organizational knowledge resources and capabilities (Gold, Malhotra, & Segars, 2001; Hammer & Champy, 1993; Maier & Remus, 2001). Generating performance improvements via a KM initiative thus requires a deep understanding of how process work is organized, what knowledge is inherent to and derived from it, what factors influence knowledge workers, and how all of these factors relate to an organization's business environment (Massey & Montoya-Weiss, 2002).

In this chapter, we offer a framework that provides a multi-level view of the performance environment surrounding organizational knowledge work. The framework provides a useful means by which to identify, define, analyze, and address knowledge-based problems or opportunities relative to multi-level (business, process, and knowledge worker) performance goals and requirements. Our perspective responds to a current call in the literature for KM frameworks that take a systems-oriented perspective by considering problems and opportunities in their entirety (Rubenstein-Montano et al., 2001; Senge, 1990). We draw from and integrate literature concerned with approaches to dealing with complexity and purposeful (i.e., performance-oriented) systems (Checkland & Howell, 1998), business process reengineering (Hammer & Champy, 1993), and human performance (Stolovich & Keeps, 1999).

Rather than suggesting that KM requires a whole new perspective with its own special laws, our framework purports that KM sits well within our current understanding of what drives performance (Soo, Devinney, Midgley, & Deering, 2002).

We illustrate the efficacy of our framework to KM using case studies conducted at IBM and Nortel Networks. In addition, based on the framework and the insights we drew from our case studies, we offer a series of steps that can help to direct organizations as they undertake KM initiatives. Finally, we illustrate the generalizability of these steps by demonstrating them in context of the software development process, using insights gained from a study with a software development firm. We conclude our chapter with a discussion of broader implications for organizational practice and directions for future research.

Background and Motivation

The general goal of KM is to capitalize on knowledge assets in order to achieve maximum attainable business performance (Barney, 1991; Becerra-Fernandez & Sabherwal, 2001; Davenport & Prusak, 1998). Organizations are faced with two key questions: What should an organization consider before undertaking a KM initiative? How can KM become a strategic asset?

In a review of existing KM frameworks, Rubenstein-Montano et al. (2001) suggest that most frameworks to date have been prescriptive and focused primarily on knowledge flows. As such, they do not provide a comprehensive, holistic approach to integrate KM practices with strategic goals of the organization to realize potential for improving performance. Moreover, they do not consider nontask-oriented aspects that ultimately influence knowledge workers as they carry out business process activities. A further review of the literature suggests that KM has considered a broad array of issues and approaches, addressing things such as capturing and sharing best practices, building databases and intranets, measuring intellectual, establishing corporate libraries, installing groupware, enacting cultural change, and fostering collaboration (Ackerman, Pipek, & Wulf, 2003; Alavi & Leidner, 1999, 2001; Fahey & Prusak, 1998; Grover & Davenport, 2001; O'Dell & Grayson, 1998; Stewart, 2001). Thus, while no generally accepted framework has been adopted, it seems that KM has involved all kinds of approaches, practical activities, measures, and technologies.

In order to make KM a strategic asset and to realize the potential for improving performance, there is a need for a unifying framework that considers KM relative to the entirety of the organizational system as well as its subcomponents (i.e., the business, its processes, and knowledge workers) (Soo et al., 2002). Such a framework should provide a general sense of direction (i.e., be prescriptive) for KM initiatives in order to ensure that the same general requirements are addressed across the organization, but it also should be descriptive in that it considers factors that ultimately influence KM success or failure (Rubenstein-Montano et al., 2001; Tsoukas, 1996).

A systems approach to KM can ensure a holistic and purposeful (performance-oriented) consideration of the interrelationships between the business, its processes, and knowledge workers (Ackoff & Emery, 1972). The objective is to enhance understanding of and responsiveness to a problem by examining relationships among various parts of the system

(Checkland, 1981; Checkland & Howell, 1998; Gao, Li, & Nakamori, 2002). A systems approach can enhance KM initiatives by examining and depicting the complex relationships among components so that an organization can ascertain where and how KM might respond (Rubenstein-Montano et al., 2001). In the following section, we develop and offer a KM framework that considers the complex interdependencies among the business, its processes, and knowledge workers surrounding organizational knowledge work. When applied, the framework offers a systematic way to identify, define, and analyze performance problems or opportunities, their drivers, and causes at multiple levels (business, process, and individual). By doing this, desired performance outcomes at all levels can be described, and behaviors that will produce those outcomes can be identified (Gordon, 1996). With this robust understanding, organizations can more precisely specify and implement interventions in order to address problems or capitalize on opportunities and, ultimately, improve performance (Gery, 1997; Massey & Montoya-Weiss, 2002; Rosenberg, 1995; Stolovitch & Keeps, 1999).

A Multi-Level Performance Framework

In Figure 1, we offer a framework to ensure that KM initiatives and multi-level requirements are addressed in a similar vs. ad hoc fashion across the organization. The framework draws from and integrates literature concerned with approaches to dealing with complexity and purposeful (i.e., performance-oriented) systems (Checkland & Howell, 1998), business process reengineering (Hammer & Champy, 1993), and human performance (Rummler & Brache, 1992; Stolovich & Keeps, 1999). It possesses both prescriptive (task-oriented activities) and descriptive (consideration of factors that influence success or failure) elements, which, in turn, facilitate a holistic perspective. Importantly and consistent with a systems approach, the framework does not imply that the same methodologies will be used for all situations; rather, the framework facilitates a method to KM that is adaptive and responsive to different situations.

Since knowledge is context-specific and KM will be most powerful when applied to a specific domain (Sviokla, 1996), a component of our framework is its focus on core business processes. Thus, at the process level, we draw from the business process reengineering (BPR) literature, which is concerned with a fundamental rethinking of and redesign of business processes in order to achieve performance improvements (Hammer & Champy, 1993). Although BRP involves the analysis and design of workflows, it does not consider explicitly the complex environment that influences knowledge workers (Davenport & Short, 1990). Without consideration of the human element in knowledge-intensive processes, BPR rarely will be successful. Therefore, by leveraging literature concerned with human performance (Rummler & Brach, 1992; Stolovich & Keeps, 1999), our KM framework includes factors that influence individual work behaviors and performance. It is likely that a KM initiative that only considers isolated subcomponents of the overall system will not enhance performance. Rather, success will hinge on understanding how each part (i.e., strategic goals, business process, knowledge workers) influences and interacts with other parts.

As illustrated, the external environment presents an organization with opportunities, pressures, events, and resources (Holsapple & Joshi, 2000, 2002). In response, an organization

Figure 1. Knowledge management framework (Adapted from Stolovich & Keeps, 1999; Rummler & Brach, 1992)

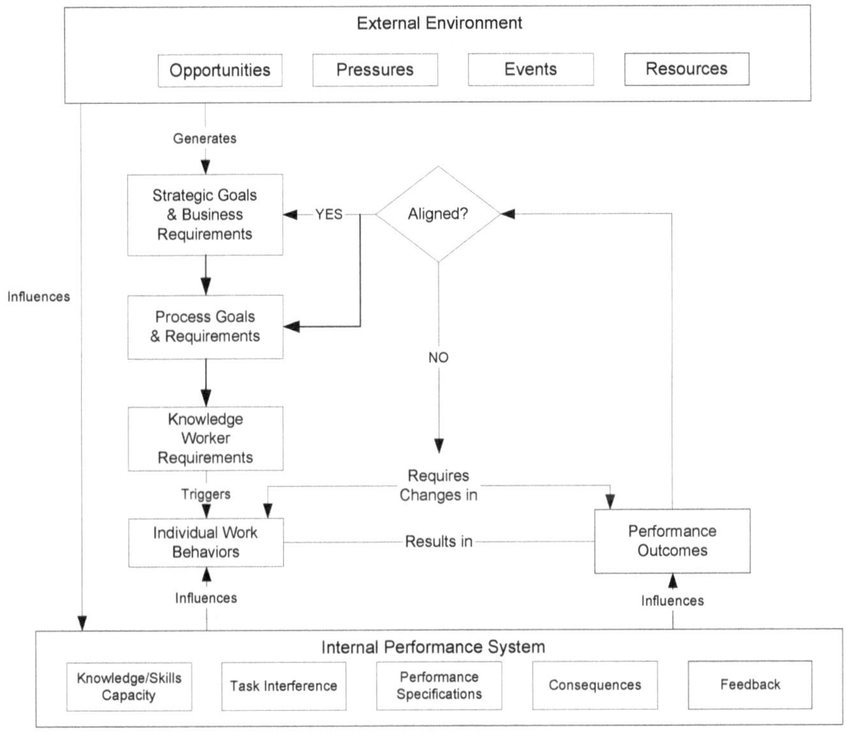

generates business and process requirements (i.e., a set of actions that allows the organization to capitalize on external opportunities and/or to respond to threats). For example, in order to remain competitive, a strategic business performance goal may be to increase market acceptance of new products (Moorman & Rust, 1999). In a software-related business, the business-level requirement may be to increase the rate of new software introduction into the marketplace. This business requirement generates process-level requirements (e.g., the new product development process must produce a stream of continuous new products or services).

Gaps between current process capabilities and defined requirement(s) may force the organization to reengineer the business process so that the process performs at the required level of performance (Davenport, 1993; Hammer & Champy, 1993; Teng, Grover, & Fiedler, 1994). Recognizing that processes are knowledge-intensive (Davenport, DeLong, & Beers, 1998; Massey & Montoya-Weiss, 2002), reengineering efforts should focus on decomposing and structuring the process so that data, information, and knowledge activities and workflows between activities are clearly defined (Davenport & Short, 1990; Hammer & Champy, 1993; Teng et al., 1994). Importantly, structuring the process and identifying knowledge exchange activities inherent to the process will assist in identifying knowledge worker requirements

(Leonard & Sensiper, 1998; O'Dell & Grayson, 1998). In particular, this involves defining what knowledge and what types of knowledge (tacit/explicit) are needed to accomplish activities. It also involves identifying who or what are the sources and receivers of knowledge (e.g., human, archives, etc.) as well as defining desired performance outcomes of process-level work. A purposeful and seamless flow of data, information, and knowledge (a defined knowledge cycle) then can occur among collaborating knowledge workers tasked with various process activities.

In addition to rethinking how process work should be done via reengineering (Davenport, 1993; Hammer, 1990), it is important to consider the knowledge worker(s) who will be tasked with carrying out process activities. Thus, in addition to specifying the knowledge cycle, we must consider factors that influence the behaviors and performance of knowledge workers at the task/activity level (Checkland, 1981; Rubenstein-Montano et al., 2001; Rummler & Brache, 1992). As shown in Figure 1, the task/activity level factors are referred to as a knowledge worker's internal performance system. Here, it becomes important to recognize that individual (or often team) performance is not simply a function of knowledge, skills, or capacity. Rather, other factors influence performance, including the nature and clarity of the business process work tasks and whether anything (e.g., lack of resources) interferes with task completion, clarity of performance specifications and goals, positive and negative work consequences, and performance feedback (Rummler & Brache, 1992). By taking a broader view of knowledge workers, cause(s) of poor performance and/or opportunities to enhance performance (beyond knowledge, skills, and capacity) can be identified.

As already described, the framework enables a holistic examination of the interrelationships between multi-level goals and requirements, which allows for the identification of problems or opportunities that should be addressed to enhance performance (Senge, 1990). In the following section, we illustrate the framework based on our work with IBM and Nortel. Our purpose is not to provide detailed case studies; rather, our intent is to illustrate key elements and interrelationships (see Massey, Montoya-Weiss, and Holcolm, 2001 and Massey, Montoya-Weiss, and O'Driscoll, 2002 for further, in-depth case studies).

Performance-Driven KM Initiatives

In the latter half of the 1990s, both IBM and Nortel Networks were facing significant external pressures. From 1986 to 1992, IBM's market share dropped from 30% to 19% with each percentage point representing $3 billion in revenues. Rather than paying attention to customer needs, IBM focused on its own financial needs and tried to reduce costs by cutting customer service staff and levels of support. In the end, customers were driven away. Thus, by the mid-1990s, the changing market environment and downsizing necessitated that IBM rethink the basic way that it serviced customers in order to reduce customer defections and increase sales. Throughout the 1980s and early 1990s, IBM's primary points of contact with its customers were through business partners, the direct catalog, and the traditional "Blue suits." Given that these points of contact were not supporting the business strategic goals and requirements to remain competitive, an internal task force was charged with reengineering IBM's customer relationship management (CRM) process.

CRM involves attracting, developing, and maintaining successful customer relationships over time (Berry & Parasuraman, 1991; Day, 1994, 2000). At the core of CRM is the development of a learning relationship that engages customers in a two-way collaborative dialogue that is effective and efficient for both customers and the firm (Peppers, Rogers, & Dorf, 1999). When effective, this knowledge-based process leads to a relationship that gets smarter and deeper through every interaction. The task force charged with addressing the business problem recognized that advanced information technology, the burgeoning Internet, and the emerging network-centric environment presented great opportunities for reengineering IBM's CRM process and leveraging its knowledge assets.

Similarly, at Nortel Networks, the Telecommunications Reform Act of 1996 produced intense competition in the telecom industry, yielding an explosion in the development of innovative telecommunications technology. The new rules of the deregulated telecommunications marketplace forced Nortel to recognize that differentiation through innovation was one of the few strategies that might allow the company to continue to succeed. Like IBM, an internal group was charged with the task of addressing this strategic business goal and requirement. After preliminary research, the group discovered that the generation and existence of innovative ideas within Nortel was not the issue. Rather, Nortel's existing new product development (NPD) process had no formal mechanism to systematically capture, develop, and manage internally generated ideas (i.e., ideas that could be developed into product or service concepts and evaluated for funding). Developing ideas and evaluating concepts is knowledge-intensive work based on the individual and collective expertise of employees. The Nortel task force set out to reengineer its NPD process in order to leverage its knowledge assets. As described, the efforts of both the IBM and Nortel initiatives were guided by strategic business goals and requirements that, in turn, led them to focus on business processes that were most relevant to achieve desired performance (Figure 1).

Core business processes like CRM and NPD represent the fundamental link between business and knowledge workers performance. The reality for both IBM and Nortel was that their respective business requirements would be achieved through processes, and both organizations were only as good as their processes, which ultimately depend on the behaviors of knowledge workers. Driven by this performance reality, IBM's reengineered CRM process was designed to enhance the customer relationship, while Nortel's reengineered NPD process was designed to produce a continuous stream of products and services. Although the specific details of the process reengineering efforts are beyond the scope of this chapter (details are available upon request), both organizations structured their new processes by decomposing the process into knowledge-based activities, simultaneously identifying the required flows of data, information, and/or knowledge between activities and knowledge workers. The analysis led to the specification of the knowledge-based drivers (types, sources, and receivers) of each activity, decision, or information flow. In their reengineering efforts, both task forces went through several process prototype iterations (Davenport & Short, 1990), simultaneously considering whether (or not) the new process, in fact, could support business goals and requirements. As one IBM task force member put it, "We had done things the same way for so long. We realized early on that any changes to our [CRM] process had to demonstrate they would, in fact, improve business performance or nobody—our reps or our customers—would buy in. We continually asked ourselves whether the new process supported our business [level] goals." This quote illustrates the strong link between business and process levels shown in Figure 1 and how important it is to not only decompose the levels but also to integrate them.

The reengineering of IBM's CRM process and Nortel's NPD process created new knowledge worker performance requirements, triggering requisite changes to individual work behaviors. As one Nortel task force member observed, "While we believed our new [NPD] process could perform as desired, we were not sure if our people could or would actually carry out the new process. We needed to gain a better understanding of their capabilities and motivations." Given this, both organizations sought to understand the internal performance system of its various knowledge workers in light of the reengineered processes and requirements (see Figure 1). Specifically, did they possess the knowledge/skills/capacity to carry out reengineered or new process activities? Did they possess and/or understand the inputs required to carry out process tasks? Did anything exist that would interfere with task requirements? Did they understand the desired performance outcomes intended to support business and process requirements? What contextual factors would motivate or demotivate knowledge workers to share knowledge and carry out the new process (i.e., consequences and feedback)?

Answers to these questions enabled a collaborative learning relationship between IBM representatives and customers and a more collaborative relationship between the engineers and managers involved in the NPD process at Nortel. For example, Nortel's NPD process called for idea generators (often engineers) to develop a raw product or service idea into a robust concept along the lines of marketing, business, technology, and human factors (areas used by managers to make funding decisions). While engineers are technically knowledgeable, they typically do not possess sufficient knowledge in the other areas required in the new NPD process. This drove Nortel's team to consider interventions to support the specific knowledge gaps of workers engaged in this process activity. Similarly, IBM's team considered the factors that would influence the behaviors of CRM knowledge workers. For example, IBM sales representatives felt threatened by the CRM reengineering effort due to their perception that customer relationships would be transferred largely from human contacts to technology. In response, IBM undertook efforts to show sales representatives that the new CRM process, in fact, would allow them to more proactively sell and market products and services.

Ultimately, both IBM and Nortel designed and implemented technology-based interventions to support the performance of knowledge workers. Drawing from the disciplines of KM and CRM, IBM developed an Internet-based system called Inside IBM. The system allowed customers to link directly to IBM's intranet and backend cross-functional knowledge-based resources. Inside IBM subsequently was adopted as a corporate standard leading to IBM's e-Services, as it is known today. Deploying artificial intelligence, information systems, and user-centered design, Inside IBM aggregated IBM's accumulated product support knowledge into a single system and enabled collection of information about its customers. IBM's efforts facilitated a collaborative and learning relationship between IBM and its customers. This led to improved decision-making for both the customer and the organization's sales and service workforce, leading to increased sales and customer loyalty. IBM estimated that $525 million of incremental revenue and $50 million of productivity savings were realized over a three-year period as a result of this initiative (see Massey et al., 2001, for further details concerning IBM's initiative).

Similarly, Nortel developed a KM system called Virtual Mentor, which supported both the performance of knowledge workers (engineers) engaged in developing raw ideas into robust concepts and decision makers (managers) tasked with making funding decisions. Virtual Mentor subsequently was integrated into a broader corporate time-to-market strategy that is in place today. Nortel's efforts led to decreased time-to-market, increased time-to-market acceptance, and improved funding decisions. Over a three-year period, Nortel's new product introduction rate increased by more than 50% (see Massey et al., 2002, for further details concerning Nortel's initiative).

Clearly, the bottom line for IBM and Nortel was to increase profitability, sales, share, and return on investment by leveraging and managing its knowledge assets. As evidenced, IBM's and Nortel's KM initiatives were guided by a holistic understanding of interdependent multi-level (business, process, knowledge worker) performance goals and requirements. This facilitated problem/opportunity identification and definition, diagnoses of the changes required, and the subsequent design of suitable interventions needed to affect the performance of knowledge workers tasked with process activities. Addressing what to do from a performance perspective drove the reengineering of two knowledge-intensive business processes. Both processes called for improvements to cross-functional coordination; collaboration and learning; and knowledge exchange in business, technology, and marketing (and other relevant areas). Considering "how to do it" and simultaneously understanding the behavioral factors that influenced knowledge workers informed the development and implementation of interventions designed to enhance performance. In the end, both IBM and Nortel were directed by a deep understanding of the complex interdependencies inherent to their organizational systems. In their respective efforts, they not only decoupled the organizational environment into its smaller parts (business, process, knowledge worker) but also continually considered how the parts were linked in hierarchies to form the whole performance environment.

In the following section, we present a series of steps that underlie our framework and provide direction for KM initiatives. We then illustrate the efficacy of our approach in a software engineering context. Our data in this context is based on interviews with managers and developers regarding KM systems currently in use at TechCo (a pseudonym), a well-known Indian software development firm that has several centers certified at Level 5 of the capability maturity model (Paulk, Weber, Curtis, & Chrissis, 1995).

Steps for KM Initiatives

- **Step 1. Select a target business process:** Once an organization has identified its business goals and requirements, a KM initiative then must identify the firm's key leverage points for achieving business results. As noted earlier and as evidenced at IBM and Nortel, since knowledge is context-specific (Sviokla, 1996), KM likely will be most powerful when it addresses a particular domain such as new product development, operations, sales, and customer service. Organizations should start where advocacy exists for doing something different. Processes such as those targeted for improvement by the organizations we studied is where work is accomplished. Once the process is identified, establish a process and project owner and ensure that the new initiative is managed as a business change project, not an information technology project (as many early KM projects were managed). In this step, it is also important to establish performance measures for the business case. Demonstrating success with a single process may lead to acceptance for other processes.

- **Step 2. Model the process:** This step requires that the inherent underlying process structure be found or defined in order for an initiative to move forward. Oftentimes, process activities and the data, information, and knowledge flows among activities are poorly defined. Clarifying activities and promoting an integrative view of the whole process is the starting point for managing knowledge and improving performance. For example, in the front-end of the NPD process, idea-to-concept development and concept selection activities often are called the *fuzzy* because they involve ill-defined activities and ad-hoc decisions carried out by multiple and diverse stakeholders (Cooper & Kleindschmidt, 1995). Via careful analysis and benchmarking, Nortel reengineered and enhanced the front end of its NPD process by defining a consistent and structured approach for developing, screening, and cataloging new product ideas.
- **Step 3. Identify activity-based knowledge exchange processes:** This step requires understanding the context of work (i.e., the knowledge needs associated with each process activity defined in Step 2). For example, in IBM's CRM process, in order for customer representatives to proactively target sales and marketing, they had to determine how to acquire knowledge concerning customer requirements. Similarly, at Nortel, different knowledge workers and functions had different pieces of data, information, and knowledge relevant to the NPD process. These pieces needed to be exchanged in order to create a common and logically organized bank of knowledge about a product or service concept. The objective of this step is to identify the knowledge exchange processes that are or must be in place to support value-creating activities.
- **Step 4. Identify desired knowledge exchange performance outcomes:** When individuals or teams exploit knowledge in a business process, it is reflected in the quality of a valued outcome that benefits the organization. This step involves specifying the performance outcomes that should be derived from the knowledge exchange processes identified in Step 3. For example, in Nortel's NPD process, one desired outcome was that a decision maker (manager) could make an informed decision regarding further funding for product development. Another was when the right combination of product-related data (e.g., marketing, business, and technology) needed to be readily accessible in the right format for different tasks and functional areas. Alternatively, in IBM's CRM process, a desired outcome was that the right people, information, and services would be readily accessible to the customer.
- **Step 5. Identify the knowledge drivers of each process activity, decision, and information flow:** This step requires the identification of the types of knowledge required, the sources of that knowledge (internal and/or external people, archived data), and the receivers of knowledge (people, other databanks). In Nortel's case, this step required identification of the specific knowledge required by an idea generator (i.e., an engineer or knowledge worker source) so that he or she could develop a raw idea into a robust concept in the areas of marketing, business, human factors, and technology. With this knowledge in hand, a raw idea could be developed into a complete and robust concept so that decision makers (i.e., manager or knowledge worker receiver) could evaluate the concept and make a funding decision.
- **Step 6. Identify and develop interventions:** In concert, Steps, 2 through 5 specify the knowledge inputs, exchange processes, sources and receivers, and desired outcomes associated with the targeted and defined business process. The factors that influence

> individual work behaviors (i.e., the internal performance system of knowledge workers) also must be considered to ensure that desired performance outcomes are achievable. With this holistic understanding in hand, an organization now can specify more precisely its KM interventions or solutions to support individual and/or teamwork. Interventions reflect both responses to identified causes of performance problems as well as opportunities for improving performance. Potential interventions could include the development of individuals or teams (e.g., training) or solutions that focus on rewarding performance (e.g., incentive/reward systems). Interventions also may include information technology-based KM systems (Alavi, 2000; Alavi & Leidner, 2001;Gery, 1997; Hinds & Pfeffer, 2003; Rosenberg, 1995). Intervention selection should be done in light of appropriateness (internally and externally), economics, feasibility (given organizational constraints or barriers to implementation), and acceptability to the organization and knowledge workers. Again, by taking a multi-level view and understanding the performance environment first rather than starting with a solution looking for a problem, one can more appropriately and precisely identify interventions.

One key issue to be considered when supporting activities within a business process context is the issue of language translation. Knowledge workers deploy local languages relative to their areas of expertise. Thus, successfully enabling the flow of data, information, and knowledge among process activities and diverse knowledge workers may require language translation. For example, as noted earlier in our Nortel case, we found that idea generators (engineers) did not speak the language of decision makers (managers). Nortel's KM solution, Virtual Mentor, thus was designed to depict and translate knowledge in forms appropriate for different audiences (engineers, managers, process owners). For example, through concept development and rating forms designed in the language of engineers, idea generators provide knowledge concerning a new concept and its potential application(s). Virtual Mentor then translates the contextual structure of this concept information into a form so that decision makers can conduct a SWOT (strengths, weaknesses, opportunities, threats) analysis. Virtual Mentor enables collaboration by supporting the local languages of disparate knowledge workers who must exchange knowledge to improve decision making. Another issue that needs to be dealt with is the differential navigational needs of the various stakeholders. As an example, the navigation needs of a customer representative in the CRM process seeking to acquire customer requirements differs significantly from the needs of a customer seeking information.

An Applications of the Steps for KM Initiatives

In this section, we provide further evidence that demonstrates the validity of our multi-level performance framework and underlying steps identified in the previous section. Here, we describe the path taken by TechCo in arriving at the KM solutions in use today. TechCo is one of the leading software services and consulting organization in Southeast Asia, providing systems development and integration services to Global Fortune 500 clients.

During the, 1990s, TechCo saw a significant increase in competition in the offshore software development arena. With a business goal of maintaining its position as one of the market leaders in this arena, TechCo sought to gain a competitive edge by focusing its efforts on improving the quality of its core software development processes. This effort was very similar to Nortel's efforts described earlier, which focused on enhancing its NPD process.

Software development by its very nature is a knowledge-intensive process that involves many people working on several different activities and phases (Rus & Lindvall, 2002; Ward & Aurum, 2004). Success hinges on the creation, acquisition, identification, adaptation, organization, distribution, and application of knowledge within and among projects. It is also a dynamic process, evolving with technology, organizational culture, and development practices (Ward & Aurum, 2004). Inherent to software development is knowledge embedded in products and meta-knowledge concerning not only the products but also development processes (Rus, Lindvall, & Sinha, 2001). While individuals engaged in software development projects make decisions based on personal knowledge, the sharing of this knowledge historically has been limited to informal means (Rus & Lindvall, 2002) (see Rus et al., 2001, for a review of KM and software engineering).

TechCo's focus on process improvement initiatives was driven by a desire to provide a measure of control and accountability within complex software development projects. Example processes that could be targeted in a software lifecycle context include the requirements analysis, software development, and software maintenance processes, as well as more managerial processes such as the project management or change management process (Jalote, 2000; Rus et al., 2001). TechCo sought to address several of these processes through its efforts to achieve the Carnegie Mellon Institute's Capability Maturity Model Level 5 certification (Paulk et al., 1995).

Having identified a set of target processes to reengineer (Step 1 of our checklist), TechCo began to specify and document the standard activities and information flows for each major process in the software development life cycle. This activity (a requirement in order to be certified at Level 3 of the CMM) helped TechCo to achieve the objectives stated in Step 2 of our KM checklist. Each process was broken down into stages that consisted of activities, which, in turn, were divided into subactivities. Key participants for each stage also were identified as part of the process definition. For example, the requirements analysis process was divided into the activities of preparation, eliciting requirements, analyzing requirements, and so forth. Examples of subactivities that were identified for the requirements analysis activity included the creation of logical data models and process models.

Steps 3 and 4 of our KM initiative checklist were achieved as a natural consequence of TechCo's efforts to detail the activities that comprised each process. TechCo used the ETVX (Entry, Task, Verification, and eXit) model (Radice, Roth, O'Hara, & Ciarfella, 1985) to define the details of each stage in a process. The entry criteria and input specification together defined the primary knowledge inputs to each activity, while the exit criteria and associated metrics defined the knowledge exchange outcomes associated with each activity.

Step 5 of our checklist deals with the identification of the knowledge sources and receivers for each activity. At TechCo, the knowledge sources and receivers for each activity were defined in the process definition handbooks. These handbooks contained generic guidelines for performing activities such as group reviews, defect prevention, and so forth, as well as detailed checklists for accomplishing activities such as high-level design, functional design, code review, and so forth. In addition, TechCo created a series of templates for producing various types of documents generated during the software development process (e.g., requirements specification, unit test plan, and acceptance test documents). Specifying these items to a sufficient level of detail so that every project could follow the guidelines as well as produce documents in a standardized fashion was a key step in helping TechCo to achieve Level 5 certification. These templates represented a codification of knowledge that then could be exchanged with the various sources and receivers.

Having defined in detail its software engineering processes, TechCo began to examine the best mechanism by which it could support the activities of the knowledge workers executing these processes (Step 6 of our checklist). It is well-known that software development requires coordination and collaboration among various stakeholders (Kraut & Streeter, 1995) (i.e., project leaders, module leaders, analysts, and developers, as well as members of quality assurance groups). Armed with an understanding of the individual tasks performed by each knowledge worker, the types of knowledge exchanged among the various stakeholders and the coordination and communication needs of knowledge workers during each phase of the life cycle, TechCo was able to design a project-level KM system: the Project Reporting and Management System (PRMS). PRMS facilitates efficient knowledge sharing among workers by providing (a) support for essential collaborative activities such as configuration management of work products (e.g., documents and code); (b) division, scheduling, and assignment of subactivities to various knowledge workers; (c) support for testing and problem reporting; and (d) change management. In addition, PRMS captures various metrics relating to defects per stage, effort spent per stage, and so forth. In essence, PRMS is a project-level KM that serves as a one-stop shop for sharing key knowledge related to a given project, including informal knowledge generated during the course of the project.

By achieving the high-level of process maturity and control over its software development processes and the use of tools such as PRMS, TechCo was able to maintain its competitive edge in the marketplace. However, it still did not have any organizationwide mechanism in place to facilitate knowledge sharing across various projects. This often resulted in wasted effort and costly mistakes in personnel and time estimation. For example, there was no easy mechanism to solve the problems related to the "who-knows-what" issue that plagues large organizations. Moreover, no mechanisms for sharing knowledge regarding best practices and processes were in place.

To address these problems, TechCo developed and deployed an organization-level KM system in the form of an electronic knowledge asset library (KAL). This system serves as a repository for knowledge about its software development process (i.e., the guidelines, checklists, templates, etc.). TechCo organized knowledge generated from prior projects based on two criteria: industry vertical (e.g., manufacturing, pharmaceutical) and technology characteristics (e.g., languages, tools, databases). Detailed knowledge from each project (captured in the PRMS) in the form of all final documents produced during the various phases of the lifecycle (e.g., requirements documents, high-level design documents, program code, and records of quality assurance reviews) is stored in this system. Furthermore, because of its highly mature processes, TechCo also is able to capture quantitative information (e.g., effort and defects per stage) in the system. This system also serves as a forum for posting white papers and tutorials on emerging technology topics. Each knowledge item in the system has associated with it a contact person's information, thus creating knowledge about where expertise resides within the organization.

The knowledge captured in KAL is accessible to all users in the organization. Access to the library is provided through a groupware system based on Lotus Notes® technology. Common navigation functionality, such as ability to search projects based on keywords and other criteria, is provided. Thus, using this system, a project leader initiating a project using J2EE technology in the financial industry can retrieve documents related to prior J2EE projects in the financial industry and use the knowledge in the system to estimate the manpower and time needed to execute the new project successfully. The project leader is able to find and communicate with other project leaders with experience in that domain and make requests for software engineers who have performed well in a specific domain. At the same time, a developer can read tips on how to develop wireless applications using J2ME. Similar systems, specifically the process asset database (PDB) and Knowledge Map are in use at Infosys, one of TechCo's chief competitors (Ramasubramanian & Jagadeesan, 2002). To encourage the sharing of knowledge via these KM systems, TechCo linked knowledge worker financial incentives to systems use, which TechCo believes has led to performance improvements.

Through the use of these KM systems, TechCo has been able to deliver consistently high-quality software products by reducing the barriers of time and space associated with virtual software development (Carmel & Agarwal, 2001). It is worth noting that the process of accomplishing the six steps has taken more than five years. TechCo's efforts were spent on defining and refining the details of the software development processes (Steps 1 through 5) and the needs and motivations of its knowledge workers prior to considering and designing the subsequent technology-based KM system interventions. In the end, TechCo's efforts reflect its response to external competitive pressures and desire to improve interdependent, multi-level (business, process, and knowledge worker) performance (Figure 1). The fact that TechCo has been able to maintain its leadership position in an extremely competitive IT outsourcing/offshoring space provides evidence of the value of the KM initiatives. Thus, the TechCo case reiterates the importance of taking a multi-level, performance-centric view of KM.

Implications for Practice and Research

Successful organizations are searching for ways to improve performance by leveraging knowledge assets more effectively. New products, services, and customer relationships are key drivers of growth for sales and profitability, particularly for firms facing intense competition and rapid technological change (Alavi, 2000; Huber, 2001). Viability often hinges directly upon the competitive quality and exploitation of a firm's underlying knowledge base. Relative to their own environment, every organization will respond differently to the fundamental question posed earlier in this chapter: What performance goal(s) is the organization trying to achieve by managing its knowledge assets? While KM cannot be applied generically, we have provided an overseeing framework and underlying steps that may assist organizations to address this question (Rubenstein-Montano et al., 2001; Tsoukas, 1996).

For practice, our perspective is both adaptive and responsive to different situations. Importantly, our approach considers the entire KM process—strategic objectives, operational factors, the role of technology, and people/culture—as well as underlying knowledge types, flows, tasks, and learning that must to be considered when considering the fit of a KM initiative to a particular organization. As evidenced in our cases, any KM initiative must be aligned with the existing strategic environment (Liebowitz & Beckman, 1998). An organization should assess the relationship of the initiative to current value chain processes, the level of change and resources required to implement the envisioned solution, and the level of senior management support. Senior level support establishes an appreciation of knowledge assets and is essential for the ongoing funding and investment for necessary human and technical resources (Holsapple & Joshi, 2000). A KM initiative must fit with the operational environment. Interventions may change workflow and interpersonal relationships and, thus, may necessitate new roles and/or skills for knowledge workers. Deploying information technology in the form of a KM system also requires consideration of the existing technical environment (Flanagin, 2002; Holsapple & Joshi, 2002; Huber, 2001). The solution must be compatible with networks and platforms, and the organization must be ready to deal with the level of investment and change necessary to implement desired technical functionality. Perhaps the most challenging issue is the assessment of the fit between a KM initiative and the cultural environment. Creating a culture of knowledge sharing is critical to success (Davenport et al., 1998; Fahey & Prusak, 1998; Grover & Davenport, 2001). Given this, an organization needs to assess incentive and reward systems and identify internal inconsistencies. Understanding the internal performance system of a knowledge worker will assist in identifying factors that positively or negatively influence the behaviors of knowledge workers.

For researchers, while we recognize the limits of a case study approach to generalizability, we maintain that the very nature of our framework requires study of its application. This suggests a need for additional qualitative case studies conducted in collaboration with organizations that have engaged in or that are considering KM initiatives. It is only when a sufficient amount of systematic qualitative case study research has been conducted that themes and relationships inherent to our framework can be validated further via quantitative research methods.

Conclusion

A KM strategy entails developing a portfolio of strategically focused initiatives required to achieve business results. Organizations must prioritize these initiatives based on business value, enterprise support, and funding. As such, holistically and systematically understanding the performance environment surrounding organizational knowledge work takes on heightened importance (Massey & Montoya-Weiss, 2002; Rubenstein-Montano et al., 2001). With both prescriptive and descriptive elements, the framework and associated steps developed and offered in this chapter should guide future research and assist organizations that are interested in undertaking and leading KM initiatives.

References

Ackerman, M., Pipek, V., & Wulf, V. (2003). *Sharing expertise: Beyond knowledge management*. Cambridge, MA: The MIT Press.

Ackoff, R. L., & Emery, F. E. (1972). *On purposeful systems.* Chicago: Aldine Atherton.

Alavi, M. (2000). Managing organizational knowledge. In R. W. Zmud (Ed.), *Framing the domains of IT management research: Glimpsing the future through the past* (pp. 15-28). Cincinnati, OH: Pinnaflex Educational Resources.

Alavi, M., & Leidner, D. (1999). Knowledge management systems: Issues, challenges, and benefits. *Communications of the Association for Information Systems, 1.* Retrieved from http://cais.isworld.org/articles/1-7/article.htm

Alavi, M., & Leidner, D. (2001). Review: Knowledge management and knowledge management systems: Conceptual foundations and research issues. *MIS Quarterly, 25*(1), 107-136.

Barney, J. B. (1991). Firm resources and sustained competitive advantage. *Journal of Management, 17*(1), 99-120.

Becerra-Fernandez, I., & Sabherwal, R. (2001). Organizational knowledge management: A contingency perspective. *Journal of Management Information Systems, 18*(1), 23-55.

Berry, L. L., & Parasuraman, A. (1991). *Marketing services.* New York: Free Press.

Carmel, E., & Agarwal, R. (2001). Tactical approaches to alleviating distance in global software development. *IEEE Software, 18*(2), 23-29.

Checkland, P. B. (1981). *Systems thinking, systems practice.* Chichester: Wiley.

Checkland, P. B., & Howell, S. (1998). *Information, systems and information systems: Making sense of the field.* New York: Wiley & Sons.

Cooper, R., & Kleindschmidt, E. (1995). An investigation into the NPD process: Steps, deficiencies, impact. *Journal of Product Innovation Management, 12,* 374-391.

Davenport, T. H. (1993). *Process innovation: Reengineering work through information technology.* Cambridge, MA: Harvard Business School Press.

Davenport, T. H., DeLong, D. W., & Beers, M. C. (1998). Successful knowledge management projects. *Sloan Management Review, 39*(2), 43-57.

Davenport, T. H., & Prusak, L. (1998). *Working knowledge: How organizations manage what they know.* Boston: Harvard Business School Press.

Davenport, T. H., & Short, J. E. (1990). The new industrial engineering: Information technology and business process redesign. *Sloan Management Review, 31*(4), 11-27.

Day, G. S. (1994). The capabilities of market-driven organizations, *Journal of Marketing, 58*(4), 37-52.

Day, G. S. (2000). Managing marketing relationships. *Journal of the Academy of Marketing Science, 28*(1), 24-31.

Dyer, G., & McDonough, B. (2001, May). The state of knowledge management. *Knowledge Management, 4*(5), 31-36.

Fahey, L., & Prusak, L. (1998). The eleven deadliest sins of knowledge management. *California Management Review, 40*(3), 265-276.

Flanagin, A. J. (2002). The elusive benefits of the technology support of knowledge management. *Management Communication Quarterly, 16*(2), 242-248.

Gao, F., Li, M., & Nakamori, Y. (2002). Systems thinking on knowledge and its management: Systems methodology for knowledge management. *Journal of Knowledge Management, 6*(1), 7-17.

Gery, G. (1997). Granting three wishes through performance-centered design. *Communications of the ACM, 40*(7), 54-59.

Gold, A. H., Malhotra, A., & Segars, A. H. (2001). Knowledge management: An organizational capabilities perspective. *Journal of Management Information Systems, 18*(1), 185-214.

Gordon, J. (1996). Performance technology. In D. Zielinski (Ed.), *The effective performance consultant* (pp. 1-7). Minneapolis: Lakewood Publications.

Grover, V., & Davenport, T. H. (2001). General perspectives on knowledge management: Fostering a research agenda. *Journal of Management Information Systems, 18*(1), 5-21.

Hammer, M. (1990). Reengineer work: Don't automate, obliterate. *Harvard Business Review, 67*(4), 104-112.

Hammer, M., & Champy, J. (1993). *Reengineering the corporation: A manifesto for business revolution.* New York: Harper Collins.

Hinds, P. J., & Pfeffer, J. (2003). Why organizations don't "know what they know": Cognitive and motivational factors affecting the transfer of expertise. In M. Ackerman, V. Pipek, & V. Wulf (Eds.), *Sharing expertise: Beyond knowledge management* (pp. 3-26). Cambridge, MA: The MIT Press.

Holsapple, C. W., & Joshi, K. D. (2000). An investigation of factors that influence the management of knowledge in organizations. *Journal of Strategic Information Systems, 9*, 235-261.

Holsapple, C. W., & Joshi, K. D. (2002). Knowledge management: A three-fold framework. *The Information Society, 18*(1), 47-64.

Huber, G. P. (2001). Transfer of knowledge in knowledge management systems: Unexplored issues and suggested studies. *European Journal of Information Systems, 10*(2), 72-79.

Jalote, P. (2000). *CMM in practice: Processes for executing software projects at Infosys.* Reading, MA: Addison-Wesley.

Kraut, R. E., & Streeter, L. (1995). Coordination in software development. *Communications of the ACM, 38*(3), 69-81.

Leonard, D., & Sensiper, S. (1998). The role of tacit knowledge in group innovation. *California Management Review, 40*(3), 112-132.

Liebowitz, J., & Beckman, T. (1998). *Knowledge organizations: What every manager should know*. Boca Raton, FL: St. Lucie/CRC Press.

Maier, R., & Remus, U. (2001). Towards a framework for knowledge management strategies: Process orientation as strategic starting point. In *Proceedings of the 34th Hawaii International Conference on System Sciences.*

Massey, A. P., & Montoya-Weiss, M. (2002). Performance-centered design of knowledge intensive processes. *Journal of Management Information Systems, 18*(4), 37-58.

Massey, A. P., Montoya-Weiss, M., & Holcom, K. (2001). Re-engineering the customer relationship: Leveraging knowledge assets at IBM. *Decision Support Systems, 32*, 155-170.

Massey, A. P., Montoya-Weiss, M., & O'Driscoll, T. (2002). Knowledge management in pursuit of performance: Insights from Nortel Networks. *MIS Quarterly, 26*(3), 269-289.

Moorman, C., & Rust, R. T. (1999). The role of marketing. *Journal of Marketing, 63*, 180-197.

O'Dell, C., & Grayson, C. J. (1998). If only we knew what we know: Identification and transfer of internal best practices. *California Management Review, 40*(3), 154-174.

Paulk, M., Weber, C. W., Curtis, B., & Chrissis, M. B. (1995). *The capability maturity model for software: Guidelines for improving the software process*. Reading, MA: Addison-Wesley.

Peppers, D., Rogers, M., & Dorf, R. (1999). *The one-to-one fieldbook*. New York: Currency and Doubleday.

Radice, R. A., Roth, N. K., O'Hara, A. C., & Ciarfella, W. A. (1985). A programming process architecture. *IBM Systems Journal, 24*(2), 79-90.

Ramasubramanian, S., & Jagadeesan, G. (2002). Knowledge management at Infosys. *IEEE Software, 19*(3), 53-55.

Rosenberg, M. (1995). Performance technology, performance support, and the future of training. *Performance Improvement Quarterly, 8*(1), 12-20.

Rubenstein-Montano, B., Liebowitz, J., Buchwalter, J., McCaw, D., Newman, B., Rebeck, K., et al. (2001). A systems thinking framework for knowledge management. *Decision Support Systems, 31*(1), 5-16.

Rummler, G., & Brache, A. (1992). Transforming organizations through human performance technology. In H. D. Stolovitch & E. J. Keeps (Eds.), *Handbook of human performance technology: A comprehensive guide for analyzing and solving performance problems in organizations* (pp. 32-49). San Francisco: Jossey-Bass.

Rus, I., & Lindvall, M. (2002). Knowledge management in software engineering. *IEEE Software, 19*(3), 26-38.

Rus, I., Lindvall, M., & Sinha, S. S. (2001). Knowledge management in software engineering. DACS state of the art report (SOAR). Retrieved April 10, 2005, from http:www.dacs.dtic.mil/techs/kmse/kmse.html

Senge, P. M. (1990). *The fifth discipline*. New York: Doubleday.

Soo, C., Devinney, T., Midgley, D., & Deering, A. (2002). Knowledge management: Philosophy, processes, and pitfalls. *California Management Review, 44*(4), 129-150.

Stewart, T. (2001). *The wealth of knowledge: Intellectual capital and the twenty-first century organization.* New York: Doubleday.

Stolovitch, H. D., & Keeps, E. J. (1999). What is human performance technology? In H. D. Stolovitch & E. J. Keeps (Eds.), *Handbook of human performance technology: A comprehensive guide for analyzing and solving performance problems in organizations* (pp. 3-23). San Francisco: Jossey-Bass.

Sviokla, J. J. (1996). Knowledge workers and radically new technology. *Sloan Management Review, 37*(4), 25-40.

Teng, J. T. C., Grover, V., & Fiedler, K. D. (1994). Business process reengineering: Charting a strategic path for the information age. *California Management Review, 36*(3), 9-31.

Tsoukas, H. (1996). The firm as a distributed knowledge system: A constructionist approach. *Strategic Management Journal, 17,* 11-25.

Ward, J., & Aurum, A. (2004). Knowledge management in software engineering: Describing the process. In *Proceedings of the 2004 Australian Software Engineering Conference, (IEEE Computer Society.* Retrieved April 10, 2005, from http://csdl.computer.org/comp/proceedings/aswec/2004/2089/00/20890137abs.htm

Chapter VII

The Influence of Organizational Trust on the Use of KM Systems and on the Success of KM Initiatives

Vincent Ribière, New York Institute of Technology, USA

Francis Tuggle, Chapman University, USA

Abstract

While the discipline of knowledge management (KM) is no longer emerging, some organizations are still struggling to find the right approach that will allow them to fully take advantage of their intellectual assets. Having the proper organizational culture remains an important barrier to knowledge management success. This empirical research project, conducted with data from 97 organizations involved in KM, explores relationships between the level of organizational trust and the use of KM methodologies, in particular the use of codification KM methodologies and personalization KM methodologies. The presence of trust also can be used as an indicator of KM initiative success. The contribution of this research may help organizations seeking to launch or adapt a KM initiative to choose which KM tools and technologies to deploy in order to maximize their chances of success. Finally, a rank-ordered list of KM methodologies in descending order of usefulness is reported.

Introduction

Knowledge management (KM) initiatives are expanding across all types of organizations and companies worldwide. Many benefits resulting from the successful implementation of KM have been demonstrated and published, but unfortunately, not all KM initiatives are successful. Studies and surveys looking at some of the causes of KM initiative failure (Barth, 2000; Chua & Lam, 2005; *Knowledge Management Review*, 2001; KPMG Consulting, 2000; Microsoft, 1999; Pauleen & Mason, 2002) all come to the same conclusion: Organizational culture is an important barrier to KM success and is an important precondition for KM success (Tuggle & Shaw, 2000). After having primarily focused KM efforts on information technology (IT), practitioners now are realizing the importance of the "soft" aspect of KM initiatives. It seems that the IT tools designed to facilitate knowledge creation, capture, representation, storage, and sharing are now available, but their efficient use and acceptance by knowledge workers remains constrained by organizational culture.

There is general agreement that a knowledge-sharing organizational culture must be present or nurtured in order to succeed with KM (see, for example, Alavi and Leidner [2001] and Jennex and Olfman [2001]). However, King (2006) presented a model wherein two types of organizational cultures (supervisory control and organizational support) are assessed based on their influence on knowledge-sharing frequency and sharing effort. The findings of this study show that the presence of a knowledge-sharing culture is not always important for some types of organizations. We agree with the fact that you can force people to share and obtain a larger amount of knowledge collected than in a natural knowledge-sharing culture, but an approach is neither enduring nor desirable. The quality of the knowledge shared is critical. If knowledge workers are forced to share their knowledge, the quality of the knowledge shared will not be as high as when shared for the good of the company or for other self-motivated reasons. So we stand on our position that culture is critical and cannot be ignored.

However, the current literature is weak regarding the identification of the critical cultural success factors that lie behind the term *culture*. Often mentioned as important cultural factors are altruism, reciprocity, trust, repute, openness, solidarity, sociability, motivation, commitment, tie strength, and others (Davenport & Prusak, 1998; Levin & Cross, 2004; Lucas, 2005; Malhotra & Galletta, 2003; Park, Ribière, & Schulte, 2004; Yih-Tong Sun & Scott, 2005). Jennex and Olfman (2004) identify 14 KMS success factors, one of which (SF5) specifically states, "An organizational culture that supports learning and the sharing and use of knowledge." Based on a literature review, trust seems to be one of the most critical factors among the ones listed previously. If knowledge workers are reluctant to trust each other, they are not likely to share and exchange knowledge. Without the presence of trust, the benefits of reciprocity, repute, openness, solidarity, and sociability will not occur, and levels of motivation and commitment might be seriously affected (negatively). "Without trust, knowledge management will fail, regardless of how thoroughly it is supported by technology and rhetoric and even if the survival of the organization depends on effective knowledge transfer" (Davenport & Prusak, 1998). Does having a low trust organizational culture mean that a company should not consider launching a KM initiative? Since culture is difficult to change in the short term, are there some KM approaches/tools that might succeed and might also help culture to evolve in the right direction? This chapter will attempt to shed light on these issues.

Statement of the Problem

As Bell DeTienne, Dyer, Hoopes, and Harris (2004) point out, even though studies have been conducted to better understand the concept of trust, very few studies have looked at how trust functions in a KM environment and how trust correlates with knowledge-sharing factors. This study attempts to better understand these phenomena as well as to measure how organizational trust affects the choice and use of KM tools and technology and the resulting success of the organization's KM initiative or lack thereof. Our main research question is as follows:

> *Does the level of organizational trust influence which KM tools and technology will be successful?*

In order to study this research question, the level of organizational trust is assessed through a questionnaire distributed to knowledge workers from various organizations involved in KM. Second, the types of KM tools and technology implemented and used in these organizations were evaluated. Finally, the level of success achieved was assessed. The research project reported on herein is a straightforward extension of these pilot activities. The next sections define these aforementioned variables.

Organizational Trust

Considerable research has been conducted concerning the concept of trust, both interpersonal trust and organizational trust. As with the concept of organizational culture, organizational trust has been defined somewhat differently in the literature by numerous authors (Carnevale & Wechsler, 1992; Culbert & McDonough, 1986; Griffin, 1967; Luhmann, 1979; Matthai, 1989; McKnight & Chervany, 2000). The definitions of trust are numerous and sometimes confusing, mainly due to each discipline viewing trust from its own perspective. Two definitions of trust were selected:

> *Trust consists of a willingness to increase your vulnerability to another person whose behavior you cannot control, in a situation in which your potential benefit is much less than your potential loss if the other person abuses your vulnerability.* (Zand, 1997)

and

> *Belief that those on whom we depend will meet our expectations of them.* (Shaw, 1997)

Research on trust often is associated with research on organizational commitment and work attitudes (Cook & Wall, 1980; Mowday, Steers, & Porter, 1979; Nyhan, 1999). Research conducted by Daley and Vasu (1998), examining employee attitudes of organizational trust toward those in top management positions, demonstrated that demographic controls (education, pay level, race, and gender) exhibited no substantive effect. Attitudes assessing internal job characteristics (benefits, extrinsic rewards, and work environment) demonstrated a relationship in fostering trust. External work characteristics (job satisfaction, supervisory evaluation, and political interference) also emerged as determinants of organizational trust (Daley & Vasu, 1998). Trust often is categorized in two forms (Levin, Cross, & Abrams, 2002a, 2002b; McAllister, 1995): cognition-based trust and affective-based trust. The cognition-based dimension of trust is associated with beliefs about competence, integrity, responsibility, credibility, reliability, and dependability. It is mainly task-oriented. The affective-based dimension of trust is based on beliefs about reciprocated care and concern, benevolence, altruism, commitment, and mutual respect. It is relationship-oriented. In organizational settings, the cognition-based form of trust is more central, since it impacts more particularly reliability and dependability (Cook & Wall, 1980). This dimension of trust will be assessed and used for this study.

In addition to the many definitions of trust, many tools also have been created to assess its level in an organization. Among them is the survey tool designed by Cook and Wall (1980) that since has been extended by Wilson (1993). Wilson (1993) developed a heuristic conceptualization in the form of an influence diagram that can be used by managers in assessing the level of organizational trust. Cummings and Bromiley (1996) designed a survey tool called the organizational trust inventory (OTI). This tool is intended to measure the degree of trust among units of an organization or among organizations. Their questionnaire is based on a three-by-three definitional matrix of trust as a belief in which three dimensions of trust (keeps commitments, negotiates honestly, and avoids taking excessive advantage) are mapped against three components of belief (affective state [feel], cognition [think], and intended behavior). Nyhan and Marlowe (1997) developed a 12-item scale to measure an individual's level of trust in his or her supervisor and his or her work organization as a whole. Two books on trust also offered assessment tools. Ciancutti and Steding (2000) offer an audit questionnaire based on 21 questions as well as six open-ended questions. This questionnaire is designed to detect both the overall level of trust and the type of issues in which closure is a concern. The second book by Lewis (1999) is oriented more toward how companies build mutual trust and how interpersonal relationships are a critical component. The tool presented in this book is defined as a yardstick for measuring how close your company is to building high trust. A set of 21 trust practices is listed, and for each of these, a low trust behavior as well as a high trust behavior is listed. "Where you and your partner fall in the continuum between high and low trust determines your ability to rely on each other to reach a common objective" (Lewis, 1999). Five trust factors defined by De Furia (1996, 1997) were determined to be most relevant to our research: (1) sharing relevant information; (2) reducing controls; (3) allowing mutual influences; (4) clarifying mutual expectations; and (5) meeting expectations. These factors are described in more detail in the following section of this chapter. Very often, people think that an organizational culture with a high level of sociability also implies a high level of trust. That is not always true. Consider the example of a parent-child relationship: parents love their children, but it does not imply that they trust them (e.g., they will not leave their children by themselves). The opposite is also true: you might trust someone, but you might not necessarily like that person (e.g., an airplane pilot).

One also needs to remember that trustworthiness takes a long time to build, and yet trust can be destroyed in an instant. These examples show the complexity and fragility associated with trust. The benefits of high trust include the following (De Furia, 1997):

- Stimulates innovation,
- Leads to greater emotional stability,
- Facilitates acceptance and openness of expression, and
- Encourages risk taking.

Consequences of low trust include the following (De Furia, 1997):

- Values, motives of others are misperceived,
- Less accurate communication, poor reception,
- Diminished ability to recognize and accept good ideas,
- Increased attempt to obtain relevant information (grapevine),
- Increased control mechanisms,
- Self control replaced by external controls,
- Delayed implementation of actions and projects,
- Increased rejection, defensiveness, hostility, and
- Win-lose mentality replaces win-win.

> *Trust is the one essential lubricant to any and all social activities. Allowing people to work and live together without generating a constant, wasteful flurry of conflict and negotiations.* (Cohen & Prusak, 2001)

Knowledge Management Tools and Technologies

Numerous publications present knowledge management practice/tool/technology frameworks. Among them, the knowledge management spectrum presented by Binney (2001) offers a good overview of various KM tools and practices that are offered to organizations in order to better manage their knowledge. The tools and practices are organized into six categories: transactional, analytical, asset management, process, developmental, and innovation and creation. Most of them are IT-oriented, since IT is the main enabler for KM. Nevertheless, other KM practices that are not driven by IT also must be taken into consideration in order to fully understand the KM strategy of an organization.

Two main KM strategies or approaches emerged: codification vs. personalization. Hansen, Nohria, and Tierney (1999) describe how different companies focus on different practices and strategies in order to manage their knowledge. Additional reasons for this particular categorization of KMS approaches are offered by Jennex and Olfman (2003). Dennis and Vessey

(2005) also used these two strategies as the bedrock for their three knowledge management systems: knowledge hierarchies (in which knowledge is viewed as a formal organizational resource), knowledge markets (in which knowledge is treated as an individual resource), and knowledge community (in which knowledge is viewed as a communal resource).

The Codification Approach

The first strategy identified by Hansen et al. (1999) is called *codification*, which relies heavily on IT. One of the benefits of the codification approach is the reuse of knowledge. "Knowledge is codified and stored in databases, where it can be accessed and used easily by anyone in the company. Knowledge is codified using a people-to-documents approach: it is extracted from the person who developed it, made independent of that person, and reused for various purposes" (Hansen et al., 1999). It has been named and described differently by other authors: the cognitive network model (Swan, Newell, Scarbrough, & Hislop, 1999); the collecting dimension (Denning, 1998); the product view approach (Know-Net, 2000); the transformation model (Natarajan & Shekhar, 2000); distributive applications (Zack & Michael, 1998); and the document-centered approach and the technological approach (Wick, 2000). After a close analysis of these different portrayals, one can conclude that all of these descriptions and definitions are very similar and depict the same type of practices and tools (Ribière, 2001).

The Personalization Approach

The personalization approach (Hansen et al., 1999) focuses on developing networks for linking people so that tacit knowledge can be shared. It invests moderately in IT. This approach focuses on dialogue between individuals, not knowledge in a database. "Knowledge that has not been codified—and probably couldn't be—is transferred in brainstorming sessions and one-on-one conversations" (Hansen et al., 1999). An investment is made in building networks of people in which knowledge is shared not only face-to-face but also over the telephone, by e-mail, and via videoconference. All the previously cited authors who defined the codification approach also came up with their own definitions for this approach: the

Figure 1. The codification approach (Adapted from Zack and Michael, 1996)

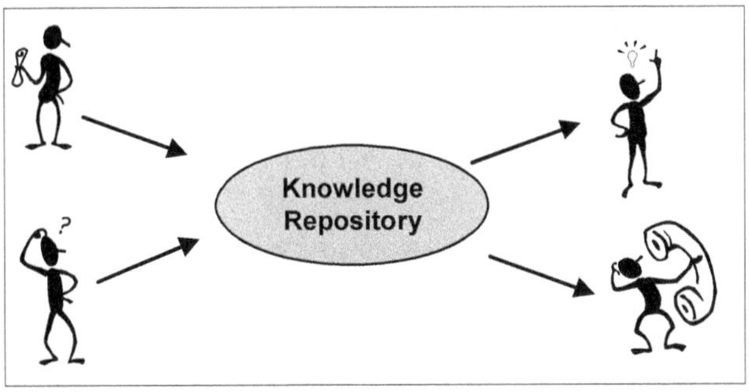

Figure 2. The personalization approach (Adapted from Tiwana, 2002)

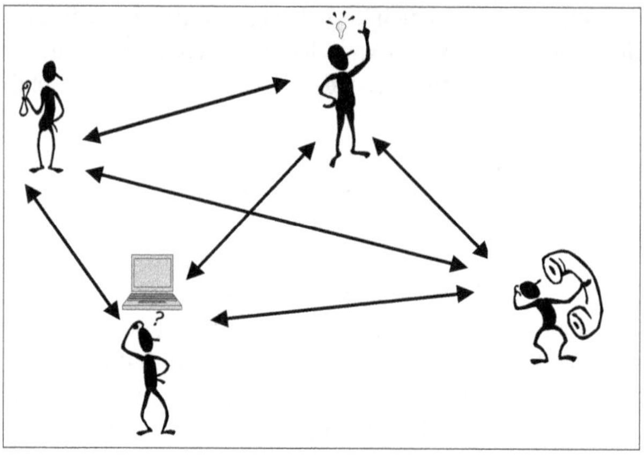

community networking model (Swan et al., 1999); the connecting dimension (Denning, 1998); the process-centered approach (Know-Net, 2000); the independent model (Natarajan & Shekhar, 2000); the collaborative approach (Zack & Michael, 1998); and socio-organizational knowledge management (Wick, 2000).

KM Initiative Success

It is always difficult and open to controversy to define and measure success. Various metrics (qualitative and quantitative) can be used to measure success. For example, Jennex and Olfman (2004) offer a success model based upon the Delone and McLean (1992) IS success model and discuss four models of KM success: (1) the knowledge value chain (Bots & Bruiin, 2002); (2) the KM success model (Massey, Montoya-Weiss, & Driscoll, 2002); (3) the KM effectiveness model (Lindsey, 2002); and (4) the KMS success model (Jennex & Olfman, 2003). Four main indicators defined and used by Davenport et al. in their publication concerning successful knowledge management projects were adopted (Davenport, De Long, & Beers, 1998):

1. Growth in the *volume* of knowledge available since the KM initiative has been launched (e.g., number of documents available)
2. Growth in the *usage* of knowledge available since the KM initiative has been launched (accesses to repositories or the number of participants for discussion-oriented projects)
3. The likelihood that the project would survive without the support of a particular individual or two; that is, the project is an organizational initiative, not an individual project
4. Growth in the resources (e.g., people, money) attached to KM initiatives

Success was measured, based on two dimensions. Since the main purpose of a KMS is to facilitate the flow and dissemination of knowledge, an important dimension for success is the fact that different employees use the system. Success factors #1 and #2 were used to measure this dimension of success. The second dimension of success is based on the robustness of the KM initiative. If KM is given the resources and if there is a clear commitment from senior management to make it happen, then robustness is a success factor. Success factors #3 and #4 were used to measure this second dimension of success.

We believed that it also would be relevant to check if the expected benefits of the KM initiative were achieved and, if yes, to what degree. To do so, we used a questionnaire developed by KPMG (2000). Fifteen main benefits often expected after KM implementation were used (KPMG, 2000).

Additional success factors could have been used, such as the 12 KMS success factors presented by Jennex and Olfman (2004), but it was easier to work with a smaller number of core variables. The average of all the success factors was used to obtain the success level score. This score helped us to differentiate highly successful KM initiatives from less successful ones.

Research Design and Methodology

Research Hypotheses

Relationships Between Organizational Trust and the Dominant KM Approach

We previously introduced the two main KM approaches: codification vs. personalization. In their original paper, Hansen, Nohria, and Tierney (1999) defended the idea that effective firms excel by emphasizing one of the approaches and using another in a supporting role. For companies adopting this strategy, we will call the main approach the *dominant approach.* For companies focusing simultaneously and equally on both approaches we will call it a *balanced approach.*

Table 1. Fifteen common KM benefits

Better decision making	Sharing best practice
Better customer handling	Reduced costs
Faster response to key business issues	New ways of working
Improved employee skills	Increased market share
Improved productivity	Create additional business opportunities
Increased profits	Improved new product development
Increased innovation	Staff attraction/retention
Increased share price	

Early in the 1990s, Jack Welsh had already underlined the important role of trust:

> *Trust is enormously powerful in a corporation. People won't do their best unless they believe they'll be treated fairly—that there's no cronyism and everybody has a real shot. The only way I know to create that kind of trust is by laying out your values and then walking the talk. You've got to do what you say you'll do, consistently and over time.* (Welch, 1993)

The early KM efforts conducted by Buckman laboratories have been coroneted with success, and once again, trust was mentioned as a critical component:

> *It is important to create a climate of continuity and trust so that we may have proactive knowledge sharing across time and space. Organizational culture must change from a state of hoarding knowledge to gain power to one of sharing knowledge to gain power.* (Davenport & Prusak, 1998)

To a large degree, trust requires time to build. Through interaction, open communication, loyalty, reciprocity, and competence, the level of trust increases over time. Without trust, individuals are not likely to share and collaborate in knowledge exchanges. Among the studies conducted to demonstrate the relationship between trust and knowledge sharing, we cite the works of Nelson and Cooprider (1996) and Politis (2003).

Many other studies have been conducted to demonstrate the impact of other variables (e.g., motivation, social capital, communication, etc.) on knowledge sharing, and trust often is mentioned as a precondition for knowledge sharing. As we presented earlier, the personalization approach consists of connecting people so they can exchange tacit knowledge. This interaction takes place on a face-to-face basis, or the interaction can be assisted by IT. In both situations, knowledge workers will not be likely to share their knowledge (or will share it only partially) if they don't trust the person. We can suppose that in an atmosphere of low trust, people will be more likely to use a knowledge management system (codification tools and practices) to look for information rather than directly asking a colleague they do not trust. When the level of trust is high, people are more open, more honest, and more likely to collaborate. We then can suppose that they will be more likely to use personalization tools and practices. Based on the previous discussion, we will postulate the following four hypotheses:

H_1: Organizations with a low level of organizational trust are more likely to adopt a codification-dominant approach than organizations with a high level of organizational trust.

H_2: Organizations with a high level of organizational trust are more likely to adopt a personalization-dominant approach than organizations with a low level of organizational trust.

H_3: Organizations with a high level of organizational trust are more likely to adopt balanced approaches than organizations with a low level of organizational trust.

H_4: Organizations with a high level of organizational trust have a higher usage level of personalization tools and practices than organizations with a low level of organizational trust.

Relationship between Organizational Trust and the Success of a KM Initiative

Since the success of a KM initiative relies heavily on people sharing knowledge, we can assume that if the level of organizational trust is high, then people will have fewer barriers to share knowledge and, consequently, the level of success of the KM initiative should be higher. Based on this assumption, we postulate the following hypothesis:

H_5: Organizations with a high level of organizational trust are more successful in their KM initiatives than organizations with a low level of organizational trust.

Relationships between the Dominant KM Approach and the Success of a KM Initiative

Hansen, Nohria, and Tierney (1999) defend the idea that a 20-80 split between codification and personalization (or vice versa) is a proper strategy for a firm to follow. They postulate that companies trying to excel at both approaches risk failing at both. The 20-80 split raised much discussion in the *Harvard Business Review* forum that referred to this article (1999). Denning mentioned that organizations that focus entirely on a personalization approach, with little or no attempt at codification, can be very inefficient (Denning, 1998). Koening (2004) affirms, based on a case study in the pharmaceutical industry, that a 50-50 balance also could be a good strategy. The findings of a study conducted by Choi and Lee (2002) show that companies having a dynamic style (focusing simultaneously on managing tacit and explicit knowledge) obtain the highest performance score. This style is equivalent to our balanced strategy. The scores obtained by the system-oriented style (strong focus on managing explicit knowledge) and by the human-oriented style (strong on managing tacit knowledge) were lower and almost equal. These two styles correspond to our dominant codification strategy and to our dominant personalization strategy. Other authors defend the idea that only the personalization approach can be used to leverage knowledge and improve creativity, and can bring the most benefits (Delmonte & Aronson, 2004; McDermott, 1999). They do not deny the usefulness of the codification approach, but they see it as secondary. Based on the result of previous studies, we postulate the following three hypotheses:

H_6: Organizations with a balanced approach are more successful in their KM initiative than organizations with a codification dominant approach.

H_7: Organizations with a balanced approach are more successful in their KM initiative than organizations with a personalization dominant approach.

H_8: Organizations with a personalization dominant approach are more successful in their KM initiative than organizations with a codification dominant approach.

Assessment of Variables

A survey tool (a questionnaire) was developed in order to assess:

1. The level of organizational trust
2. The level of use of different KM tools and technologies deployed in each organization (codification emphasis vs. personalization emphasis)
3. The perceived success of the KM initiative

Assessing Organizational Trust

The tool selected, the Organizational Trust Survey (OTS), was developed and validated by De Furia (1996, 1997) in which trustworthiness (TW) is based on five behaviors:

TW = SI + RC + AI + CE + ME

- **Sharing relevant information (SI):** Refers to the behaviors whereby one individual transmits information to another person.
- **Reducing controls (RC):** Refers to the behaviors affecting the processes, procedures, or activities with which one individual (1) establishes the performance criteria or rules for others, (2) monitors the performance of another person, (3) adjusts the conditions under which performance is achieved, or (4) adjusts the consequences of performance (i.e., positive or negative reinforcements).
- **Allowing for mutual influences (AI):** Occurs when one person makes a decision that affects both individuals. Mutual influence means that both individuals have approximately an equal number of occurrences of convincing the other or making the decision for both individuals.
- **Clarifying mutual expectations (CE):** Refers to those behaviors wherein one person clarifies what is expected of both parties in the relationship. It involves sharing information about mutual performance expectations.
- **Meeting expectations (ME):** Involves any behaviors in which one individual fulfills the behavioral expectations of another person. It is related closely to confidence, reliability, and predictability.

The OTS allows organizations to measure the trust-related behaviors of various categories of people within the organization—upper managers, first line supervisors, and coworkers—in relation to how employees' trust-related expectations are being met. It also measures trust-related behaviors between organizational units and the perceived impacts of organizational policies and values on trust-related behaviors. This tool (questionnaire) is based on 50 questions (10 questions for each of the five factors). Because of the existence of a pretested questionnaire with a small number of variables, necessary because of the somewhat limited size of our dataset, the OTS was used.

Table 2. Codification and personalization KM tools and technologies

	KM Tools and Technologies
Codification	E-mail and Listserv
	Corporate Intranet—Extranet—Internet
	Database management systems
	Search engines—Intelligent agents
	Data warehouses—Data marts
	Web-based training—E-learning
	Help-desk applications
	DMS
	Multimedia repositories
	DSS and expert systems
	Data mining—Knowledge discovery
	Knowledge mapping
Personalization	Expertise locators—Corporate Yellow pages—Who's who
	Communities of practice (interests in the same topic, field)
	Communities of purpose (project, task-oriented)
	Groupware
	Teleconferencing (shared applications, whiteboards)
	Best practices repository
	Videoconferencing (using audio and/or video)
	Mentoring—Tutoring
	Storytelling
	Desktop computer conferencing
	Online chat and instant messaging

Assessing the Use of KM Tools and Technologies and the Dominant KM Approach Adopted

For this section of the questionnaire, an assessment tool was developed. The most common tools and technologies used for knowledge management initiatives were listed, based on a literature review. These technologies cover the six categories of the knowledge management spectrum presented by Binney (2001).

Respondents were asked to list the KM tools and technologies used at the organizational level (see Table 2). A sense of the degree of use or utilization ranging from *most used* to *least used* was employed to enrich this insight. It might be argued that some of the personalization tools (e.g., corporate yellow pages), in fact, are examples of codified knowledge; the critical delineator is *how* the tools are used in practice. For example, the crucial fact about corporate yellow pages is not that it is a knowledge repository, but that employees use it to connect to experts.

The level of usage of each KM approach was calculated to determine the dominant approach (codification or personalization). We estimated that in order to qualify an approach as a dominant one, its score should be at least 10% higher than the score of the other one. This means that the dominant strategy should attain a usage level of at least 55% of the overall usage level. Organizations obtaining usage scores similar for both approaches (in the 10% bracket) will be considered to have a balanced strategy (no dominant one).

KM Initiative's Success

As mentioned earlier in this chapter, four main indicators were used to assess the level of success as well as 15 expected benefits. Respondents were asked to assess on a five-point Likert scale to what degree they believed that the following statements corresponded to the current success status of their organizational KM initiative:

Strongly Agree	Agree	Neither Agree nor Disagree	Disagree	Strongly Disagree
2	1	0	-1	-2

- I have noticed a significant growth in the *volume* of knowledge available since the KM initiative has been launched (number of documents available).

  2 1 0 -1 -2

- I have noticed a significant growth in the *usage* of knowledge available since the KM initiative has been launched (accesses to repositories and number of participants for discussion-oriented projects).

  2 1 0 -1 -2

- I believe that the project would survive without the support of a particular individual or two.

  2 1 0 -1 -2

- I believe that resources (e.g., people, money) attached to KM initiatives are going to grow.

  2 1 0 -1 -2

Regarding the 15 KM benefits expected and achieved, the respondents were asked to assess on a five=point Likert scale to what degree they believed that the benefits were achieved (if expected):

To a very high extent	To a high extent	To some extent	To a little extent	To a very little extent
2	1	0	-1	-2

Benefits	Expected	Achieved
Better decision making	Yes / No	2 · 1 · 0 · -1 · -2
Better customer handling	Yes / No	2 · 1 · 0 · -1 · -2

Validity and Reliability of the Survey Instrument

Due to the space limitation of this chapter, we only will provide a summarized version of the results of the different tests that were conducted to verify the level of validity and reliability of our instrument (Ribière, 2005). In order to test the internal validity of the different dimensions assessed, we performed a Cronbach alpha test (Table 3). The results demonstrate an acceptable level of internal validity. Some items were removed from the instrument due to their low level of correlation with the other items composing the construct.

We conducted a factor analysis to test the validity of each construct. For the codification and personalization constructs, some items had to be removed due to their low loading on

Table 3. Cronbach alpha test

Construct (number of items remaining)	α
Organizational Trust (24)	0.94
Codification (7)	0.801
Personalization (7)	0.827
KM Success (19)	0.951

the factors. For the other constructs, all the items were retained. Overall, we consider that the levels of validity and of reliability of the assessment tool were acceptable.

Data Collection and Analysis

Data were collected through two main mechanisms: an online version of the questionnaire posted on the Web and a paper version. Most of the responses received (98%) came from the online version. The target population was chief knowledge officers (CKOs), managers, and other employees involved in knowledge management initiatives at any level in an organization. A total of 1,050 e-mails asking for participation were sent out to targeted people involved with KM (members of KM groups and associations). A total of 129 responses were received. This represents a response rate of 12%. A fundamental premise of the research was that targeted organizations must have had experience with KM initiatives. Of the 129 questionnaires received, only 97 were complete and were representative of organizations involved in KM.

Organizations that participated were predominantly large organizations in the consulting and IT and telecommunications fields as well as agencies in the federal government. Respondents mainly were service-oriented, offering both standardized and customized products/services, and were predominantly located in the US.

Data Analysis

Hypotheses Related to the Relationships between Organizational Trust and the Dominant KM Approach

In order to test $\mathbf{H_1}$, $\mathbf{H_2}$, and $\mathbf{H_3}$, we performed a Chi-square test () with our two variables (trust level and dominant approach) being discrete.

The value of the Chi-square (p=.008) is statistically significant (<.05), meaning that we can reject the null hypothesis that our two variables are independent. A dominant codification approach is present in 54.1% of cases (first column) for organizations with a low level of trust and present in 45.9% of cases for organizations with a high level of trust. This finding supports $\mathbf{H_1}$. A dominant personalization approach is present in 73.7% of cases (third column) for organizations with a high level of trust and present only for 26.3% of cases for organizations with a low trust level. This finding allows us to accept $\mathbf{H_2}$. Finally, a balanced approach is used 78% of the time for organizations with a high trust level (second column) compared to only 22% of the time for organization with a low level of trust. This finding allows us to accept $\mathbf{H_3}$.

Table 5. Results of Chi-Square Test between the trust *and* dominant *approach variables*

	Cases					
	Valid		Missing		Total	
	N	Percent	N	Percent	N	Percent
Trust level * Dominant approach	97	96.0%	4	4.0%	101	100.0%

Organizational trust level * Dominant approach

			Dominant approach			
			Codification	Balanced	Personalization	Total
Level of trust	High	Count	17	32	14	63
		Expected count	24.0	26.6	12.3	63.0
		% Trust level	27.0%	50.8%	22.2%	100.0%
		% KM Approach	45.9%	**78.0%**	**73.7%**	64.9%
		Residuals	-7.0	5.4	1.7	
	Low	Count	20	9	5	34
		Expected count	13.0	14.4	6.7	34.0
		% Trust level	58.8%	26.5%	14.7%	100.0%
		% KM Approach	**54.1%**	22.0%	26.3%	35.1%
		Residuals	7.0	-5.4	-1.7	
Total		Count	37	41	19	97
		Expected count	37.0	41.0	19.0	97.0
		% Trust level	38.1%	42.3%	19.6%	100.0%
		% KM Approach	100.0%	100.0%	100.0%	100.0%

Chi-Square test

	Valeur	dl	Asymp. Sig. (2-sided)
Pearson Chi Square	9.596[a]	2	.008
Likelihood Ratio	9.561	2	.008
N of Valid Cases	97		

Note: [a] *0 cells (0.0%) have expected count less than 0.5. The minimum expected count is 6.66.*

In order to test the remaining hypotheses, an independent sample tailed t-test was used to analyze the differences of means between the companies with low and high trust and with a particular dominant KM approach, which provides the results of these tests.

The personalization factor score variable was measured on an interval/ratio scale of values ranging from 0 to 55 (mid-point 27.5). The codification factor score variable was measured on an interval/ratio scale of values ranging from 0 to 60. The level of trust factor score variable was measured on an interval/ratio scale of values ranging from 25 to 125. However, a cutoff point of 75 (the mid-point in the range) was used to divide the variable into two sets. Organizations that obtain trust factor scores greater than 75 were categorized as having a high trust culture, while organizations with scores less than or equal to 75 were categorized as having a low trust culture. The level of success score variable was measured on an interval/ratio scale of values ranging from 1 to 5.

Table 7. Descriptive statistics and results of independent sample tests

Hypoth.	Mean	SD	N	Mean	SD	N	t	p	F/A*
	High Trust			**Low Trust**					
H4	30.73	11.49	63	20.23	9.61	34	4.53	<.0005	**A**
H5	3.32	0.71	51	2.50	0.73	17	3.99	<.0001	**A**
	Codification dominant			**Balanced approach**					
H6	2.95	0.86	29	3.16	0.82	26	-0.92	<.1815	**F**
	Personalization domin.			***Balanced approach***					
H7	3.41	0.57	12	3.16	0.82	26	-0.96	<.1713	**F**
	Codification dominant			**Personalization domin.**					
H8	2.95	0.86	29	3.41	0.57	12	-1.70	<.048	**A**

Note: * *F = Fail to reject the alternative hypothesis; A = Accept the alternative hypothesis*

From the above table, one can see that **H**$_4$, **H**$_5$, and **H**$_8$, each cannot be rejected from our data, but **H**$_6$ and **H**$_7$ both fail to be accepted.

Discussion

The previous findings help us to validate some assumptions made regarding the critical role of organizational trust in organizations involved in KM. **H**$_1$, **H**$_2$, and **H**$_3$ show that organizations with a low level of trust have a tendency to adopt a codification-dominant approach, while the ones with a high level of trust are more likely to adopt a dominant approach of personalization or a balanced approach. In the presence of a trusting culture, knowledge workers are more likely to use personalization tools in order to contact, assist, and share knowledge with their trusted co-workers. If organizational trust is low, knowledge workers are suspicious and don't want to increase their vulnerability by sharing their knowledge with co-workers through one-to-one interactions/communications (physical, vocal, or virtual). They then will rely mainly on codification tools and technologies to manage knowledge. The efficiency of the codification approach also might be affected by lack of trust. Actually, if knowledge workers don't trust their peers and superiors and if they don't feel inclined to use personalization tools, they likely will not trust codified documents either. Since the main benefit of the personalization approach is to leverage individual knowledge, one can suspect that the level of creativity in low trust organizations also might be affected. Such types of organization certainly will encounter problems to become a learning organization.

We have to keep in mind that adopting a dominant personalization approach does not exclude the use of codification tools and practices, but they are used at a lower level. We cannot imagine, and we did not find any organizations focusing 100% on a particular approach. We did not even encounter any organizations that adopted an 80-20 split, as recommended by Hansen, Nohria, and Tierney (1999). This finding is important, since it can be used as a guide for organizations trying to decide on which dominant approach to focus. After having assessed their level of organizational trust, they can decide which approach is more likely to be accepted or adopted by knowledge workers.

H$_5$ demonstrates that organizations with a high level of trust are more successful in their KM initiatives than organizations with a low level of trust. This is probably the most important finding of this research. We present strong statistical evidence that trust plays a role in the success of a KM initiative.

We failed to validate **H**$_6$ or **H**$_7$ trying to demonstrate that a balanced approach could bring more success than a dominant codification or personalization approaches. This shows that the balance of the two KM approaches can be difficult to generalize. Each organization might have different needs or a different culture, and we cannot suppose that a balanced approach will work for all. But we were able to show some evidence (**H**$_8$) that organizations that adopt a personalization dominant approach are more successful in their KM initiative than organizations that focus on a codification dominant approach. Since the main benefit of the personalization approach is to leverage individual knowledge, one can suspect that the level of creativity in low trust organizations also might be affected. Such types of organizations certainly will encounter problems becoming a learning organization.

Table 8. KM tools and technologies most frequently used for knowledge exchanges

Ranking	Technology	KM Tools and Technologies	Average Score (1-5)
1	C	E-mail and Listserv	4.83
2	P	Expertise locators—Corporate Yellow pages—Who's Who	4.47
3	C	Corporate Intranet—Extranet—Internet	4.36
4	C	Database management systems	4.08
5	C	Search engines—Intelligent agents	3.95
6	C	Data warehouses—Data marts	3.84
7	C	Web-based training—E-learning	3.82
8	P	Communities of practice (interests in the same topic, field)	3.78
9	P	Communities of purpose (project, task-oriented)	3.73
10	C	Help-desk applications	3.70
11	P	Groupware	3.68
12	P	Teleconferencing (shared applications, whiteboards)	3.62
13	C	DMS	3.59
14	C	Multimedia repositories	3.58
15	P	Best practices repository	3.51
16	P	Videoconferencing (using audio and/or video)	3.47
17	C	DSS and expert systems	3.33
18	C	Data mining—Knowledge discovery	3.25
19	P	Mentoring—Tutoring	3.24
20	P	Storytelling	3.00
21	C	Knowledge mapping	2.91
22	P	Desktop computer conferencing	2.91
23	P	Online chat and instant messaging	2.89

Note: C = Codification; P = Personalization

It is important to keep in mind that our sample is composed predominantly of large organizations in the consulting and IT or telecommunications fields as well as agencies in the federal government. Respondents were mainly service-oriented, offering both standardized and customized products or services, and were located predominantly in the U.S. This small and limited number of samples may have biased our findings. More data need to be collected and further research needs to be conducted in the future. Similar research should be conducted in manufacturing firms, retailing firms, financial firms, and nonprofit/not-for-profit organizations.

Finally, the most frequently used KM tools and technology for the 97 respondents of our study were examined. They are ordered by the average score obtained (see Table 8).

Note that e-mails are by far the most frequently used tool to transfer knowledge. This can be explained by the fact that this is a technology that has been around for a long time and also by the fact that a large majority of knowledge workers uses them every day for various purposes. People are more likely to use tools with which they are familiar in order to exchange knowledge rather than learning or using new ones. If knowledge workers are very familiar with a technology, one should not force them to use different tools, but instead, one should find ways to build on the familiar technology in order to turn it into a more powerful KM tool (e.g., indexing of e-mail content, expertise profile created based on e-mail content, etc.). This table can be interpreted by looking at the ranking of all the factors but also by looking independently at the most frequently used codification tools and personalization tools. Expertise locator tools (personalization) take the second general position and the first position of personalization tool used. An expertise locator (who is who and who knows what in your organization) is often one of the first KM tools implemented by organizations. They are easy to build, simple to use, and provide rapid benefits. Communities of practice and purpose take the 8th and 9th general position and are in the 2nd and 3rd personalization position. During the past years, a strong emphasis on communities and their benefits has emerged in the KM literature. Their implementation is also quite simple, but one needs to be aware that knowledge worker participation relies heavily on organizational culture. Once again, assessing the level of organizational trust might be useful before engaging in such practices.

Conclusion

If an organization is going to be enduringly successful (Collins, 2001), considerable empirical research clearly indicates that such an organization needs, among other attributes, a single culture that aims all employees at disciplined thought and disciplined action. The bedrock of such a success culture is that it must characterize a high-trust organization.

The motive behind this research project is to begin testing empirically the proposition that KM success likewise relies upon the KMS being implemented in an organization with a trusting culture. Our research begins to shed some light on that phenomenon. In particular, this study indicates that organizations with a higher level of trust are more successful in their use of KM than those organizations with a lower level of trust. (Whether or not those higher level of trust organizations are also more successful in the marketplace is an intriguing research question left for the future.) Additionally, our work indicates an interesting interaction between the type of KMS used (codification vs. personalization) and the level of trust in the organization. Specifically, in low trusting organizations, personalization KM tools tend to not be used—and why should they be used, since co-workers have little faith in one another's reliability?

KM is an IT practice that is implemented in the faith that doing so will lead to higher levels of organizational performance. Our empirical research study begins to establish some of the parameter settings in the domain of characteristics of organizational culture that are necessary for KM success. Clearly, much more research is needed. Data were drawn from one set of organizations, primarily IT consulting, telecommunications, and the federal government. Our results might be a spurious outcome from peculiar characteristics of this set of organizations. Similar studies restricted to manufacturing firms, retailing firms, financial firms, or nonprofits/not-for-profits might yield quite different results. Also, our firms in the sample had existing KM systems. Firms developing KMS might simultaneously take steps to alter organizational policy, procedures, and, yes, even culture, to ensure that those investments in KM would enjoy high payoffs—there are no data to report on that front. And our organizations all had but a few years' experience with their KMS—the results might be quite different after firms have dozens of years' worth of experience with their KMS.

An organization, of course, is a legal fiction—at heart, an organization consists of a set of people who agree to work together for some vaguely common end. KMS is alleged to assist groups of people into behaving more effectively and more efficiently as they pursue those agreed-upon targets. This study helps shed some light on conditions that make that claim true. However, future KMS work might be directed at other issues relating to KMS success and usefulness, such as: How do leadership styles in the organization impact KMS success? How do the presence or absence of particular group norms impact KMS success? When a KMS is successful, what radiates from that project success elsewhere in the organization? When a KMS is not successful, what impact does that have on other IT-related projects? What are the causes of KMS failure, and are those failures rectifiable through greater funding, technological advances, culture change, better training, stronger leadership, or some other vehicle? To what extent is KMS success a function of the educational skill level of employees that are expected to use it? To what extent is KMS success a function of the cluster of attitudes possessed by users individually and collectively?

KM is a technology that is still not completely mature. KM technology holds great promise for organizational rationalization, but there are clearly many issues remaining surrounding KMS that need to be studied. Empirical research of all stripes—laboratory studies, field research, case studies, and so forth—will help scientists and managers to put KMS to best use. This only can benefit all of us.

References

Alavi, M., & Leidner, D. E. (2001). Review: Knowledge management and knowledge management systems: Conceptual foundations and research issues. *MIS Quaterly, 25*, 107-136.

Barth, S. (2000, October). KM horror stories. *Knowledge Management Magazine, 3,* 37-40.

Bell DeTienne, K., Dyer, G., Hoopes, C., & Harris, S. (2004). Toward a model of effective knowledge management and directions for future research: Culture, leadership and CKOs. *Journal of Leadership and Organizational Studies, 10*(4), 26-43.

Binney, D. (2001). The knowledge management spectrum—Understanding the KM landscape. *Journal of Knowledge Management, 5*(1), 33-42.

Bots, P.W. G., & Bruiin, H. (2002). Effective knowledge management in professional organizations: Going by the rules.In *Proceedings of the 35th Hawaii International Conference on System Sciences.*

Carnevale, D. G., & Wechsler, B. (1992). Trust in the public sector. *Administration and Society, 23*, 471-494.

Choi, B., & Lee, H. (2002). An empirical investigation of KM styles and their effect on corporate performance. *Information & Management, 40.*

Chua, A., & Lam, W. (2005). Why KM projects fail: A multi-case analysis. *Journal of Knowledge Management, 9*(3), 6-17.

Ciancutti, A., & Steding, T. L. (2000). *Built on trust. Gaining competitive advantage in any organization.* Contemporary Books.

Cohen, D., & Prusak, L. (2001). *In good company. How social capital makes organizations work.* Harvard Business School Press.

Collins, J. (2001). *Good to great—Why some companies make the leap ... and others don't.* Harper Business.

Cook, J., & Wall, T. (1980). New work attitude measures of trust, organizational commitment and personal need non-fulfillment. *Journal of Occupational Psychology, 53*, 39-52.

Culbert, S. A., & McDonough, J. J. (1986). The politics of trust and organizational empowerment. *Public Administration Quarterly, 10*, 171-188.

Cummings, L. L., & Bromiley, P. (1996). The organizational trust inventory (OTI). In R. M. Kramer, & T. R. Tyler (Eds.), *Trust in organizations; frontiers of theory and research.* Thousand Oaks, CA: Sage Publications.

Daley, D. M., & Vasu, M. L. (1998). Fostering organizational trust in North Carolina. The pivotal role of administrators and political leaders. *Administration and Society, 30*(1), 62-84.

Davenport, T., De Long, D. W., & Beers, C. M. (1998). Successful knowledge management projects. *Sloan Management Review*, 43-57.

Davenport, T., & Prusak, L. (1998). Working knowledge. How organizations manage what they know. *Harvard Business School Press.*

De Furia, G. L. (1996). *A behavioral model of interpersonal trust* [unpublished doctoral dissertation]. Springfield, LA: St. John's University.

De Furia, G. L. (1997). *Facilitator's guide to the interpersonal trust surveys.* Pfeiffer & Co.

Delmonte, A. J., & Aronson, J. E. (2004). The relationship between social interaction and knowledge management system success. *Journal of Knowledge Management Practice, 5.*

Delone, W. H., & McLean, E. R. (1992). Information systems success: The quest for the dependent variable. *Information Systems Research, 3*, 60-95.

Denning, S. (1998). *What is knowledge management?* Retrieved from http://www.worldbank.org/ks/index.html

Dennis, A. R., & Vessey, I. (2005). Three knowledge management strategies: Knowledge hierarchies, knowledge markets, and knowledge communities. *MIS Quarterly Executive, 4*(4), 399-412.

Griffin, K. (1967). The contribution of studies of source credibility to a theory of interpersonal trust in the communication process. *Psychological Bulletin, 68*, 104-120.

Hansen, M. T., Nohria, N., & Tierney, T. (1999). What's your strategy for managing knowledge? *Harvard Business Review*, 106-116.

HBR Forum. (1999). *HBR Forum.* Retrieved from http://www.hbr.org/forum

Jennex, M. E., & Olfman, L. (2001). Development recommendations for knowledge management/organizational memory systems. In M. K. Sein et al. (Eds.), *Contemporary trends in IS development* (pp. 209-222): Kluwer Academic.

Jennex, M. E., & Olfman, L. (2003). A knowledge management success model: An extension of DeLone and McLean's IS success model. In *Proceedings of the Ninth Americas Conference on Information Systems.*

Jennex, M. E., & Olfman, L. (2004). Assessing knowledge management success/effectiveness models. In *Proceedings of the 37th Hawaii International Conference on System Sciences.*

King, W. R. (2006). Maybe a "knowledge culture" isn't always so important after all! *Information Systems Management, 23*(1), 88-89.

Knowledge Management Review. (2001, November/December). KM Review survey reveals the challenges faced by practitioners. *Knowledge Management Review, 4,* 8-9.

Know-Net. (2000). *The approach.* Retrieved from http://www.know-net.org

Koenig, M. E. D. (2004). Knowledge management strategy. Codification versus personalization (a false dichotomy). In M. E. D. Koenig & K. T. Srikantaiah (Eds.), *Knowledge management lessons learned. What works and what doesn't.* Medford, NJ: Information Today.

KPMG Consulting. (2000). *Knowledge management research report.*

Levin, D. Z., & Cross, R. (2004). The strength of weak ties you can trust: The mediating role of trust in effective knowledge transfer. *Management Science, 50*(11), 1477-1490.

Levin, D. Z., Cross, R., & Abrams, L. C. (2002a). *Trust and knowledge sharing: A critical combination.* IBM Institute for Knowledge-Based Organizations.

Levin, D. Z., Cross, R., & Abrams, L. C. (2002b). *Why should I trust you?* (White paper presented at 2002 Academy of Management meetings).

Lewis, J. D. (1999). *Trusted partners. How companies build mutual trust and win together.* New York: The Free Press.

Lindsey, K. (2002). Measuring knowledge management effectiveness: A task-contingent organizational capabilities perspective. In *Proceedings of the Eighth Americas Conference on Information Systems.*

Lucas, L. M. (2005). The impact of trust and reputation on the transfer of best practices. *Journal of Knowledge Management, 9*(4), 87-101.

Luhmann, N. (1979). *Trust and power.* New York: John Wiley.

Malhotra, Y., & Galletta, D. F. (2003). Role of commitment and motivation in knowledge management systems implementation: Theory, conceptualization, and measurement of antecedents of success. In *Proceedings of the Hawaii International Conference on System Sciences (HICSS 36),* Hawaii.

Massey, A. P., Montoya-Weiss, M. M., & O'Driscoll, T. M. (2002). Knowledge management in pursuit of performance: Insights from Nortel Networks. *MIS Quarterly, 26*(3), 269-289.

Matthai, J. M. (1989). *Employee perceptions of trust, satisfaction, and commitment as predictors of turnover intentions in a mental health setting* [unpublished doctoral dissertation]. Vanderbilt University.

McAllister, D. J. (1995). Affect and cognition-based trust as foundations for interpersonal cooperation in organizations. *Academy of Management Journal, 38*(1), 24-59.

McDermott, R. (1999). Why information technology inspired but cannot deliver knowledge management. *California Management Review, 41*(4), 103-117.

McKnight, H. D., & Chervany, N. L. (2000). What is trust? A conceptual analysis and an interdisciplinary model. In *Proceedings of the Americas Conference on Information Systems (AMCIS), Long Beach, California.*

Microsoft. (1999). *Practicing knowledge management.*

Mowday, R. D., Steers, R. M., & Porter, L. W. (1979). The measurement of organizational commitment. *Journal of Vocational Behavior, 14*(2), 224-247.

Natarajan, G., & Shekhar, S. (2000). *Knowledge management: Enabling business growth.* New Delhi: Tata McGraw-Hill.

Nelson, K. M., & Cooprider, J. G. (1996). The contribution of shared knowledge to IS group performance. *MIS Quarterly, 20*(4), 409-432.

Nyhan, R. C. (1999). Increasing affective organizational commitment in public organizations. The key role of interpersonal trust. *Review of Public Personnel Administration,* 58-70.

Nyhan, R. C., & Marlowe, H. A. J. (1997). Development and psychometric properties of the organizational trust inventory. *Evaluation Review, 21*(5), 614-635.

Park, H., Ribière, V., & Schulte, W. D. (2004). Critical attributes of organizational culture promoting knowledge sharing & technology implementation successes. *Journal of Knowledge Management, 8*(3), 106-117.

Pauleen, D., & Mason, D. (2002). *New Zealand knowledge management survey: Barriers and drivers of KM uptake.* Retrieved January 10, 2004, from http://www.nzkm.net/mainsite/NewZealandKnowledgeManagementSurveyBarriersandDriv.html

Politis, J. D. (2003). The connection between trust and knowledge management: What are its implications for team performance. *Journal of Knowledge Management, 7*(5), 55-66.

Ribière, V. (2001). *Assessing knowledge management initiative successes as a function of organizational culture* [unpublished DSc dissertation]. Washington, DC: George Washington University.

Ribière, V. (2005). *Le rôle primordial de la confiance dans les démarches de gestion du savoir* [unpublished PhD dissertation]. Aix en Provence, France: Université Paul Cézanne.

Shaw, R. B. (1997). *Trust in the balance. Building successful organizations on results, integrity, and concern.* San Francisco: Jossey-Bass.

Swan, J., Newell, S., Scarbrough, H., & Hislop, D. (1999). Knowledge management and innovation: Networks and networking. *Journal of Knowledge Management, 3*(4), 262-275.

Tiwana, A. (2002). *The knowledge management toolkit, orchestrating it, strategy, and knowledge platforms* (2nd ed.). Upper Saddle River, NJ: Prentice Hall.

Tuggle, F. D., & Shaw, N. C. (2000). The effect of organizational culture on the implementation of knowledge management. In *Proceedings of the Florida Artificial Intelligence Research Symposium (FLAIRS),* Orlando, Florida.

Welch, J. (1993, January 25). Jack Welch's lessons for success. *Fortune, 127,* 86.

Wick, C. (2000). Knowledge management and leadership opportunities for technical communicators. *Technical Communications.*

Wilson, M. B. (1993). *A new method for assessing Cook and Wall's informal theory of organizational trust: A Coast Guard sample* [unpublished doctoral dissertation]. Washington, DC: George Washington University.

Yih-Tong Sun, P., & Scott, J. L. (2005). An investigation of barriers to knowledge transfer. *Journal of Knowledge Management, 9*(2), 75–90.

Zack, M. H., & Michael, S. (1996). *Knowledge management and collaboration technologies* (white paper). The Lotus Institute, Lotus Development Corporation.

Zack, M. H., & Michael, S. (1998). *Knowledge management and collaboration technologies*. Retrieved from http://www.lotus.com/services/institute.nsf/550137bfe37d25a18525653a005e8462/000021ca

Zand, D. E. (1997). *The leadership triad—Knowledge, trust, and power*. Oxford University Press.

Chapter VIII

Knowledge Management's Impact on Organizational Performance

Vittal Anantatmula, Western Carolina University, USA

Abstract

Establishing criteria for knowledge management (KM) is important, because criteria help to establish a basis for assessing the value and evaluating its results. More importantly, the criteria will tell us what the KM outcomes are and their relevance to organizational performance. The literature review has revealed that widely accepted criteria and performance measures have not been developed for KM. Delphi Technique and survey-based research using a questionnaire targeting KM professionals as respondents were aimed at establishing criteria for assessing KM success for different types of organizations. The results show what organizations consider important outcomes of a KM initiative. Contributions from this research effort should support government, nonprofit, and for-profit organizations in making decisions about KM initiatives and measuring KM efforts in terms of its relevance to the performance of organizations. Future research efforts can focus on developing these KM outcomes into detailed measures.

Introduction

Knowledge is recognized as a key economic resource, and obviously, organizations must possess the right knowledge in the desired form and context under all circumstances in order to be successful. Specifically, knowledge sharing and resultant knowledge creation are critical in order for organizations to gain competitiveness and to remain competitive. Knowledge is considered important for sustaining competitive advantage.

The continuous progression of civilization is a testimony to its ability to develop, learn, and share knowledge. Recent advances in information and communication technologies have made it easy to develop, store, and transfer knowledge. Globalization, increasing international competition, and a free market philosophy are driving forces for these advances in technology, and many organizations have realized that the creation, transfer, and management of knowledge are critical for success today.

The increasing gap between the book value and the market value of some business entities indicates the increasing importance of knowledge-based intangible assets (Marr, 2003) and knowledge management (KM). However, the dimension of KM has not received adequate attention (Holsapple & Joshi, 1999). Also, the KM concept is still understood as information management and is associated with technological solutions, such as intranet and databases (Marr, 2003).

Several organizations are attempting to use KM to improve organizational performance, but commonly accepted KM principles are yet to be developed. KM's lack of focus (Fairchild, 2002) and absence of commonly accepted KM principles (Stankosky & Baldanza, 2001) are some of the gaps in this discipline. Among the commonly accepted KM principles or references that are missing are the criteria for measuring success associated with KM. In this chapter, a research effort is presented to address this knowledge gap from the practitioners' point of view and leading to identifying expected outcomes of a KM initiative in organizations.

Definitions

Knowledge is derived from thinking and is a combination of information, experience, and insight. Deriving knowledge from information requires human judgment and is based on context and experience. Knowledge categories—tacit and explicit—can be found in different forms. While explicit knowledge can be found in articulated and documented forms, tacit knowledge, which is personal and specific to a social, organizational, or interpersonal context, does not always acquire physical form and can be found in people's actions and interpersonal communications. Much of the tacit knowledge—a greater component of organizational knowledge—is found in social interactions, and different social contexts facilitate different modes of knowledge integration.

It should be understood that the primary focus of KM is to utilize information technology and tools, business processes, best practices, and culture to develop and share knowledge within an organization and to connect those who possess knowledge to those who need the knowledge. Ultimately, leveraging relevant knowledge assets to improve organizational performance is what knowledge management is all about.

Background: KM and Organizational Performance

While many organizations have implemented knowledge management (KM) initiatives, it remains unclear the extent to which they are successful in delivering the anticipated outcomes, and why. Research studies show that it is difficult to assess return on investment of KM. Improving organizational performance by using a KM initiative is an investment decision, and we, therefore, must have an understanding of its outcomes. While discussing approaches to building KM systems (KMS), Jennex and Olfman (2004) contended that the measurement of a KMS is crucial to understanding how these systems should be developed and implemented. They cite several reasons for measuring success of a KMS, including three from Turban and Aronson (2001): to provide a basis for valuation, to stimulate management's focus on what is important, and to justify investments.

However, inherent intangible characteristics of knowledge assets make them difficult to measure (Ahn & Chang, 2002). Unlike materials or equipment, the core competencies and distinctive abilities of employees are not listed on balance sheets (Austin & Larkey, 2002). As a result, factors that contribute substantially to a firm's success elude traditional means of quantification, thereby presenting significant challenges to KM performance measurement.

Bassi and Van Buren (1999) suggest that the lack of understanding of how to measure and evaluate impacts of intellectual capital is a major obstacle to turning investments toward promoting intellectual capital into a source of competitive advantage. Similarly, Ernst & Young's Center for Business Innovation survey identified measuring the value and performance of knowledge assets as the second most important challenge faced by companies behind the challenge of changing people's behaviors (Van Buren, 1999).

Instead of trying to measure knowledge directly, which may not be possible, a different approach is to measure its contribution to business performance, which still is considered a major research agenda (Ahn & Chang, 2002). Major consulting organizations agree that measuring KM effectiveness and contributions is a key concern for consulting organizations (Wikramasinghe, 2002).

Some studies have suggested nontraditional KM measurements. A survey of 100 FTSE (index used by London Stock Exchange and Financial Times) companies attempted to establish levels of engagement with KM, the organizational implications, and evidence of impact on performance (Longbottom & Chourides, 2001). The survey results suggest that performance measures are not well developed and that these measures should be linked to balance scorecard frameworks. According to Fairchild (2002), KM activities are considered integral to other management activities and processes; measuring KM is about how and when KM is integrated into organizational activities, which can be measured. Thus, it is important to identify these activities and to determine KM contributions to these activities.

The study suggests that organizations should require less precision and exact figures and more interest in trends using a balance scorecard approach, such as customer and employee satisfaction and intellectual capital.

These research findings lead to the conclusion that KM results are difficult to measure and that commonly accepted outcomes of a KM initiative are not yet established. This research effort is aimed to address this knowledge gap in order to develop an understanding of the relevance of KM to organizational performance. This chapter uses a literature review to identify a number of KM outcomes at the organizational level that then are translated into KM criteria. Based on this literature review, a list of KM criteria and important research questions was established for the Delphi study. To support the Delphi findings, a survey consisting of the same list of criteria and questions was distributed. Based on these research findings, expected KM initiative outcomes from the practitioner's point of view were established. Finally, limitations of the study and future research opportunities are discussed.

Literature Review

Research related to KM success can be classified into two focus areas: KM success factors and KM outcomes. KM success factors can be viewed as facilitating factors for a KM initiative. Though the main focus of this chapter is on outcomes of KM initiatives, a brief discussion on success factors is relevant for this study in order to understand the distinction between the two.

There have been efforts to identify organizational factors for successful KM initiatives (Chourides, Longbottom, & Murphy, 2003; Jennex & Olfman, 2004). While discussing KMS frameworks, Jennex and Olfman (2004) recommend that developing a successful KMS would involve designing a technical infrastructure, incorporating KM into processes, developing a secured KMS and knowledge structure for the enterprise, gaining senior management support, and building motivational factors into the system. Other research indicates that leadership, investing in people, and developing supporting organizational conditions are critical to achieving success in a KM program (Chourides et al., 2003). Similar success factors were suggested, based on a study of several projects (Davenport, DeLong, & Beers, 1998). While these are facilitating factors for a KM initiative, outcomes or results of a KM initiative are different.

A conference in London, Measuring Knowledge Value 2002, addressed the knowledge measurement issue from both macro and micro perspectives (Perkmann, 2002). The macro perspective focused on quantifying intangible assets to capture the value of human capital, competencies, customer relationships, employee collaborations, and so forth, which are not purely financial measures and emphasize the importance of intangible assets. The micro perspective addressed the issue of quantifying the impact of individual knowledge projects. While analyzing the 2002 London conference proceedings, Perkmann (2002) supported the idea of case studies and anecdotal evidence by illustrating that ROI can capture only a part of the project's impact (efficiency and productivity concerns) and that projects always have unintended consequences or effects (competency development and learning), negative or positive, that cannot be captured easily in quantitative or financial terms. However, anecdotal

Table 1. KM benefits (KPMG, 1999)

• Better decision making	• Reduced costs
• Better customer handling	• New or better ways of working
• Improved employee skills	• Increased market share
• Faster response to key business issues	• Create additional business opportunities.
• Improved productivity	• Improved new product development
• Increased profits	• Staff attraction/retention
• Sharing best practices	• Increase share price

evidence and case studies are context-specific artifacts that may not reflect overall reality and may not be commonly accepted. In addition, they do not meet some of the desired characteristics of measures, such as reliability, applicability, and transferability.

In their case study of professional service firms creating competitive advantage through KM, Ofek and Sarvary (2001) identified reducing costs, enhancing product or service quality, or creating value to customers as business strategies for designing and implementing KM in order to create competitive advantage.

KPMG International, UK (1999) produced a report based on a survey of 423 organizations from Europe and the U.S. In their survey, KPMG identified the following as expected KM benefits (see Table 1).

A 1997 survey of 431 business organizations in the U.S. and Europe identified four KM application areas (Ruggles, 1998): creation of intranets; data warehousing and knowledge repositories; implementing decision-support tools; and implementing groupware to support collaboration. These application areas focus on knowledge transfer, knowledge retention, better decision making, and support collaboration.

According to a benchmarking study by APQC (Elliott & O'Dell, 1999), the most common reason for managing and sharing knowledge is the transfer of best or exemplary practices within the organization, followed by increasing employee capabilities and providing customer or market information.

Successful KM programs achieve competitive advantage, customer focus, employee relations and development, innovation, and lower costs (Skyrme, 1997). Though KM promotes development and application of knowledge to attain enterprise's ultimate goal of profitability, the implicit purpose of KM is to empower knowledgeable individuals with intellectual tasks in order to promote learning (Wiig, 1999).

Based on the previous discussions and several other references, 26 factors were identified to be included in the list of outcomes (Anantarmula, 2004). All of them have direct references, not necessarily as outcomes but under different terms such as *benefits*, *impact*, *focus*, *performance factors*, *metrics*, *results*, *strategies*, and *value*. Table 2 presents a summary of literature review consisting of KM outcomes and important sources.

Table 2. Summary of KM outcomes

KM Outcome	Source
Better decision making	KPMG (1999); Ruggles (1998)
Better customer handling	KPMG (1999); Skyrme (1997); Kelly (2003); Van Buren (1999); Longbottom et al. (2001)
Faster response to key business issues	KPMG (1999); Van Buren (1999); BP Amoco (2001); Longbottom et al. (2001)
Improved employee skills	Elliott & O'Dell (1999); KPMG (1999); Skyrme (1997); Perkman (2002); Van Buren (1999)
Improved productivity	KPMG (1999); Perkman (2002); Kelly (2003); Van Buren (1999); BP Amoco (2001)
Increased profits	KPMG (1999); Van Buren (1999); Wiig (2000)
Sharing best practices	KPMG (1999); Perkman (2002); Van Buren (1999); BP Amoco (2001); Longbottom et al. (2001); Ruggles (1998); Allee (1997)
Reduced costs	KPMG (1999); Skyrme (1997); Wiig (1993); Ofek & Sarvary (2001); BP Amoco (2001); Longbottom et al. (2001)
New or better ways of working	KPMG (1999); Perkman (2002); BP Amoco (2001); Longbottom et al. (2001); Ruggles (1998); Allee (1997
Increased market share	Elliott & O'Dell (1999); KPMG (1999); Wiig (2000); Kelly (2003); BP Amoco (2001)
Creation of new business opportunities	Elliott & O'Dell (1999); KPMG (1999); Wiig (2000); Longbottom et al. (2001)
Improved new product development	KPMG (1999); Wiig (2000); BP Amoco (2001); Longbottom et al. (2001)
Better staff attraction/retention	KPMG (1999); Skyrme (1997); Kelly (2003); BP Amoco (2001)
Increased share price	KPMG (1999); BP Amoco (2001)
Enhanced product or service quality	Skyrme (1997); Wiig (2000); Kelly (2003); Ofek & Sarvary (2001); Van Buren (1999); Longbottom et al. (2001)
Creation of more value to customers	Elliott & O'Dell (1999); Skyrme (1997); Wiig (2000); Kelly (2003); Ofek & Sarvary (2001); Van Buren (1999); Longbottom et al. (2001)
Enhanced intellectual capital	Allee (1997)
Improved communication	BP Amoco (2001); Longbottom et al. (2001); Allee (1997)
Increased innovation	Skyrme (1997); Perkman (2002); Allee (1997)
Improved learning/adaptation capability	Skyrme (1997); Perkman (2002); Kelly (2003); Van Buren (1999); BP Amoco (2001); Ruggles (1998)
Return on investment of KM efforts	Van Buren (1999); BP Amoco (2001)
Increased market size	Wiig (2000); Kelly (2003); BP Amoco (2001)
Entry to different market type	BP Amoco (2001); Ruggles (1998)
Increased empowerment of employees	Skyrme (1997); Wiig (2000); BP Amoco (2001)
Enhanced collaboration	BP Amoco (2001); Jennex & Olfman (2002); Perkman (2002); Ruggles (1998)
Improved business processes	Jennex & Olfman (2002); Elliott & O'Dell (1999)

The Table 2 list of outcomes is used in the Delphi as a reference list for its participants and in the survey questionnaire to establish priority.

Research Questions

The main research objective is to establish the criteria for measuring KM success. Consequently, it led to understanding the relevance of KM to organizational performance. Since a criterion can be considered as a standard on which a judgment may be based, establishing criteria and using them to evaluate KM initiatives will lead to knowing expected outcomes of KM initiatives. Thus, the main research question is:

What should be the criteria for measuring KM success?

Though KM principles are similar, irrespective of the type of organization, criteria and consequent KM outcomes could be different for different types of organizations for two reasons. First, an organization's reason to invest in a KM initiative could be business-specific and, thus, could be different. Second, this initiative is driven by what the organization's goals and objectives are, and each type of an organization may have different objectives and goals.

Many research studies support the contention that KM initiatives should be aligned with organizational goals and objectives. A poll of executives from 80 large companies in the U.S., such as BP Amoco, Chemical Bank, Hewlett-Packard, and Kodak, indicated that 80% believed that managing knowledge of their organization should be an essential or important part of business strategy (Takeuchi, 1998). Strategic goals and business requirements drive process requirements, which, in turn, determine knowledge requirements (Massey & Montaya-Weiss, 2002). Massey and Montaya-Weiss (2002) contend that KM initiatives will be effective when they are aligned with the performance goals and requirements of a business, its processes, and its people. Citing that KM is about creating synergy in organizations, Davenport and Probst (2001) (as reported by Chourides et al., 2003) contend that such action translates into aligning individual goals with organizational goals. In other words, aligning KM practices with organization goals is a desired way to implement a KM initiative. Thus, an extension of the main research question is:

Are the criteria for measuring KM success different for different types of organizations?

The second research question focuses on establishing the criteria or outcomes for different types of organizations.

Research Methodology

Literature review findings and research questions discussed in previous sections suggest that those who are well-versed in KM theory and practice can better address these issues. For this reason, this research effort uses the Delphi Technique with occasional use of in-depth interviewing and personal discussions. The Delphi Technique uses a group of experts to deliberate a research issue or a problem anonymously (i.e., without having direct interaction among group members and without knowing who the other members of the group are). The Delphi Technique does not involve face-to-face group discussion (Anderson, 1990), and it does not have the disadvantages of conventional groups, because it provides anonymity and controlled feedback. However, the Delphi Technique has certain disadvantages. It is time-consuming (Anderson, 1990), and swiftness of the decision making process is controlled by participating individuals to some extent. As results are limited by the number of experts and the number of organizations they represent (Anantarmula, 2004), the Delphi Technique research effort is supplemented by a survey questionnaire, which also helped to set up priority among the established criteria. The survey was aimed only at KM professionals for the same reasons the Delphi was chosen for this research effort.

The Delphi Technique

The Delphi Technique was developed by RAND Corporation in the 1960s to forecast purposes and was later enhanced by the U.S. government for group decision making (Cline, 2000). It is a technique that also has been used to develop lists of objectives or indicators of successful programs (Abramson, Tittle, & Cohen, 1979). In the KM discipline, the Delphi Technique was used for identifying attribute dimensions to characterize knowledge (Holsapple & Joshi, 2001).

The Delphi Technique was used to get responses to important research questions by choosing KM professionals and researchers. The Delphi Technique involved the following steps:

- Test research questions.
- Select members of the expert group.
- Have each member respond to questions.
- Summarize responses and distribute the summary.
- Seek second responses.
- Summarize and distribute responses again.
- Continue the process until an agreement emerges.

The Delphi Technique addressed the following questions to the select group of KM experts:

1. If an organization wishes to measure success of its KM program, what should be the criteria for measuring KM success?
2. Are the criteria for measuring KM success different for government, nonprofit, and for-profit organizations? If different, what should be the criteria for different types of organizations?

The chosen participants of the Delphi Technique were academicians and senior-level KM professionals in organizations such as the George Washington University, Gartner Research Group, Xerox Corporation, U.S. Federal Government, and independent consultants.

The Survey

The initial survey instrument was generated from Table 2 and then modified based on pilot test feedback (the pilot test checked the survey's reliability and validity). The final survey was designed with the following features:

- There are 19 questions divided in three parts, making the questionnaire short and less time-consuming.
- A brief note about the purpose of the questionnaire is provided in the cover letter.
- Definitions are provided for important terms used in the questionnaire.
- An information sheet is included to obtain consent and to ensure confidentiality of the responses.
- Instructions and questions are simple and easy to understand.
- Respondents are given an option to receive research findings.

The survey consisted of 17 close-ended and two open-ended questions. The primary research question has three parts:

- Identify the criteria that are used to measure KM success
- Importance of each criterion
- Effectiveness of each criterion

The importance of the criterion gives the evidence of significance or consequence, whereas effectiveness denotes the capability of being used to a purpose. A criterion that is important may or may not be effective. If a criterion is chosen as both important and effective, it is considered useful.

The questionnaire was aimed at a target population of KM professionals and practitioners. Those surveyed were from government, nonprofit, and for-profit organizations. The survey instrument was used to solicit responses from a number of KM professionals around the world. A total of 152 valid responses were received. Statistical analysis of the results presents the most useful criteria for measuring efforts associated with KM efforts.

Findings

Final Delphi Technique responses are summarized next.

1. If an organization wishes to measure success of its KM program, what should be the criteria for measuring KM success?

All the participants agreed on knowledge sharing and learning and organization performance as the criteria for KM. Table 3 lists the criteria identified in response to the question.

All Table 3 criteria can be found in Table 2, either with identical phrases or with similar meanings. However, this list could not be shortened, as no further agreement could be reached among the Delphi participants.

Table 3. KM criteria identified by the Delphi respondents

• improve efficiency and effectiveness • improve innovation • make faster, better decisions • improve processes • better business practices • improve capacity (learning and adaptation) • improve collaboration • bottom line (return on investment, increase profits)	• increase intellectual capital • improve knowledge sharing and learning • improve schedule • reduce cost • improve performance • improve customer satisfaction • improve employee satisfaction

Table 4. KM criteria for different types of organizations identified by the Delphi

Government	Nonprofit	For-Profit
• cost (budget limitations and fair value) • schedule (milestone requirements) • performance (constituent satisfaction) • customer service • fulfilling government mandate • public service • mission accomplishment • mission results	• public service • mission results • mission accomplishment	• business cost (profit and loss) • schedule (cost reduction and meet customer needs) • performance (quality and customer satisfaction) • business results • customer satisfaction • financial results • innovation • customer loyalty • quality

2. Are the criteria for measuring KM success different for government, nonprofit, and for-profit organizations? If different, what should be the criteria for different types of organizations?

KM criteria for nonprofit organizations could include criteria from those of both government and for-profit organizations and are influenced by government. One participant noted that the mission and objectives of each department or agency are different, and hence, measures are different for each agency or department. Another participant felt that except for the bottom-line argument, the criteria would be the same for all.

Similar to the literature review findings, the Delphi Technique results revealed a list of 15 criteria, making it necessary to use a survey to determine the most useful criteria.

Survey Results

All survey respondents have some KM experience; 79% of respondents have more than three years experience, and 42% have six or more years experience. When asked to rate themselves on KM expertise, only 3.3% of respondents rated themselves as novice, and more than two-thirds considered themselves to be either experts or close to being experts. The majority of those surveyed hold positions such as chairman/CEO, president/CEO, founder/CEO, chief knowledge officer, managing director, director (KM), director, KM architect, KM consultant, senior knowledge strategist, principal, and principal strategy officer. Through descriptive statistical analysis, it was evident that respondents have KM experience, consider themselves fairly knowledgeable about KM, and are involved in KM initiative decision making. Finally, their roles and responsibilities appear to be consistent with their organizational profiles.

Of the organizations represented, 31.5% have equal to or fewer than 100 employees, 36.2% have 2,500 to 10,000 employees, and 24.8% have 10,000 or more employees. In terms of revenue, 47% of organizations have more than $1 million revenue with 21% having more than $1 billion. Revenue is not applicable for 37% of organizations. Since all the respondents indicated that they have KM experience and that they have answered KM-related questions, we can assume that most of these organizations are involved in implementing KM.

The mean value and standard deviation of importance and effectiveness of each criterion are computed and compiled in Table 5 with the criteria listed in order of most useful to least useful. Both importance and effectiveness have been used for this purpose.

Both the importance and effectiveness measures have identical scales with *5* representing *very high* and *1* representing *very low*, thus values closer to *5* indicate higher importance or effectiveness. Table 6 lists the most and least useful criteria based on quartile values, with the first quartile representing high importance—high effectiveness of criteria—and the last quartile representing low importance—low effectiveness of criteria.

It is interesting to note that the least useful criteria can be quantified and easily measured, whereas the most useful criteria are difficult to measure and cannot be tied easily to bottom

Table 5. Descriptive statistics of KM criteria (N = 152)

Criteria	Importance		Effectiveness	
	Mean	SD	Mean	SD
Better decision making	4.303	0.828	3.746	0.962
Better customer handling	4.135	0.936	3.727	0.971
Faster response to key business issues	4.023	0.919	3.563	1.005
Improved employee skills	4.094	0.926	3.829	0.985
Improved productivity	4.109	0.981	3.739	0.974
Increased profits	3.458	1.297	3.103	1.267
Sharing best practices	4.068	0.906	3.802	0.988
Reduced costs	3.708	1.069	3.388	1.100
New or better ways of working	3.992	0.897	3.692	1.083
Increased market share	3.067	1.241	2.790	1.081
Creation of new business opportunities	3.815	1.137	3.416	1.194
Improved new product development	3.769	1.151	3.542	1.142
Better staff attraction/retention	3.604	1.048	3.316	1.118
Increased share price	2.593	1.412	2.493	1.298
Enhanced product or service quality	4.110	0.959	3.743	1.022
Creation of more value to customers	4.065	1.058	3.582	1.102
Enhanced intellectual capital	3.992	1.073	3.761	1.008
Improved communication	4.244	0.842	3.992	0.929
Increased innovation	3.875	1.050	3.627	1.148
Improved business process	3.974	1.021	3.819	0.998
Improved learning/adaptation capability	3.975	0.987	3.761	1.017
Return on investment of KM efforts	3.644	1.299	3.268	1.229
Increased market size	2.933	1.321	2.843	1.254
Entry to different market type	3.128	1.387	3.088	1.389
Increased empowerment of employees	3.844	1.132	3.653	1.108
Enhanced collaboration within organization	4.346	0.794	3.976	0.911

Table 6. Survey Results of KM Criteria.

Most Useful Criteria	Least Useful Criteria
• Enhanced collaboration	• Increased share price
• Improved communication	• Increased market size
• Improved employee skills	• Increased market share
• Improved productivity	• Entry into different market type
• Better decision making	• Increased profits
	• Better staff attraction/retention
	• Return on investment of KM efforts

line results. It can be concluded that KM efforts have internal focus and may have indirect impact on business results, specifically market performance.

Other criteria that are associated with business results—increased profits, reduced costs, improved new product development, return on investment of KM efforts, and enhanced product or service quality—are not among the most or least useful criteria. It is important to understand that KM efforts also can lead to results associated with the least useful criteria.

It was possible that respondents may employ one or more criteria, which are not listed among the 26 criteria presented in the survey. To address this issue, respondents were asked to provide any other criteria that they consider important for measuring KM success. Only 28% of respondents answered this question, which implies that the majority of the respondents (72%) found their useful criteria in the Table 2 list of criteria. Since there were no omissions in the criteria list and not a single or multiple criteria are mentioned more than twice as a response to this open-ended question, they were not included in the results of the study.

KM Criteria and Types of Organizations

While a majority of the respondents agreed that the criteria for measuring KM success are based on an organization's mission, objectives, and goals, the pairwise correlation analysis indicated that the aligned criteria are not necessarily the most useful criteria. Some of the criteria related to business performance and growth are easily measurable and aligned with the mission, objectives, and goals of an organization.

Pairwise correlation analysis suggested that top management support is aligned with factors relating to business performance and the delegation of power. Participation of functional managers in KM efforts is aligned with many criteria effectiveness, which signifies its value to KM efforts. Respondents were asked to identify their organization from the following options; the percentage of responses for each is summarized below.

- Federal or state government: 21%
- Nonprofit organization: 26%

Table 7. ANOVA—(a) criteria importance, (b) criteria effectiveness

Criteria	Government		Nonprofit		For-Profit		Probability
	size	mean	Size	mean	size	mean	
Increased profits	16	2.250	20	3.000	54	3.963	<0.0001
Increased market share	15	1.667	21	2.953	49	3.429	<0.0001
creation of new business	18	3.055	28	3.536	58	4.156	0.0004
Increased share price	15	1.800	19	2.211	43	2.954	0.0116
Increased market size	16	2.062	25	2.800	44	3.205	0.0106
Entry to different market	16	2.250	22	3.046	44	3.364	0.0224

(a)

Criteria	Government		Nonprofit		For-Profit		Probability
	size	Mean	Size	mean	size	mean	
Increased profits	15	3.667	19	3.210	49	3.368	0.009
Increased market share	13	1.693	20	2.850	45	3.089	0.000
creation of new business	17	2.923	27	3.186	54	3.704	0.013
increased empowerment	22	3.318	27	3.334	48	3.917	0.035

(b)

Table 8. KM criteria for different types of organizations

Government	Nonprofit	For-Profit
• Improved communication • Improved productivity	• Improved communication • Enhanced collaboration within organization • Improved learning, adaptation capability	• Enhanced collaboration within organization • Improved employee skills • Improved communication • Enhanced product or service quality • Sharing best practices • Better customer handling • Better decision making • Creation of more value to customers

- For-profit organization: 48%
- Other, please specify: 5%

Responses to option 4 (Other) were eliminated for the analysis as they are not relevant to the study, and a one-way ANOVA was attempted to determine whether the sample mean values of criteria (both importance and effectiveness) are different for each type of organization. Results showed significant difference in the sample mean value for only 6 and 4 variables of criteria importance and criteria effectiveness, respectively (Tables 7a and 7b). The differences of mean value are not significant for 20 and 22 (out of 26) criteria importance and effectiveness, respectively, for different types of organizations.

While the differences of mean value are significant, their mean values are not. Except for the criterion, *creation of new business opportunities*, which has a mean value of 4.156 for for-profit organizations, all others have a lower mean value. Table 8, which uses a similar methodology to that used for Table 6 to identify the most useful criteria for all organizations, was used to identify the most useful criteria for each type of organization.

These results indicate that criteria for measuring KM efforts are different for different types of organizations. While government and nonprofit organizations focus on internal performance only, for-profit organizations focus on both internal and external performance.

Figure 1. Significant KM outcomes

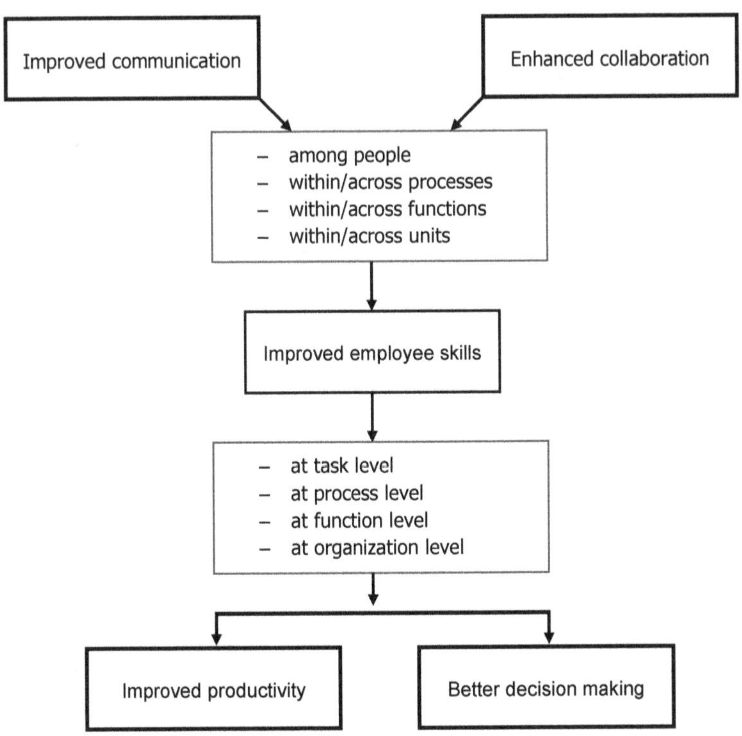

Measures of the Most Useful KM Outcomes

Through this research, both the most useful and the least useful criteria are established for outcomes of KM initiatives. While the most useful outcomes are difficult to measure, the least useful outcomes can be quantified and are easily measurable. The research identified enhanced collaboration within organization, improved communication, and improved employee skills as the top three outcomes overall, followed by improved productivity and better decision making. All of them contribute to organizational performance. However, as these criteria are difficult to measure, it is important that they need to be broken down further to different levels in order to develop detailed measures, as shown in Figure 1.

Developing measures for improved communication and enhanced collaboration requires critical thinking. Some of the suggested methods to develop measures for these criteria are discussed next.

Improved Communication

- A gap survey to determine communication effectiveness is a first step. Once these gaps are identified, organizations should address these issues through KM initiatives and tools, which could be organization-specific. Impact of these actions can be measured by using controlled groups or by conducting a similar gap survey later. Some of the tools to address communication gap are newsletters, kiosks, newsdesks, and so forth. However, these tools should be designed to communicate key business knowledge with a focus on improving organizational performance.
- Constant and continuous transformation of individual learning to organizational learning and vice-versa is a source of effective communication. To facilitate this transformation, organizations must encourage both formal and informal communication channels and monitor their performances. Communities-of-practice, electronic yellow pages, intranet, and best-practice database systems should help this transformation, and their effectiveness measures in terms of business performance can be developed.
- Organizations must develop skill development workshops and employee training development seminars to improve communication and to ensure transformation of explicit knowledge to tacit knowledge and vice-versa. Effectiveness of these workshops and seminars can be developed easily.
- Quantification of organizational or explicit knowledge and developing measures for their usefulness is another way of measuring communication. For example, a number of documented and well-defined processes, project management practices, and decision making procedures can be developed; frequency of reference and number of revisions could be guiding factors for measuring the effectiveness of these communication tools.

Enhanced Collaboration

- Opportunities for individuals to participate in management activities such as decision making should be designed to improve collaboration, and it should be aimed at improving organizational performance. The effectiveness of these actions can be measured by controlled group studies.
- Formation of committees at several levels of management to develop new problem-solving methods and resolve management problems will ensure enhanced collaboration. It is relatively easy to measure the effectiveness and contribution of these committees.
- Constant and continuous transformation of individual learning to organizational learning and vice-versa is a source of effective communication and enhanced collaboration as well. Some of the measures discussed previously are also applicable for enhanced collaboration.
- Delegation of authority and accountability to encourage individuals would result in greater collaboration at lower levels of management. It would lead to increased employee morale, motivation, and individual performance, which can be measured.

Improving employee skills can be achieved by improved communication and enhanced collaboration. In addition, recognition programs, such as employee of the month, years of service, and attendance awards, would create incentives for employees to improve their individual performance. There are many ways to measure improved productivity and better decision making, which can be judged by results associated with decisions.

To summarize these research results, knowledge as a source of competitive advantage will continue to gain strategic importance, and organizations will be compelled to implement KM initiatives to improve organizational performance. KM will continue to evolve to develop industry and organization-specific systems and processes.

Limitations of the Study

As mentioned earlier, the number of experts and the number of organizations they represent limits Delphi Technique results. The survey questionnaire is similar to a one-time case study in which all the respondents were asked to respond to the questionnaire only once.

Of the internal and external validity factors (Campbell & Stanley, 1963), only statistical regression and biases are relevant to the survey and the Delphi Technique. Others are relevant for controlled experimental studies. External validity factors are no threat to the research study. Statistical regression, which is concerned with selection of groups on the basis of their extreme scores, is part of the research design, as responses from KM professionals and experts are sought for this research and its findings are limited to KM initiatives only.

Bias, which results in differential selection of respondents for the comparison groups, is not directly related to the current research, as there are no comparison groups. However, selection bias is a possibility for the Delphi Technique. To avoid this, the Delphi Technique group was selected by a leading KM expert and professor, who was not directly involved with the study.

The study grouped organizations into government, nonprofit, and for-profit organizations, and research data were separated using these groups to examine hypotheses using statistical analysis. The underlying assumption is that within each group, organizations have similar KM purpose and focus, which may or may not be true.

Incomplete responses were ignored for the statistical analysis. Of the valid responses received, 48% of responses represent for-profit organizations, whereas responses from government and nonprofit organizations constitute 21% and 26%, respectively. Due to this imbalance in organizational representations, the results may be biased toward for-profit organizations.

Some respondents chose not to answer questions associated with importance and effectiveness of the criteria that they did not use. As a result, responses to criteria importance and effectiveness varied from 75 (criteria effectiveness—*increase in share price*) to 135 (criteria importance—*sharing best practices*). However, more than 80% of responses were in the range of 105 to 135. Since not all respondents provided their contact information, it was not possible to do a follow-up mailing or phone call to get full responses.

Research findings and conclusions of this research must be seen in the context of the profiles of the respondents and organizations they represent. Specifically, these findings are not tied to any specific geographical region.

Suggestions for Future Research

Statistical analysis and research findings helped to identify the criteria for measuring KM efforts, which, in turn, can be described as desired outcomes. The research study also helped to identify new areas of interest for further research. Some of these gray areas and new areas of interest are as follows:

- The most useful criteria identified through this research can be developed further into detailed measures of KM success, as discussed briefly in the previous section. The research questions—What are the detailed measures for enhanced collaboration within an organization? What are they for: improved communication and improved employee skills?—are required to be answered in this effort. The research effort would entail establishing detailed measures for each useful criterion, validating their relation to the criteria and validating their effectiveness through research.
- Based on geographical location as well as industry type, the differences in KM criteria can be analyzed using multiple factors. The research questions—What are the differences in KM criteria based on geographical location? Are they industry specific?—will have to be addressed in a follow-up research effort. However, by using data similar to those obtained from this research, the differences in KM criteria can be examined for Europe and the U.S.

Relationships among all 26 criteria can be explored to establish associations and classifications among these criteria by factorial analysis. KM criteria and outcomes may be classified based on business results, market results, customer service, and internal performance.

References

Abramson, T., Tittle, C., & Cohen, L. (1979). *Handbook of vocational education evaluation.* Beverly Hills, CA: Sage.

Ahn, J., & Chang, S. (2002). Valuation of knowledge: A business performance-oriented methodology. In *Proceedings of the 35th Hawaii International Conference on System Sciences, HICSS35, IEEE Computer Society.*

Allee, V. (1997). *The knowledge evolution: Expanding organizational intelligence.* Newton, MA: Butterworth-Heinemann.

Anantarmula, V. (2004). *Criteria for measuring knowledge management efforts in organizations.* UMI Dissertation Services (3123064): ProQuest.

Anderson, D. R. (1990). Increased productivity via group decision making. *SuperVision, 52*(9), 6.

Austin, R., & Larkey, P. (2002). The future of performance measurement: Measuring knowledge work. In A. Neely (Ed.), *Business Performance Measurement.*

Baldanza, C., & Stankosky, M. (2000). Knowledge management: An evolutionary architecture toward enterprise engineering. *INCOSE,* 13.2-1-13.2-8.

Bassi, L., & Van Buren, M. (1999). Valuing investments in intellectual capital. *International Journal of Technology Management, 18*(5,6,7,8), 414-432.

BP Amoco. (2001). Retrieved April 2001, from http://bp.com

Campbell, D., & Stanley, J. (1963). *Experimental and quasi-experimental designs for research.* Boston: Houghton Mifflin.

Chourides, P., Longbottom, D., & Murphy, W. (2003). Excellence in knowledge management: An empirical study to identify critical factors and performance measures. *Measuring Business Excellence, 7*(2), 29-45.

Cline, A. (2000). *Prioritization process using Delphi technique* [white paper]. Carolla Development. Retrieved June 2001, from http://www.carolla.com/wp-delph.htm

Davenport, T., & Probst, G. (2001). *Knowledge management case book—Siemens best practice.* Germany: MCD Verlag and Willey and Sons.

Davenport, T. H., DeLong, D. W., & Beers, M. C. (1998). Successful knowledge management projects. *Sloan Management Review, 39*(2).

The Delphi Method: Definition and Historical Background. (n.d.). Retrieved July 2001, from http://www.carolla.com/wp-delph.htm

Drucker, P. (1994). *Post-capitalist society.* New York: Harper Business.

Elliott, S., & O'Dell, C. (1999). Sharing knowledge & best practices: The hows and whys of tapping your organization's hidden reservoirs of knowledge. *Health Forum Journal, 42*(3), 34-37.

Ellis, K. (2003). K-span: Building a bridge between learning and knowledge management. *Training, 40*(10), 46.

Fairchild, A. (2002). Knowledge management metrics via a balanced scorecard methodology. In *Proceedings of the 35th Hawaii International Conference on System Sciences–2002,* Hawaii.

Fowles. (1978). The Delphi method. *Department of Civil and Architectural Engineering, Illinois Institute of Technology*. Retrieved June 2001, from http:///www.iit.edu/~it/delphi.html

Griffin, H., & Houston, A. (1980). Self-development for managers: Making most of existing resources. *Personnel Management, London, 12*(9), 46.

Gubbins, M. (2003, November 27). Enterprise—Conference call—Reap the long-term rewards of knowledge management. *Computing, 20.*

Holsapple, C. W., & Joshi, K. D. (1999). Description and analysis of existing knowledge management frameworks. In *Proceedings of the 32nd Hawaii International Conference on System Sciences. HICSS32, IEEE Computer Society,* Hawaii.

Holsapple, C. W., & Joshi, K. D. (2001). Organizational knowledge resources. *Decision Support Systems, Amsterdam, 31*(1), 39.

Jennex, M. E., & Olfman, L. (2002). Organizational memory/knowledge effects on productivity, a longitudinal study. In *Proceedings of the 35th Hawaii International Conference on System Sciences. HICSS35, IEEE Computer Society,* Hawaii.

Jennex, M. E., & Olfman L. (2004). Accessing knowledge management success/effectiveness models. In *Proceedings of the 37th Hawaii International Conference on System Sciences. HICSS37, IEEE Computer Society,* Hawaii.

Jennex, M. E., Olfman, L., & Addo, T. B. A. (2003). The need for an organizational knowledge management strategy. In *Proceedings of the 36th Hawaii International Conference on System Sciences. HICSS36, IEEE Computer Society,* Hawaii.

Jiang, J., & Klein, G. (1999). Project selection criteria by strategic orientation. *Information & Management, Amsterdam, 36*(2), 63-75.

Kelley, J. (2003). *Strategic premises and propositions for I-value growth.* Retrieved September 2003, from http://www.transformpartners.com

KPMG International, UK. (1999, November). *Knowledge management research report 2000.* UK: KPMG Consulting.

Longbottom, D., & Chourides, P. (2001). Knowledge management: A survey of leading UK companies. In *Proceedings of the 2nd MAAQE International Conference,* Versailles, France, 113-126.

Marr, B. (2003, February). Known quantities. *Financial Management*, 25-27.

Massey, A. P., & Montaya-Weiss, M. (2002). A performance environment perspective of knowledge management. In *Proceedings of the 35th Hawaii International Conference on System Sciences. HICSS35, IEEE Computer Society,* Hawaii.

O'Dell, C., Jackson, C., & Grayson. (1998). If only we knew what we know: Identification and transfer of internal best practices. *California Management Review, 40*(3), 154-174.

Ofek, E., & Sarvary, M. (2001). Leveraging the customer base: Creating competitive advantage through knowledge management. *Management Science INFORMS, 47*(11), 1441-1456.

Perkmann, M. (2002, July). Measuring knowledge value? Evaluating the impact of knowledge projects. KIN Brief #7. *Warwick Business School, Leicester University, UK*. Retrieved January 2003, from http://www.ki-network.org

Ruggles, R. (1998). The state of notion: Knowledge management in practice. *California Management Review, 40*(3), 80-89.

Simpson, J. A., & Weiner, E. S. C. (1989). *The Oxford English dictionary* (2nd ed., Vols. 1-20). Oxford, UK: Oxford University Press.

Skyrme, D. (1997). *Knowledge networking: Creating the collaborative enterprise.* Boston: Butterworth-Heinemann.

Stankosky, M., & Baldanza, C. (2001). *A systems approach to engineering a KM system* [unpublished manuscript].

Sveiby, K. (1997). *The new organisational wealth: Managing & measuring knowledge-based assets.* San Francisco: Berrett- Koehler.

Takeuchi, H. (1998, June). *Beyond knowledge management: Lesson from Japan.* Retrieved December 2001, from http://www.sveiby.com.au/LessonsJapan.html

Turban, E., & Aronson, J. E. (2001). *Decision support systems and intelligent systems* (6th ed.). Prentice Hall.

Van Buren, M. (1999). A yardstick for knowledge management. *Training Development, Alexandria, 53*(5), 71-77.

Wiig, K. M. (1993). *Knowledge management foundations: Thinking about thinking—How people and organizations create, represent, and use knowledge.* Arlington, TX: Schema Press.

Wiig, K. M. (1999). What future knowledge management users may expect. *Journal of Knowledge Management, 3*(2), 155.

Wikramasinghe, N. (2002). Practising what we preach: Are knowledge management systems in practice really knowledge management systems. In *Proceedings of the 35th Hawaii International Conference on System Sciences-2002,* Hawaii.

Chapter IX

Factors that Contribute to the Success of Knowledge Management Communities of Practice

Eric W. Stein, Pennsylvania State University, USA

Abstract

A community of practice (CoP) is an organizational form that promotes sense making, knowledge management, and learning. It is important to understand how and why these communities form and grow over time. These questions are explored in a qualitative analysis of a knowledge management (KM) community of practice. This case study includes a description of how the organization formed, survived, grew, and matured over a five-year period (1999-2004). Several practices and structures related to CoP development are identified: operations, roles and responsibilities, communications, subgroup structures, use of information technologies, and other aspects of organizing. Using data from several sources (e.g., membership surveys, interviews with key informants, document analysis), four sets of critical success factors are identified: Individual factors, content factors, meeting factors, and organizational factors. These factors are arranged into a descriptive model of the function and structure of CoPs over the life cycle. This work also sheds light on how to set up and successfully grow a community of practice.

Overview and Objectives

A successful community of practice (CoP) has the ability to sustain and renew itself over time (Barab & Duffy, 2000). This observation raises several questions. How are CoPs formed? Why do some survive? What is inherent in the structures and operations of successful CoPs that allow them to stay in existence? What other critical success factors are required, such as intrinsic or extrinsic rewards for members? The purpose of this study is to better understand communities of practice, how and why they form, and what sustains them over time.

The answers can shed light on loosely structured extraorganizational and intraorganizational forms and the factors that lead to their success over the life cycle. We define CoP success here as effectively forming, being in existence for a significant period of time, and showing continued signs of growth and development. This work also provides insight for individuals who wish to set up a successful, long-term CoP within their organizations as part of a broad KM strategy. For those specifically interested in developing KM-centered communities of practice, this study provides insights into the formation, survival, and growth of such structures.

Review: Toward a Shared Definition of Communities of Practice

This work is grounded in the literature on communities of practice (CoPs), organizational memory (OM), and knowledge management (KM). The concept of a community of practice has emerged as a useful construct to describe a social form that has been in existence for centuries (e.g., guilds) but recently has been rediscovered in the context of corporations and applications in knowledge management. The concept owes its early modern formulation to the works of Lave (1988), Wenger (1998), Lave and Wenger (1991), and Brown and Duguid (1991, 2001). The initial works focused on the shared meaning and knowledge that developed in occupational groups such as midwives and butchers (Buysse, Sparkman, & Wesley, 2003) and repair specialists (Iverson & McPhee, 2002). It is now applied to any knowledge-sharing group within and between organizations (Brown & Duguid, 2001; Swan, Scarbrough, & Robertson, 2002) and is viewed as a nontechnical component of many knowledge management strategies.

Definition

The definition of a community of practice has evolved over time. Wenger, McDermott, and Snyder (2002) see a CoP as a set of people who "share a concern, a set of problems, or a passion about a topic, who deepen their knowledge and expertise in this area by interacting on an on-going basis" (p. 4). Buysse, Sparkman, and Wesley (2003) observe, "[A] community of practice generally can be defined as a group of professionals and other stakeholders in pursuit of a shared learning enterprise, commonly focused on a particular topic" (p. 4).

Swan, Scarbrough, and Robertson (2002) define a CoP as "an activity system about which participants share understandings concerning what they are doing and what that means in their lives and for their community. Thus they are united in both action and in the meaning that that action has, both for themselves, and for the larger collective" (p. 2). Brown and Duguid (1998) observe that "collective practice leads to forms of collective knowledge, shared sense-making, and distributed understanding that doesn't reduce to the content of individual heads. A group (in which) such know-how and sense-making are shared ... has been called a 'community of practice" (p. 5).

Characteristics

Wenger's (1998) work specifies three characteristics of CoPs: mutual engagement (i.e., interaction among the members), negotiation of joint enterprise (i.e., enacting meaning and significance; defining goals and priorities), and shared repertoire (i.e., the stories, methods, tools, and theories used by members). The works of Buysse, Sparkman, and Wesley (2003) and Barab and Duffy (2000) define three essential characteristics of communities of practice: (a) shared cultural and historical heritage, as well as shared goals and meanings; (b) interdependent participants who are part of a larger social system; and (c) a reproductive cycle whereby older members leave and new members enter the community. Wenger's most recent work (2004) identifies three characteristics of a community of practice that differ from his earlier works: (a) Domain: the area of knowledge that brings the community together; (b) Community: the group of people for whom the domain is relevant; and (c) Practice: the body of knowledge, methods, tools, stories, cases, documents which members share and develop together.

A careful reading of the prior works suggests that there are five distinct aspects of communities of practice that identify them as follows:

1. A knowledge domain of interest
2. A set of interested and interconnected participants
3. Opportunities for ongoing processes of sense making, knowledge sharing, and discovery within the domain of interest
4. A set of resources related to the domain of interest, including methods, tools, theories, practices, and so forth, that are acquired, retained, and accessible by the community
5. Processes by which the community maintains and refreshes its membership

These five criteria are useful in distinguishing between communities of practice and other similar but more narrow forms of organization, such as communities of interest (CoI), communities of learners, and learning communities (Barab & Duffy 2000; Buysse, Sparkman, & Wesley, 2003). For instance, communities of interest share some but not all of the characteristics of CoPs. According to Walters and Clark (1996), "Communities of interest or electronic communities ... offer people who live in the same locale and share common interests and concerns a virtual place to exchange information" (p. 1). The distinct focus of a CoI is on electronic information exchange as opposed to the far-richer concept of sense making and enacted meaning found in CoPs (i.e., sense making is seen as a key distinguishing feature of communities of practice).

The importance of sense making to organizing is found in the management literature in several works. Sense making in organizations has been seen as a critical factor related to crisis management (Weick, 1979, 1995), strategic management (Gioia & Chittipeddi, 1991; Gioia & Mehra, 1996), organizational learning, knowledge management, and performance (Thomas, Clark, & Gioia, 1993; Thomas, Sussman, & Henderson, 2001), creativity in organizations (Drazin, Glynn, & Kazanjian 1999), product innovation (Dougherty, Borrelli, Munir, & O'Sullivan, 2000), managing organizational complexity (Moss, 2001), and information technology use (Griffith, 1999).

Another defining characteristic of communities of practice is their ability to process knowledge through knowledge discovery, retention, and use. This function finds support in the literature on organizational memory (Ackerman & Halverson, 2000; Anand, Manz, & Glick, 1998; Casey, 1997; Cross & Baird, 2000; Croasdell, 2001; Moorman & Minor, 1997, 1998; Nissley & Casey, 2002; Rulke & Rau, 2000; Stein, 1995; Stein & Zwass, 1995; Walsh & Ungson, 1991; Wijnhoven, 1999; Wishart, Elam, & Robey, 1996) and on knowledge management (KM). According to OM research, organizations intentionally construct, acquire, retain, and retrieve organizational memories to support organizational activities (e.g., positive memories) and adapt to cope with the effects of outdated knowledge and information (e.g., negative memories).

Knowledge management has been defined in the works of Nonaka (1994), Nonaka and Takeuchi (1995, 2001) and others as inclusive of the processes of knowledge creation, transfer, and use. Earl (2001) segments KM interests and application into three areas. The first area (Technocratic) focuses on the use of information technologies to achieve KM goals such as knowledge retention and retrieval. The second area (Economic) pertains to how organizations use knowledge for competitive advantage and to formulate business strategies (Davenport & Prusak, 1998). The third area (Behavioral) is most closely aligned with this work, which "describes the use of organizational structures, or networks, to share or pool knowledge. Often described as 'knowledge communities' the archetypal organizational arrangement is a group of people with a common interest, problem, or experience" (Earl, 2001, p. 6). In a similar vein, Demarest (1997) divides the KM world into two parts: the one that makes knowledge explicit and stores it using information technologies and the one that facilitates the growth and enrichment of knowledge through the formation of social networks bound by communication and learning processes. Communities of practice, it can be argued (Iverson & McPhee, 2002) offer "a theoretical construct for understanding the interactive roles of information systems and people and also as a model for understanding how KM is negotiated communicatively between people" (p. 1). Brown and Duguid (1998) also recognize the importance of CoPs to KM but caution that they "can easily be blinkered by limitations of their own world view" (p. 5). Less circumspect, Wenger (2004) argues that "communities of practice are the cornerstones of knowledge management" (p. 1). At the least, communities of practice can be viewed as a means of implementing KM in organizations. For example, organizations that are reluctant to commit resources to KM information technologies often start with CoPs because they are perceived as a low-cost alternative. Others simply want to test the KM waters and view CoPs as a means of doing so. To those that embrace the human side of organization, communities of practice are considered the primary way to implement KM.

In summary, communities of practice represent a social form that has been in existence for many years. Studies of these systems have illuminated much about their functional aspects (i.e., ability to process knowledge and information, usefulness to its members or to the organizations that support them). On the other hand, less is known about their actual structures, how they come into existence, and how they grow and are self-sustaining over time. The goal of this work is to explore the evolution of communities of practice and the organizing activities that contribute to longevity and success.

Research Questions and Propositions

The following research questions and propositions (see Table 1) were framed based on the central questions articulated here, gaps in the literature, and observations from the field. The key questions are as follows: Why (and how) are some CoPs able to move smoothly from one life cycle phase to another? Why are some CoPs in existence for several years without

Table 1. Research questions and propositions

Research Questions	• Why are CoPs formed? How? • Why do some CoPs endure for years after they are formed? • Why do people join and contribute to a knowledge-based CoP? • How is a CoP structured? How is a CoP operated? • What benefits does a CoP deliver to its members?
Research Propositions	**Formation Propositions** • CoPs can form from a single meeting or presentation. • The cost to form a CoP is low and requires a small investment in a meeting place and refreshments. • To ensure its survival, a core set of participants must commit to future meetings and establish means of communication and coordination.
	Survival and Early Growth Propositions • Having a clearly defined and executed mission is important to survival and growth. • Social networks of the members are important to survival and growth. • Knowledge assets of the members are important to survival. • Organizational champions are a key to survival. • Pooled resources are a key to survival and growth.
	Late Growth and Maturity Propositions • CoPs provide *explicit* benefits to members at meetings (e.g., topic, knowledge exchange) as well as *implicit* benefits such as recognition, affinity, expanding social networks, and emotional support. • CoPs create organizational structures, establish roles and responsibilities, and create procedures to ensure operational efficiency and to move from survival to growth. • CoPs can choose that its structures and processes remain informal and low-cost. • Information technologies and communication processes play a key role in ensuring a CoP's continuation.
	Decline or Renewal Propositions • CoPs that adapt to changing conditions will experience renewal. • CoPs that fail to adapt to changing conditions, modify their mission, or recruit new members will experience decline and cease to exist.

a formal budget or commitment of resources? In short, we are interested in identifying the organizational factors that have led to success. These include motivational factors (e.g., what motivates organizational leaders and members to contribute) and factors related to context, knowledge content, organization, and structure. The study also addresses several how-to questions related to CoP formation and development (e.g., operations, roles and responsibilities, meeting management, etc.) that are of particular relevance to practitioners.

The propositions are grouped according to the life-cycle phases of organizations (i.e., from formation to maturity). The concept of the organizational life cycle is not new, appearing first in the works of Haire (1959) and Chandler (1962) and later in several works in organization theory. Gupta and Chin's (1994) extensive review identifies five areas of impact on the management literature: organizational effectiveness (Quinn & Cameron, 1983); entrepreneurship (Smith & Minor, 1983); strategy making (Gupta & Chin, 1992); organizational power (Mintzberg, 1984); and organizational politics (Gray & Ariss, 1985). The concept, which typically is applied to formal organizations such as corporations, contends that organizations go through various stages throughout their development, pursuing different ends and exhibiting different characteristics over time. Most works specify a four-stage model: formation, early growth, maturity, and decline or renewal.

The formation phase is characterized by the development and implementation of a plan or vision, the acquisition of resources, and the fulfillment of customer needs (Jawahar & McLaughlin, 2001). In the survival and early growth phases, the organization stabilizes its position and pursues opportunities for expansion. Most organizations focus on product or service reliability and work to formalize organizational structures (Dodge & Robbins, 1992, Jawahar & McLaughlin, 2001). In the mature phase, the organization becomes more confident in its abilities but also more risk-adverse. The rate of growth has slowed by this time, and there is some uncertainty regarding new opportunities (Jawahar & McLaughlin, 2001). Finally, the organization faces a crisis: it reinvents itself and transitions to a new cycle or it goes into a decline phase that leads to its eventual termination.

These stage distinctions are useful in grouping the research propositions with the understanding that CoPs differ from formal organizations in important ways yet share some of the same evolutionary life-cycle characteristics.

Research Design and Methods

A single case study design was chosen for this research. Case studies offer an opportunity to examine an organization in depth through the development of rich description.

Case Selection

A community of practice for the exchange of information on best practices in knowledge management was selected for study. The case organization was formed in 1999 and remains active as of this writing. This organization met the five criteria identified for CoPs as noted

Table 2. Characteristics of study case as community of practice

Characteristic	Value
A knowledge domain of interest?	Yes. Knowledge management theory and practice.
A set of interested and interconnected participants?	Yes. Drawn from area businesses and universities. Bound by strong to weak ties.
Opportunities for ongoing processes of sense making, knowledge sharing, and discovery within the domain of interest?	Yes. Monthly meetings and Executive Committee meetings. Threaded discussions online.
A set of resources related to the domain of interest, including methods, tools, theories, and so forth?	Yes. Knowledge is retained in people's heads and the documents produced by the community (e.g., presentations from the monthly meetings made available through the organization's Web site).
Processes by which the community maintains and refreshes its membership?	Yes. The organization has grown from 10 members to more than 200 over a five-year period. New members attend and are added each month.

previously: a domain of interest, a set of interconnected participants[b], opportunities for sense making, tools and supporting resources, and mechanisms for renewal (see Table 2).

Types of Data Collected

Data for the case came from multiple sources in order to provide cross-checks of validity and to triangulate the results, as suggested by Yin (1989) and Eisenstadt (1989). In general, the overall rigor of case studies can be increased by using multiple sources, establishing a chain of evidence, and having key informants review drafts of the work (Yin, 1989). These methods were employed to the greatest extent possible in this study. The work is considered exploratory in nature and accompanied by the caveats that apply to such research.

Data sources included documents, interviews, surveys, and participant observation by the author (the latter has been an executive member of the organization since its inception). Documents from 1999 to 2004 were collected from the organization, including schedules, Web materials, lists, membership databases, minutes of meetings, postings of threaded discussions, and e-mails. Interviews and surveys were conducted with present executive committee members to test assumptions and propositions regarding the organization. Surveys also were conducted with the entire membership. In addition to these data sources, data regarding the organization were collected by the author as a participant-observer within the organizational setting. All data were collected using instruments and methods approved by the university's Office for Research Protections, which ensures the protection of human subjects according to federal guidelines.

Form of the Report

These data were compiled into three sets of findings (sections 5-7) and a case discussion section (section 8). The case discussion ties the findings back to the research questions and propositions noted in Table 1. Preliminary models are provided of the factors that have contributed to the success of the organization. The insights of the discussion section are grouped according to the life-cycle stages noted previously.

Findings 1: Characteristics of the Case Organization

These findings are based on the collection, compilation, and analysis of existing documents, informal discussions with members, and participant observation.

Formation and History

The organization chosen for study was the Knowledge Management Group of Philadelphia. Data were collected over a five-year period from its inception in April 1999 to April 2004. This is the third longest continuously running KM group of this kind in the U.S. The organization has been through the formation, survival, and growth phases and is in or moving toward a state of maturity[c]. The group began from a meeting held in March 1999. The original meeting described knowledge management (KM) strategies and practices at Hewlett-Packard, with particular emphasis on its consulting division. Based on the interest generated by this meeting, a meeting sponsored by the local area Chamber of Commerce was held in April 1999 to explore starting a KM learning community. About 50 people attended this meeting, representing industry and academia. At the meeting, a handful of people willing to lead future meetings was identified. These eight people represented several industries, including education, engineering, energy, government, and consulting. Of the eight, five became active members of an executive (steering) committee (EC). The role of the executive committee was to coordinate and host meetings and to set the direction for the organization. The group met once a month in the morning throughout the calendar year (after the first year, meetings in July and August were not scheduled due to vacations and work slowdowns). By the end of the study period, the group successfully concluded its fifth year of activity, having held more than 52 meetings. The group has more than 240 registered members.

Goals and Objectives

The stated aims of the KM Group are as follows:

> *The Knowledge Management Group (KMG) was formed to address the needs of area organizations in managing knowledge assets. Knowledge assets include intellectual capital (e.g., what employees know, patents), procedural knowledge contained in documents and administrative structures, and knowledge embedded in information systems.*
>
> *Knowledge management includes activities related to the creation, capture, organization, maintenance, retrieval, and use of organizational knowledge to promote improved decision-making and performance. KMG's goal is to promote the sharing of KM best practices, to provide a forum for group problem solving on KM problems, and to encourage networking and professional collaboration in the area of KM.* (Organization's Web site)

Thus, the explicit goals of the group are (a) to promote networking and (b) to promote learning and shared understanding about knowledge management. The group promotes networking between professionals in the area who work for knowledge-intensive organizations. These include pharmaceutical firms, consulting firms, software developers, manufacturers, and academic institutions. The main learning objective is to share the latest theoretical and practical knowledge among the members. Each meeting is structured around a theme or idea, and presentations are made by member companies or by outside experts. The meetings provide a forum to challenge assumptions, foster sense making, absorb new ideas, and develop a shared understanding of the domain.

Meetings

Meetings are scheduled on the second Wednesday of the month from 7:45 to 9:45 a.m. to allow members to attend and then return to work. The first half hour is devoted to networking. At 8:15, announcements are read, an introduction is provided to KMG, and (time permitting) members introduce themselves. The main part of the meeting is about one hour followed by Questions/Answers. People leave around 9:45 a.m. or stay to discuss issues informally. About two-thirds of the meetings are located in the suburbs, and one-third takes place in the city to encourage a broad cross-section of attendees. Topics range from presentations of theory and frameworks (e.g., teams and the social construction of knowledge) to case studies (e.g., KM at the DuPont Company) to experiential exercises (e.g., knowledge acquisition and retention) to sessions about KM tools and strategies (e.g., communities of practice, IT support for KM)[d]. The meetings range in size from 30 to 50 people.

Membership

The organization began with a core group of about 20 members. After five years, the database of members included more than 240 members from several industries and organizations. Statistics summarizing the industries and positions held by the membership are presented in Table 3. About 40% of the members in the database consistently attend meetings. The largest percentage of organizations comes from the information technologies sector (18%)

Table 3. KMG-P membership by organization type, job title, and geography

Organization Type	Count	%
Consulting	22	9%
Engineering	9	4%
Finance	5	2%
Government	9	4%
Insurance	8	3%
Information Technologies	43	18%
Manufacturing	16	7%
Other	39	16%
Pharmaceutical	29	12%
University	14	6%
NA	47	20%

TOTAL 241 100%

Job Title/Type	Count	%
Consultant	13	5%
Professor	11	5%
Information Professional	35	15%
Knowledge Mgt Professional	24	10%
Library Professional	3	1%
Manager	32	13%
Senior Manager	48	20%
NA	75	31%

TOTAL 241 100%

Location	Count	%
City	33	14%
Suburbs	157	65%
NA	51	21%

TOTAL 241 100%

Notes:

1. *Organization Type table does not include 47 unidentified cases (Total N=241).*
2. *Job Type table does not include 75 unidentified cases (Total N=241). Classes based on job title.*
3. *NA = data not available or in database.*

followed by pharmaceuticals (12%) and consulting (9%). Managers and senior managers from these companies make up almost 33% of the attendees, followed by information technology (15%) and KM professionals (10%). The majority of the firms (> 65%) are located outside the city in its suburbs or neighboring states.

Administrative Structures

The primary administrative structure of the organization is the executive committee (EC). The purpose of the committee is to select topics for future meetings, find host sites, set policy and procedures, and guide the group. Members of the EC serve on a voluntary basis. The committee is composed of members of large and small organizations, consultants, and academics. The board strives for a balanced representation of these three groups. Two of the current members, including the author, have been in the organization since its start in 1999. The EC members have both breadth and depth of knowledge about its activities.

Table 4. Requirements for application and roles and responsibilities of EC members

Requirements for Application to the EC
• Attend and participate in KMG meetings for six months or more prior to application • Assist in the planning and management of one or more KMG meetings prior to application • Contribute to the intellectual capital of the KMG meetings for six months or more • Be nominated for participation on the board by two or more EC members • Commit to regularly attend and to contribute intellectual assets and time to monthly main and EC meetings if elected to EC

Role*	Responsibility	Tasks
Meeting Coordinator	Take responsibility for coordinating the hosting of one or more meetings per year	• Find venue • Arrange for refreshments at meeting • Manage logistics at meeting
Speaker Coordinator	Take responsibility for obtaining a speaker for one or more meetings per year	• Find and manage speaker • Introduce speaker at meeting
Resources Coordinator	Solicit and identify resources on behalf of the organization	• Obtain in-kind services, host locations, and so forth
Management Coordinator	Manage meeting and participate in the internal running of the organization	• Select topics for calendar year • Set agenda for EC meetings • Run meetings
Communications Coordinator	Manage communications of the organization	• Communicate announcements of meetings to members via e-mail • Manage Web site • Manage Yahoo groups
Membership Coordinator	Maintain and grow membership	• Solicit new members • Manage member database

Note: Some roles are shared among one or more members

The EC meets the Friday following the monthly meetings and conducts an after-action review of the event. These members are responsible for the running of the organization and have the highest levels of participation among the membership. From 2003 to 2005, the executive committee was composed of eight members. Membership on the EC requires taking responsibility for the welfare of the group. These requirements are noted on the organization's Web site, which evolved over the five-year period and were not codified until the fifth year.

Several roles and responsibilities evolved over time within the EC group to meet the needs of running the organization. The most important logistical issue to deal with each month is to select a topic, get a speaker, and find a venue to host the meeting. Ongoing activities include managing the membership list of names, the Web site, and the threaded discussion group on Yahoo. See Table 4 for a complete list of the requirements, roles, and responsibilities of EC members.

IT and Communications

Information technology and related procedures have played a role in the development of the organization. The member database was one of the most important.[e] A database (e.g., in MS Word or Excel) of members was set up early in the first year. The database was populated with information obtained at monthly meetings, through referrals, and through the organization's Web site. At the monthly meetings, attendees sign in or verify contact information on a list. This information is used to augment or keep up-to-date the member database. The member database is central to the notification process of upcoming meetings. For instance, an e-mail[f] is sent out about two weeks prior to a meeting, announcing its whereabouts and the topic. The mailing list is not used for commercial purposes.

Notice of upcoming meetings also is posted on the organization's Web site. The Web site was started in the second year of the organization and includes a meeting calendar, links to other KM sites, a list of Executive Committee members and sponsors, contact information, document archives, and member signup. The Web site was designed, maintained, and hosted by one of the EC members[g] for the first five years. More recently, it is hosted using an inexpensive ISP and is maintained jointly by two EC members.[h] Calendars for each year are archived, and presentations are made available through the site. Member communications are facilitated via e-mail and through Yahoo-Groups. At Yahoo-Groups, members can engage in threaded discussions and post documents of interest.[i]

Findings 2: Membership Survey

Results of the Survey of the Membership

The following questions were posed to the general membership:

Table 5. Results of the survey of the KMG membership December 2003

How did you first learn of the existence of the Knowledge Management Group? (check all that apply)	**Number of Responses**	**Response Ratio**
Colleague	72	74%
Web search	9	9%
Other	20	21%
Why did you join the Knowledge Management Group? (check all that apply)	**Number of Responses**	**Response Ratio**
Required by job	0	0%
Career advancement	13	13%
Interested in topic	81	84%
To network with other professionals	73	75%
To learn practical methods and techniques of KM	60	62%
To be in a supportive environment	20	21%
To look for new business	18	19%
Other	6	6%
What do you like most about the organization? (check all that apply)	**Number of Responses**	**Response Ratio**
Quality of speakers	55	57%
Choice of topics	63	65%
The way the meetings are run	38	39%
The people you interact with at meetings	57	59%
The times of the meetings	26	27%
Communications following the meeting	13	13%
Friendships	11	11%
Networking	47	48%
Intellectual stimulation	64	66%
Other	3	3%
Would you call this organization a success?	**Number of Responses**	**Response Ratio**
Yes	96	99%
No	1	1%
Do you think it is likely that the organization will continue to exist for another five years?	**Number of Responses**	**Response Ratio**
Yes	90	93%
No	7	7%

- How did you first learn of the existence of the Knowledge Management Group?
- Why did you join the Knowledge Management Group?
- What do you like most about the organization?
- Would you call this organization a success?
- Do think it is likely that the organization will continue to exist for another five years?

An anonymous online survey[j] was conducted over a three-week period, which concluded the first week of December 2003. E-mail invitations to participate in the study were sent to 241 members. Of these, 37 were returned due to incorrect e-mail addresses, leaving 204 invitations. Of these, 97 completed the five-question Web-based survey, which yielded a response rate of 48%. The assumption was made that respondents were representative of the group, given the relatively high response rate.[k]

The results of the survey, provided in Table 5, are as follows:[l] 74% of the respondents indicated that they learned of the group from a colleague, 9% through a Web search, and 21% through other means. These data reinforce the networking and word-of-mouth aspects of communications regarding the group. The primary reasons that people joined the group were that they were interested in the topic (84%), they wanted to network (75%), and they wanted to learn practical methods regarding KM (62%). Of lesser importance were career advancement (13%) and looking for new business (19%). More than one out of five joined the group to be in a supportive environment (21%).

The primary aspects of the organization that people liked most included intellectual stimulation (66%), the quality of the topics (65%), the quality of the people at the meetings (59%), the quality of the speakers (57%), and networking opportunities (48%). The way the meetings were run and times of the meeting were also important (39% and 27%, respectively).

Overwhelmingly, the organization was considered a success by the respondents (99%), and 93% thought it would exist for another five years.

Findings 3: EC Interview Data

Data Collection and Methods

Data were collected from the current Executive Committee (EC) members via e-mail and informal interviews. The following questions were posed:

- Why was this organization started?
- Why do you think people join and participate in the organization?
- Why do you think someone volunteers to work on the executive committee?

- Why is the organization still functioning after five years?
- What are the factors that have led to the success of this organization?

All EC members (excluding the author) participated in the data collection (n=7), yielding a response rate of 100%. The written responses were compiled and analyzed using Atlas TI, a content analysis program. After all responses to each question were compiled for all respondents, codes were assigned to the material. The coding was done in an iterative fashion. First, the program provided frequency counts of words; these lists provided candidates for content codes. Next, the program searched for sentences that contained the codes identified in the previous step, and these were auto-coded. Several reviews of the material resulted in the addition of new codes and the refinement of existing codes. Finally, the remaining text was coded manually using the master list of codes that was created. The efficacy of the codes was tested with alternate raters (two university staff members[m]), who were asked to match the codes to a sample set of statements from the primary material. The first reviews found that interrater reliability was about 72%. Subsequent refinements improved rater reliability to better than 80%, and the codes were considered to be reliable for the purposes of this work. Once the coding of the responses was complete, quotations from each member (e.g., P1, P2, P3, … Pn) were organized according to topic, and preliminary models (e.g., influence diagrams) were constructed.

Interview Results

Q1: Why was the organization started?

Three primary reasons were given: knowledge sharing, networking opportunities, and a desire to learn more about the topic. As one member put it, "(it was started) as a way for people to build a network and share practices and learnings in an informal, low-barrier setting" (P3). However, despite the clarity of its goals, it was not clear from the outset what the evolution of the organization would be and if it would exist for five years. One member put it this way:

> *The initial focus was about learning and, likely, networking. But, I doubt there were expectations about what it would become and (it was) more about let's just get together and talk about a topic of interest and maybe we can learn some more.* (P5)

The organization thus evolved in an organic way with few initial expectations about its future.

Q2: Why do people join and participate in the organization?

As noted in the previous question, knowledge sharing, networking, and learning were important reasons to join the organization. One member said this:

> *(People join) to learn what KM is all about (everything from "what it is" to "how" and "why"), and how KM can drive their organizations to greater productivity, and help themselves in their efforts to manage their information and knowledge. To gain personal, realistic and first-hand experience and learnings from others in their same industry or area about KM. To network—personally with others sharing like interests and problems in the KM arena.* (P6)

These primary variables (networking, learning, and sharing) formed a cluster that was labeled Individual Factors. In addition, two other categories of variables were identified: meeting factors (i.e., characteristics of the meetings) and content factors (i.e., the characteristics of the knowledge exchanged at monthly meetings). For instance, positive meeting factors included the informality of the meetings, meeting locations, meeting quality, and fostering a fun atmosphere. Content factors included content relevance and content value. A model of these factors is shown in Figure 1.

One member summed up the reasons for joining this way:

> *I came originally because a friend invited me. I have invited others. The loose structure makes it easy to participate; to come and go. The opportunities for networking are also valued. The communication about meetings and topics in advance of the sessions lets individuals decide when to participate. The atmosphere is inviting; questions and answers are valued.* (P4)

Figure 1. Factors related to "why join the organization?"

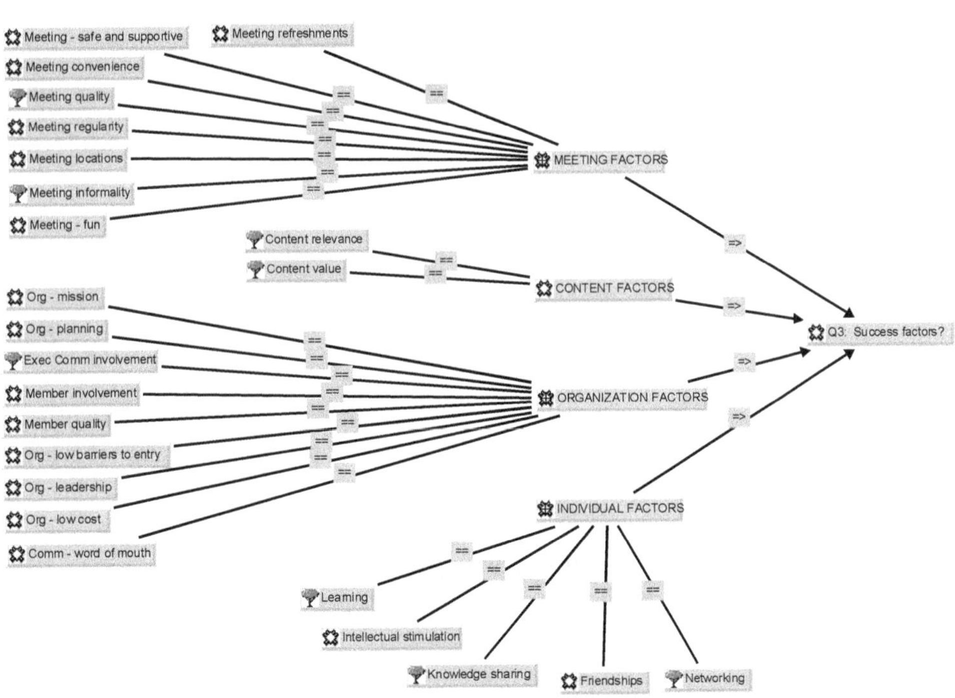

Q3: Why do you think someone volunteers to work on the executive committee?

Reasons for participating on the EC ranged from very personal ones to more pragmatic considerations.

> *I did it intuitively, because it felt right. Now I'm thinking about why it felt right, and here's what I believe. ... (I) tend to feel responsible for organizations in which I participate and ... believed it would be a good experience. ... I thought it would get me more deeply networked with other KM practitioners. ... I figured I could help the group and add value. ... I continue to do it because, in addition to what's noted above, I really like everyone on the EC, we have a lot of fun, and it's a great group of people to work with.* (P3)

Another member noted, "For me it is the desire to give back to the group and my profession; to support the continuation of the grassroots nature of the organization.". Another (P2) said that they had an "interest in KM ... (and a) desire to contribute and lead." (P4) Still another member said:

> *Various motivators: (1) Deepen their social network. Get to know colleagues in broader ways—and they get to know you. (2) Professional recognition. The membership ascribes a certain status to those guiding a functioning organization. (3) Marketing purposes (e.g., "I am on the Executive Committee of the KMG"), (which) adds a level of credibility in talking about KM. (4) Deeper learning. ... Can tap a high level of knowledge and experience. (5) Pleasure of turning an idea into reality. It has certainly given me pleasure and a sense of accomplishment to having gone from nothing to a substantial something that has impacted people's careers and businesses.* (P5)

Finally, one person saw the EC as a means to engage in sense making and deep learning as a subcommunity within the larger community of practice:

> *Learning is social—(the) EC is a smaller social group than KMG that provides opportunities (occasionally) for debating, for sharing mental models and learning.* (P7)

In summary, the factors that were most important included the quality of the experience, a desire to give back, intellectual stimulation, and professional development. See Figure 2 for a model of the motivators to join the EC.

Q4: Why is the organization still functioning after five years?

Q5: What are the factors that have led to the success of this organization?

Figure 2. Factors related to "why join the executive committee?"

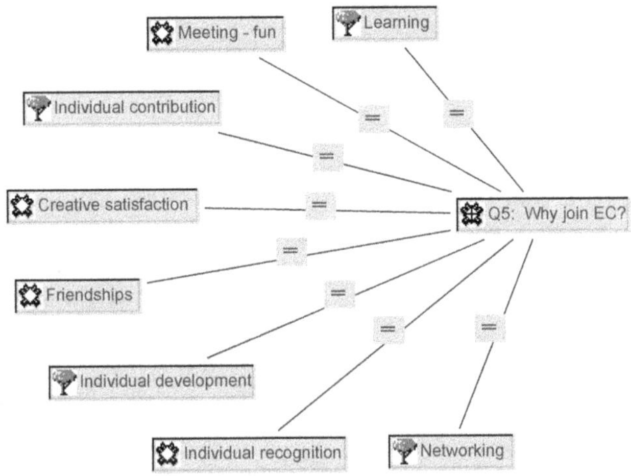

One of the key objectives of this study was to determine why the organization is still around after five years and, in this sense, is a success. The critical success factors range from personal motivators to structural ones. For example, one member commented on the community aspects of the group:

> *(It is successful) primarily because it ... meets the needs of participants as a "community." From a "community" perspective, it enables participants to be fully networked (CoP) with others in the same area, share experiences with others in like businesses or even just geographically co-located, or to participate as "interested parties" (CoI) or onlookers, feeling free to participate as interest and time permits. The personal touch, however, is (the) key to this organization's success.* (P6)

Another member identified several positive aspects of the organization:

> *Here are several factors I think are the main reasons KMG still exists after all this time. ... Above all, KMG is focused on building on and extending learning around the broad topic of KM. ... the focus of the learning has first and foremost been from a practitioner's viewpoint (e.g., practical things I can do tomorrow). This has kept the sessions very grounded. ... A handful of people have continued to step forward to make sure the session topics reflected the interests and needs of those who were interested in attending KMG sessions. ... With an emerging professional group, there is a need for affiliation. This plays out in many ways, such as, just finding like minds; looking for jobs/projects so one can actually do KM and get paid. ... KMG has evolved because of an awareness and sensitivity to what is taking place in the field. Thus, topics have been timely and current ... contributing to the value participants receive from attending meetings. No fees. No "official" membership. No administrative bureaucracy. Simply topics*

> *offered on certain dates and some minimal support infrastructure (website, Yahoo group). Essentially, nothing to interfere with the learning itself.* (P5)

Still another person said this:

> *The group provides a safe environment for people to share ideas no matter how controversial they are. People in the group have shared experiences and a history of sharing ... and networks and friendships have formed between the members. ... The topics are valuable and timely (and) focus on KM practices. The group consistently exceeds expectations of members. ... I learn a lot of stuff that is immediately applied to my current situation. It makes me think. ... It's fun. ... It's free. The EC has worked hard to make it a lasting program. Events are well planned and professional.* (P1)

The results of the content analysis show that several factors are responsible for the organization's success according to the respondents. These findings have been grouped into four categories based on an extension of the earlier categorization: individual factors, content factors, meeting factors, and organizational factors. Organizational level factors include leadership, EC involvement, barriers to entry, planning, member quality, among others. Figure 3 identifies the key variables in each category and illustrates the relationships among the variables.

Discussion and Interpretation of Findings

The results of the interviews, documents analysis, content analysis, and surveys are synthesized into this case report, and the findings are grouped according to the life-cycle phases. Support is found for many of the research propositions appearing in Table 1.

Formation

This study has shown that a community of practice can form around an idea or practice; in this case, knowledge management. The initiating factor in this case was a presentation on the topic by a sponsoring organization. The costs to sponsor were low (e.g., less than a few hundred dollars). Based on the initial meeting, a subset of people from the group agreed to meet to explore the possibility of meeting again in the future on a periodic basis. The drivers for this new organization were to promote networking, learning, and knowledge sharing. E-mail was the primary communication mechanism for the new group.

While the stated aims of the organization were to promote learning and networking, the community evolved in an organic way; its features unfolded over time. In several ways, a CoP is a good example of a self-organizing system (Contractor, 1999; Houston, 1999; Lorange

Figure 3. Factors related to "why is the organization successful?"

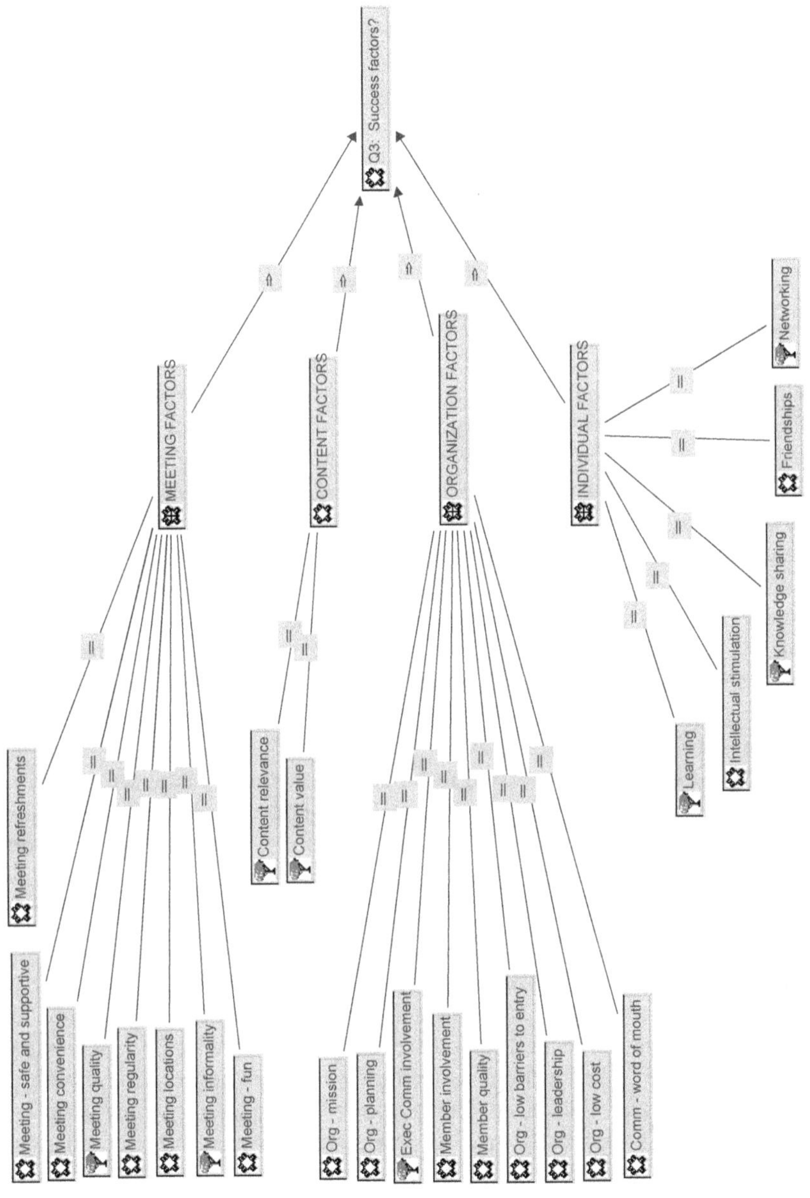

& Probst, 1987). Self-organizing systems develop patterns out of a state of disorganization (Contractor, 1999), which aptly describes a community of practice in its early stages. Its functions and structures develop over time in response to the environment, the available resources, and the goals and objectives of the members.

There are parallels here with new venture formation in the field of entrepreneurship and organizational life cycles (Cameron & Whetten, 1981, Quinn & Cameron, 1983). Like any startup, CoPs improvise procedures and policies until routines form (Nelson & Winter, 1982). In new venture organizations, structures are simple, there is little planning and control, and the primary focus is on turning ideas into reality (Dodge & Robbins, 1992; Flynn & Forman, 2001). Communities of practice share these same characteristics at this stage of development.

Survival and Early Growth

Following its inception, the KM group survived for a variety of reasons. The case study identifies several drivers, including organizational, individual, meeting, and content-related factors (see Figure 3). One of the key success factors in the organization category was having a clear sense of mission. In the case, the organization knew what it would focus on: sharing KM practices and theory and to promote networking. Another key factor was the emergence of organizational champions who brought energy, drive, and leadership to the group. They also brought assets in the form of social capital and access to resources. Without the extensive social networks of some of the key figures, the organization would not have had the connections necessary to grow and make it out of the formation stage. Furthermore, based on their positions within their respective organizations, these champions were able to get their organizations to commit modest resources in support of the meetings (i.e., by providing food and space). Finally, the organizational champions brought sufficient knowledge of the domain of interest to foster a rich and creative intellectual environment. Several of the core members of the organization volunteered to make presentations until a network of topic providers was created. This collective knowledge base was certainly a critical success factor for the organization. Other factors in this category included event communication and planning, member participation, and the low costs of membership (i.e., no costs and no dues).

In the second and third categories are grouped meeting and content-related factors. During the first year and beyond, significant time and effort were devoted to making the meetings relevant and convenient to attend. Topics were carefully selected by the core members prior to their announcement. The location of meetings was varied between the suburbs and the city to encourage diverse attendance. The general quality and consistency of the meetings was important to the group's early survival. The content was generally viewed as practical, relevant, and value-adding by the membership. Since this was the primary product offered by the CoP, we are not surprised by this result. Finally, there was a conscious desire to make the meetings informal and fun. This is consistent with research on the associations found between play and learning in work environments (Webster & Martocchio, 1992, 1993).

Finally, several individual factors were viewed as helping the organization to survive and grow. One of the most important ones was intellectual stimulation. This finding is consistent with what may be viewed as a defining feature of a community of practice; namely, the provision of an environment that allows for knowledge sharing and sense making. Making sense of the often conflicting theories and practices associated with knowledge management is done best in a group in which meaning can be tested, negotiated, and refined. While such behavior is associated with and encouraged in universities, the reverse is true in most corporate environments. In the latter, there is a regression to the mean, strong pressure put on conformity and usually little time to question corporate goals and objectives. While the learning organization (Argyris & Schon, 1978; Berends, Boersma, & Weggeman, 2003; Senge, 2003; Vera & Crossan, 2004) is an ideal for many organizations, in practice, learning based on paradigmatic change usually is discouraged in favor of getting the job done or fulfilling the client's needs. In contrast, a community of practice offers an opportunity (albeit for a brief interlude) to question assumptions, refine ideas, and sharpen vocabulary (i.e., to learn). This finding is interpreted as another reason why communities of practice have emerged as a functionally useful social form.

Late Growth and Maturity

Once the organization survived its first year of operation, several factors helped to ensure that it would continue to do so in the future. During the first year and beyond, the organization evolved key structures and processes to help it to grow. The most critical structure setup was the executive committee. This group was responsible for guiding the growth and development of the organization. Among the duties of the EC were planning, topic selection, finding host sites, review and evaluation, and setting policy and procedures. Several reasons are cited for joining the EC, including increased opportunities to learn, professional recognition, professional development, the opportunity for individual leadership, creative satisfaction, and making a contribution (see Figure 2). Also important was the fun atmosphere and friendships that developed. These factors served as potent motivators for the EC members as they took care of the numerous duties required to keep the organization functioning. Another closely related outcome among the EC members was the creation and adoption of roles in response to the evolving needs of the organization (see Table 4). This finding is consistent with what typically takes place in groups, usually in the context of formal organizations. Finally, the group established rules for admission to the EC, which were posted on the Web. For example, attendance and hosting requirements were set up for EC members. These criteria helped to distinguish between those who were really committed to the organization and those who were not. For instance, some early EC members did not show up frequently to meetings nor did they contribute to the welfare of the community significantly. By publicizing the requirements, expectations for assigned tasks and roles were clear.

Another important factor that helped the organization grow and develop was the formation of routines. These included scheduling the meeting at the same time each month (i.e., second Wednesday), holding it for the same duration (i.e., 7:45-9:45 a.m.), providing a similar structure to the meeting (e.g., first half hour devoted to networking, followed by announcements, the presentation, and a closing). Executive Committee meetings also were scheduled on the Friday immediately following the membership meeting to conduct after-action reviews and to plan future meetings. Meeting notices were sent out at approximately the same time each month. Taken together, these routines created a sense of security and permanence for the membership, despite the essentially virtual aspects of the community. In reality, the community had no physical assets, no financial resources, no space to convene, no offices, and no paid human resources. It created its own reality through the use of e-mail, a Web site, virtual group workspaces, monthly meetings, and its own routines and structures.

Although threaded discussions and other online activities were supported and encouraged by the community, for the most part, information technologies played a supportive but not dominant role in the organization's development. Information technologies served to coordinate the back-end activities in support of the human aspects of the experience (i.e., the meetings where sense-making, and learning could occur). That being said, it is unlikely that the community could have continued to exist without e-mail and Web access to facilitate the planning, communication, and coordination of meetings.

No organization continues to function unless it serves its members. The obvious benefits to the members included learning about new topics and KM practices, networking, and opportunities for self-marketing. However, there were important implicit benefits. For example, the ideas of KM suffer from ambiguity. Members found the organization a safe place to test and share ideas, such as one gets in a university classroom. In this sense, the organization provided an emotionally supportive environment for its members. Recognition also was built into the culture of this community. Iverson and McPhee (2002) observed, "Cultivation of knowledge can occur through three communicative actions: celebration, articulation, and collaboration. The general point of celebration is to recognize knowledge accomplishments and problems solved" (p. 3). It was not uncommon for members to receive public recognition at general and executive meetings. This produced the dual benefit of providing psychological reward for those doing the work and encouraging others to do the same.

Finally, there was a conscious effort on the part of the leadership of the community to choose the ways in which it would grow. Oftentimes, organizations perceive growth as a good thing and simplistically make choices without regard to the consequences; not so with this CoP. For instance, while membership growth was an important consideration, the community did not want to grow without limits. Monthly meeting size was unofficially capped at 50 people. Holding larger meetings was problematic. It put a strain on the resources of the sponsoring organization and led to discussions of fees. Paying for space would have necessitated the creation of a dues structure for the community, which would have brought a whole set of administrative tasks such as managing payments, invoicing, accounting, liability, and so forth. The community chose not to incur these costs. Larger meetings also tend to become more formal and impersonal; the organization wanted to avoid this outcome. Another conscious decision of the community leadership was to limit commercialization. For instance, members were asked to refrain from making commercials at meetings (e.g., buy my goods or services). In addition, the EC decided not to sell its mailing list to any third parties, despite several requests from various organizations. In summary, this community chose how it wanted to grow in order to preserve an informal yet rich learning and sense-making context.

Decline or Renewal?

At the close of the study in 2004, more than 93% of the respondents thought the organization would continue for another five years according to the survey. As of this writing, the community of practice is healthy and has been active for more than seven years. The members continue to experience high satisfaction with the organization. The evidence suggests that the organization will continue to adapt to changing conditions. The most dramatic change would occur if the theory and practice of knowledge management simply diminishes in value and goes away. What would the community do then? Would it continue to meet as an affinity group based on member ties, or would it organize around another practice issue and body of knowledge? Formal organizations often decline because they become rigid and inflexible. They are unable to adapt to changing conditions and are weighed down by extensive rules, policies, bureaucratic structures, and cultures that are state-maintaining. Communities of practice may suffer these same consequences, but further research is necessary to determine how and if this takes place. Perhaps what occurs with CoPs is not the result of over-structuring but rather increasing entropy and decreases in energy (i.e., a CoP ceases to exist due to lack of member interest, loss of focus, or insufficient resources).

Limitations and Next Steps

There are limitations to this study. First, this work represents a view of one organization over a five-year period. Care must be taken in generalizing these findings to other organizational forms claiming to be communities of practice. Second, since this work was exploratory in nature, the strengths of the associations found among elements were not measured. The next logical step is to replicate the results in other settings and to drill down into greater detail. Others are encouraged to test the propositions provided in this work and to use empirical methods that include hypothesis testing. One area for further research is to examine the strength of the interactions that bind the members. Weak ties suggest more of a network of practice (NoP), while strong ties suggest more of a community of practice (Brown & Duguid, 2001). Further study using social network analysis would resolve the fuzziness around the issue.

To the practice-minded, one of the findings of this study is that communities of practice are relatively easy to set up and can be fostered with limited financial resources. The challenge is to find people willing to commit the time necessary to keep it going. On the other hand, CoPs potentially can reap big rewards, especially for corporations looking for ways to preserve and grow organizational knowledge. This study hopefully provides some practical insights about how to grow and sustain a community of practice over time.

Summary and Conclusion

This research has explored the formation, survival, and growth of a CoP devoted to knowledge management. Several factors that have contributed to the success of the organization over its lifetime have been identified. Future research needs to be done across several communities of practice in order to verify these findings. Others are encouraged to frame and test hypotheses related to the strength of the associations identified in this work and to examine the life cycles of other communities of practice. In closing, all organizational forms grow and contract and change over time. Communities of practice are no different in this regard. Rather than being viewed as static and given, Communities of practice unfold in both expected and unexpected ways, like all social systems.

Acknowledgments

The author wishes to thank the following individuals for giving freely of their time in support of the preparation of this document: John Barrett, Jordan Corn, Marianne Gruber, Paul Hilt, Andrea Hornett, Dee Anne Kotzur-Cerruti, and Mary Ellen Myers. Thanks also is given to the general membership of the case organization for its contribution to the study.

References

Ackerman, M. S., & Halverson, C. A. (2000). Reexamining organizational memory. *Association for Computing Machinery. Communications of the ACM, 43*(1), 58.

Anand, V., Manz, C. C., & Glick, W. H. (1998). An organizational memory approach to information management. *The Academy of Management Review, 23*(4), 796.

Argyris, C., & Schon, D. (1978). *Organizational learning*. Reading, MA: Addison Wesley.

Barab, S. A., & Duffy, T. M. (2000). From practice fields to communities of practice. In D. H. Jonassen, & S. M. Land (Eds.), *Theoretical foundations of learning environments* (pp. 25-55). Mahwah, NJ: Lawrence Erlbaum Associates.

Berends, H., Boersma, K., & Weggeman, M. (2003). The structuration of organizational learning. *Human Relations, 56*(9), 1035.

Brown, J. S., & Duguid, P. (1991). Organizational learning and communities of practice: Toward a unified view of working, learning, and innovation. *Organization Science, 2*(1), 40-57.

Brown, J. S., & Duguid, P. (1998). Organizing knowledge. *California Management Review, 40*(3), 90.

Brown, J. S., & Duguid, P. (2001). Knowledge and organization: A social-practice perspective. *Organization Science, 12*(2), 198.

Buysse, V., Sparkman, K. L., & Wesley, P. W. (2003). Communities of practice: Connecting what we know with what we do. *Exceptional Children, 69*(3), 263.

Cameron, K. S., & Whetten, D. A. (1981). Perceptions of organizational effectiveness over organizational life cycles. *Administrative Science Quarterly, 26*(4), 525.

Casey, A. (1997). Collective memory in organizations. Advances in Strategic Management. *JAI, 14*, 111-146.

Chandler, A. D. (1962). *Strategy and structure.* Cambridge, MA: MIT Press.

Contractor, N. S. (1999). Self-organizing systems research in the social sciences: Reconciling the metaphors and the models. *Management Communication Quarterly, 13*(1), 154.

Croasdell, D. C. (2001). IT's role in organizational memory and learning. *Information Systems Management, 18*(1), 8.

Cross, R., & Baird, L. (2000). Technology is not enough: Improving performance by building organizational memory. *Sloan Management Review, 41*(3), 69.

Davenport, T. H., & Prusak, L. (1998). *Working knowledge: How organizations manage what they know*. Boston: Harvard Business School Press.

Demarest, M. (1997). Understanding knowledge management. *Long Range Planning, 30*(3), 374.

Dodge, H. R., & Robbins, J. E. (1992). An empirical investigation of the organizational life cycle. *Journal of Small Business Management, 30*(1), 27.

Dougherty, D., Borrelli, L., Munir, K., & O'Sullivan, A. (2000). Systems of organizational sensemaking for sustained product innovation. *Journal of Engineering and Technology Management, 17*(3,4), 321.

Drazin, R., Glynn, M. A., & Kazanjian, R. K. (1999). Multilevel theorizing about creativity in organizations: A sensemaking perspective. *The Academy of Management Review, 24*(2), 286.

Earl, M. (2001). Knowledge management strategies: Toward a taxonomy. *Journal of Management Information Systems, 18*(1), 215.

Eisenhardt, K. M. (1989). Building theories from case study research. *Academy of Management Review, 14*(4), 532-550.

Flynn, D., & Forman, A. M. (2001). Life cycles of new venture organizations: Different factors affecting performance. *Journal of Developmental Entrepreneurship, 6*(1), 41.

Gioia, D. A., & Chittipeddi, K. (1991). Sensemaking and sensegiving in strategic change initiation. *Strategic Management Journal, 12*(6), 433.

Gioia, D. A., & Mehra, A. (1996). Sensemaking in organizations. *The Academy of Management Review, 21*(4), 1226.

Gray, B., & Ariss, S. S. (1985). Politics and strategic change across organizational life cycles. *The Academy of Management Review, 10*(4), 707.

Griffith, T. L. (1999). Technology features as triggers for sensemaking. *The Academy of Management Review, 24*(3), 472.

Gupta, Y. P., & Chin, D. C. W. (1992). Organizational life cycle and organizational strategic orientation: An empirical examination. *International Journal of Management, 9*, 215-227.

Gupta, Y. P., & Chin, D. C. W. (1994). Organizational life cycle: A review and proposed directions. *The Mid-Atlantic Journal of Business, 30*(3), 269.

Haire, M. (1959). Biological models and empirical history of the growth of organizations. In M. Haire (Ed.), *Modern organization theory*. New York: John Wiley and Sons.

Houston, R. (1999). Self-organizing systems theory: Historical challenges to new sciences. *Management Communication Quarterly: McQ, 13*(1), 119.

Iverson, J. O., & McPhee, R. D. (2002). Knowledge management in communities of practice. *Management Communication Quarterly: McQ, 16*(2), 259.

Jawahar, I. M., & McLaughlin, G. L. (2001). Toward a descriptive stakeholder theory: An organizational life cycle approach. *The Academy of Management Review, 26*(3), 397.

Lave, J. (1988). *Cognition in practice: Mind, mathematics, and culture in everyday life.* Cambridge: Cambridge University Press.

Lave, J., & Wenger, E. (1991). *Situated learning: Legitimate peripheral participation.* Cambridge: Cambridge University Press.

Lorange, P., & Probst, G. J. B. (1987). Joint ventures as self-organizing systems: A key to successful joint venture design and implementation. *Columbia Journal of World Business, 22*(2), 71.

Mintzberg, H. (1984). Power and organizational life cycles. *The Academy of Management Review, 9*, 207–224.

Moorman, C., & Miner, A. S. (1997). The impact of organizational memory on new product performance and creativity. *Journal of Marketing Research, 34*(1), 91.

Moorman, C., & Miner, A.S. (1998). Organizational improvisation and orginizational memory. *The Academy of management review*, *23*(4), 698.

Moss, M. (2001). Sensemaking, complexity and organizational knowledge. *Knowledge and Process Management, 8*(4), 217.

Nelson, R., & Winter, S. (1982). *An evolutionary theory of economic change.* Cambridge, MA: Belknap Press of Harvard University Press.

Nissley, N., & Casey, A. (2002). The politics of the exhibition: Viewing corporate museums through the paradigmatic lens of organizational memory. *British Journal of Management, 13*, S35.

Nonaka, I. (1994). A dynamic theory of organizational knowledge creation. *Organization Science, 5*(1), 14.

Nonaka, I. O., & Nishiguchi, T. (2001). *Knowledge emergence: Social, technical, and evolutionary dimensions of knowledge creation.* Oxford: Oxford University Press.

Nonaka, I. O., & Takeuchi, H. (1995). *The knowledge-creating company: How Japanese companies create the dynamics of innovation.* New York: Oxford University Press.

Quinn, R. E., & Cameron, K. (1983). Organizational life cycles and shifting criteria of effectiveness: Some preliminary evidence. *Management Science, 29*(1), 33.

Rulke, D. L., & Rau, D. (2000). Investigating the encoding process of transactive memory development in group training. *Group & Organization Management, 25*(4), 373.

Senge, P. M. (2003). Taking personal change seriously: The impact of organizational learning on management practice. *The Academy of Management Executive, 17*(2), 47.

Smith, N. R., & Miner, J. B. (1983). Type of entrepreneur, type of firm, and managerial motivation: Implications for organizational life cycle theory. *Strategic Management Journal, 4*(4), 325.

Stein, E. W. (1995). Organizational memory: Review of concepts and recommendations for management. *International Journal of Information Management, 15*(1), 17.

Stein, E. W., & Zwass, V. (1995). Actualizing organizational memory with information systems. *Information Systems Research, 6*(2), 85.

Swan, J., Scarbrough, H., & Robertson, M. (2002). The construction of "communities of practice" in the management of innovation. *Management Learning, 33*(4), 477.

Thomas, J. B., Clark, S. M., & Gioia, D. A. (1993). Strategic sensemaking and organizational performance: Linkages among scanning, interpretation, action, and outcomes. *Academy of Management Journal, 36*(2), 239.

Thomas, J. B., Sussman, S. W., & Henderson, J. C. (2001). Understanding "strategic learning": Linking organizational learning, knowledge management, and sensemaking. *Organization Science, 12*(3), 331.

Vera, D., & Crossan, M. (2004). Strategic leadership and organizational learning. *The Academy of Management Review, 29*(2), 222.

Walsh, J. P., & Ungson, G. R. (1991). Organizational memory. *The Academy of Management Review, 16*(1), 57.

Walters, B., & Clark, J. (1996, Fall). Communities of interest. *Canadian Business*, S6.

Webster, J., & Martocchio, J. J. (1992). Microcomputer playfulness: Development of a measure with workplace implications. *MIS Quarterly, 16*(2), 201.

Webster, J., & Martocchio, J. J. (1993). Turning work into play: Implications for microcomputer software training. *Journal of Management, 19*(1), 127.

Weick, K. E. (1979). *The social psychology of organizing*. Reading, MA: Addison-Wesley.

Weick, K. E. (1995). *Sensemaking in organizations*. Thousand Oaks, CA: Sage Publications.

Wenger, E. (1998). *Communities of practice: Learning, meaning, identity*. Cambridge, MA: Cambridge University Press.

Wenger, E. (2004). Knowledge management as a doughnut: Shaping your knowledge strategy through communities of practice. *Ivey Business Journal, 68*(3).

Wenger, E., McDermott, R., & Snyder, W. M. (2002). It takes a community. *CIO, 15*(15), 1.

Wijnhoven, F. (1999). Development scenarios for organizational memory information systems. *Journal of Management Information Systems, 16*(1), 121.

Wishart, N. A., Elam, J. J., & Robey, D. (1996). Redrawing the portrait of a learning organization: Inside Knight-Ridder, Inc. *The Academy of Management Executive, 10*(1), 7.

Yin, R. K. (1989). *Case study research: Design and methods*. London: Sage Publications.

Endnotes

[a] Much of this material originally was published as "A Qualitative Study of the Characteristics of a Community of Practice for Knowledge Management and its Success

Factors" in the *International Journal for Knowledge Management, 1*(3), 2005, 1-24.

[b] Ties among members of the community ranged from strong to weak.

[c] Additional information about the group may be found in appendices that accompany this document, which is available on request from the author.

[d] See Appendix 1 for a complete list of the topics presented over a five-year period. Special thanks to P. Hilt and J. Barrett for help in preparing this appendix.

[e] Developed and maintained by P. Hilt from 1999 to 2004 and by D.-A. Kotzur-Cerruti after 2004.

[f] Invitations currently are made using eVite available at www.evite.com

[g] http://www.kmgphila.org Web site courtesy of the author

[h] J. Barrett and the author

[i] Managed by J. Barrett and M. Eichhorn

[j] Special thanks to D.-A. Kotzur-Cerruti for her help with data collection from the membership.

[k] In a future study, we would assess the characteristics of the nonrespondents.

[l] Note: Respondents were allowed to make multiple selections, and therefore, the percentages do not add up to 100%.

[m] Special thanks to Rebecca Riley and Suzanne Shaffer for assistance with coding.

Section III

Measuring Knowledge Management

Chapter X

Evaluation of Knowledge Management: A Review and Agenda for Future Research

Atreyi Kankanhalli, National University of Singapore, Republic of Singapore

Loo Geok Pee, National University of Singapore, Republic of Singapore

Bernard Cheng Yian Tan, National University of Singapore, Republic of Singapore

Abstract

Evaluation methods are essential for the advancement of research and practice in an area. In knowledge management (KM), the process of measurement, evaluation, and development of metrics is made complex by the intangible nature of the knowledge asset. Further, the lack of standards for KM business metrics and the relative infancy of research on KM evaluation point to a need for research in this area. This chapter reviews KM evaluation methods for research and practice and identifies areas in which there is a gap in our understanding. It classifies existing research based on the units of evaluation such as user of knowledge management system (KMS), KMS, project, KM process, and organization as a whole. The importance of considering differences across industries in assessing KM is also discussed. The chapter concludes by suggesting avenues for future research in KM and KMS evaluation based on the gaps identified.

Introduction

Knowledge management (KM) has become an accepted part of the business and academic agenda. Organizations have high expectations for KM to play a significant role in improving their competitive advantage (KPMG, 2000). Measuring the business value of KM initiatives has become imperative to ascertain if the expectations are realized.

Evaluation of KM involves developing measures to assess the phenomenon. Such measures are key to advancement of research and practice in an area. In research, they provide comparability of studies between individuals, time periods, organizations, industries, cultures, and geographic regions (Cook & Campbell, 1979). They also provide a basis for empirical validation of theories and relationships among concepts. Measures that are reliable and valid enable cumulation of research in a topic area and free subsequent researchers from the need to redevelop instruments (Boudreau, Gefen, & Straub, 2001).

For practitioners, evaluation measures are a way of learning what works and what does not. In fact, measuring firm performance is the focus of the entire field of management accounting. In KM, performance measures serve several objectives, including securing funding for KM implementation, providing targets and feedback on implementation, assessing implementation success, and deriving lessons for future implementation. Measures can assist in evaluating the initial investment decision and in developing benchmarks for future comparison.

Measurement is typically a complex process fraught with errors. What is easy to measure is not always important, and what is important is often difficult to measure (Schiemann & Lingle, 1998). KM measures are particularly distinct from other measures due to the intangible nature of the knowledge resource (Glazer, 1998). Something such as knowledge that is difficult to define and has multiple interpretations is likely to be difficult to value and measure. Due to such considerations and the complexity of assessing organizational initiatives in general, research (Grover & Davenport, 2001) and practice (Bontis, 2001) on the assessment of KM initiatives and knowledge management systems (KMS) are not well developed.

In light of the previously mentioned motivations, this study seeks to review KM evaluation in practice and research and to identify areas for further investigation. Previous research on measures for KM and KMS is classified based on the elements of evaluation, such as user of KMS, KMS, project, KM process, and organization as a whole. Further, the importance of considering differences across industries in assessing KM is discussed. The chapter concludes by providing avenues for future research based on the gaps identified during the review. In the next section, some basic concepts related to KM and KMS are described. This is followed by the review of practice KM measures, classification of research on KMS and KM evaluation, and finally, a discussion of areas for further investigation.

KM and KMS Basics

KM involves the basic processes of creating, storing, retrieving, transferring, and applying knowledge. The ultimate aim of KM is to avoid reinventing the wheel and to leverage cumulative organizational knowledge for more informed decision making (Alavi & Leidner, 2001). Examples of ways in which knowledge can be leveraged include transfer of best practices from one part of an organization to another part, codification of individual employee knowledge to protect against employee turnover, and bringing together knowledge from different sources to work on a specific project.

Information technology (IT) is recognized as a key enabler of KM (although there are many other factors that are necessary for KM success). Without the capabilities of IT in terms of both storage and communication, leveraging of knowledge resources hardly would be feasible (Alavi & Leidner, 2001). A variety of tools are available to organizations to facilitate the leveraging of knowledge. These tools (KMS) are defined as a class of information systems applied to managing organizational knowledge. That is, they are IT-based systems developed to support and enhance the organizational processes of knowledge creation, storage/retrieval, transfer, and application (Alavi & Leidner, 2001). Common KMS technologies include intranets and extranets, search and retrieval tools, content management and collaboration tools, data warehousing and mining tools, and groupware and artificial intelligence tools such as expert systems and knowledge-based systems.

Two models of KMS have been identified in information systems research (Alavi & Leidner, 1999), both of which may be employed by organizations to fulfill different needs. These two models correspond to two different approaches to KM (i.e., the codification approach and the personalization approach)[1] (Hansen, Nohria, & Tierney, 1999). The repository model of KMS associated with the codification approach focuses on the codification and storage of knowledge in knowledge bases. The purpose is to facilitate knowledge reuse by providing access to codified expertise. Electronic knowledge repositories (EKR) to code and share best practices exemplify this strategy (Alavi & Leidner, 2001). A related term, Organizational Memory Information System (OMIS) refers to any system that functions to provide a means by which knowledge from the past is brought to bear on the present in order to increase levels of effectiveness for the organization (Stein & Zwass, 1995).

The network model of KMS associated with the personalization approach attempts to link people to enable the transfer of knowledge. One way to do this is to provide pointers to location of expertise in the organization (i.e., who knows what and how they can be contacted). This method is exemplified by knowledge directories, commonly called yellow pages (Alavi & Leidner, 2001). It has been noted that in order to access the knowledge in an organization that remains uncodified, mapping the internal expertise is useful (Ruggles, 1998).

A second way is to link people who are interested in similar topics. The term communities of practice (COP) has come into use to describe such flexible groups of professionals informally bound by common interests, who interact to discuss topics related to these interests (Brown & Duguid, 1991). KMS that provide a common electronic forum to support COP exemplify this approach (Alavi & Leidner, 2001). The two models of KMS allow us to make sense of existing KMS measures (since measures for a particular type of KMS are similar) and to identify directions for further evaluation of KMS.

KM and KMS Evaluation in Practice

KM Measures

Most practice measures of KM initiatives focus on assessing knowledge assets or intellectual capital (IC) of a firm, assuming the outcome of a KM initiative being its impact on IC. The majority of respondents of practice surveys thinks that IC should be reported and that knowledge measurement would improve performance (Bontis, 2001). Even the process of measuring IC is considered important, whether as an internal management tool or for external communication on financial balance sheets.

Three general-purpose approaches to measuring the impact of KM initiatives include house of quality (quality function deployment, or QFD), balanced scorecard, and American Productivity Center (APQC) benchmarking approach (Tiwana, 2000). The house of quality (Hauser & Clausing, 1988) method involves the development of a metrics matrix (house). The desirable outcomes of KM initiatives are listed on the left wall of the house, the roof consists of the performance metrics, the right wall consists of the weights (relative importance of the outcomes), and the base of the house consists of targets, priorities, and benchmark values. By looking at the correlations within the body of the matrix, management can decide to focus on those areas of KM that are most likely to affect overall firm performance. A number of software tools such as QFD designer are available to automate the analysis process.

The balanced scorecard technique developed by Kaplan and Norton (1996) aims to provide a technique to balance long-term and short-term objectives, financial and nonfinancial measures, leading and lagging indicators, and internal and external perspectives. Typically four views—customer, financial, internal business, and learning and growth—are used to translate high-level strategies to real targets. Within each view, the goals, metrics, targets, and initiatives are listed. Relationships among views also must be considered. The views (dimensions) can be adapted suitably to assess the current state of KM and to evaluate the impact of initiatives in this area. Here software tools also are available, though in general, the balanced scorecard is more difficult to develop than QFD. However, it is likely to yield more balanced goals with an inbuilt consideration of the causal relationships.

The APQC process classification framework (PCF) provides a detailed taxonomy of business processes derived from the joint efforts of close to 100 U.S. businesses (APQC, 2006). The PCF can be employed to benchmark and assess impact on business processes as a result of introduction of KM initiatives. Other general measures of firm performance such as economic value added (EVA) and Tobin Q also can be used for evaluating IC (Stewart, 1997).

Three other measures specific to KM are the Skandia navigator, IC index, and intangible assets monitor. The Skandia navigator (Edvinsson & Malone, 1997) consists of 112 IC and traditional metrics (with some overlap between metrics) in five areas of focus (financial, customer, process, renewal and development, human). These areas are similar to the balanced scorecard views except for the additional human focus area in the Skandia metric (more areas also can be added in balanced scorecard, as desired, although a limit of seven areas is suggested). Out of all the indicators, the monetary indicators are combined into a single dollar value (C), while the remaining percentage completeness measures are combined into an efficiency indicator (I) that captures the firm's velocity or movement toward desired goals. The overall IC measure is a multiplication of I and C.

The IC index (Roos, Roos, Dragonetti, & Edvinsson, 1998) is an extension to the Skandia IC metric that attempts to consolidate measures into a single index and to correlate this index with changes in the market (i.e., it focuses on monitoring the dynamics of IC). It consists of monitoring both IC stock and IC flow. A third technique is the Intangible Assets Monitor (Sveiby, 1997). Intangible asset value is defined as the book value of the firm minus the tangible assets and the visible debt. Three components of intangible assets are external structure (brand, customer, and supplier relations), internal structure (management, legal, manual, attitude, and software), and individual competence (education, experience, and expertise). For each intangible asset component, three indicators focus on growth and renewal, efficiency, and stability of that component. Other KM-specific techniques include technology broker (Brooking, 1996) and citation-weighted patents (Hall, Jaffe, & Trajtenberg, 2000).

Whether it is the more general purpose or the more KM-specific techniques for firm performance evaluation, the efficacy of all techniques depends on the competence of management in applying these techniques. Although the aforementioned techniques attempt to provide systematic and comprehensive KM performance indicators, there is a number of subjective judgments to be made in applying these techniques, including determining which objectives are more important than others and which indicators need to be given greater weight. As pointed out in previous studies (Bontis, 2001), a further limitation on these techniques is that many of them use different terms to label similar measures. A lack of standards leads to proliferation of measures and difficulty in comparison. Also, since most of the evidence on KM assessment is on a case-by-case basis, there is a lack of generalizable results on this topic.

KMS Measures

Organizations employ a variety of measures to assess their KMS (Dept. of Navy, 2001). System-level measures for EKR include number of downloads, dwell time, usability surveys, number of users, and number of contributions and seeks. Measures for electronic COP include number of contributions and seeks, frequency of update, number of members, and ratio of number of members to the number of contributors. System-level measures have been used for evaluating and monitoring particular KMS implementations. Here also, the literature is mainly in the form of individual case studies (Wei, Hu, & Chen, 2002), and generalizable measurement techniques are lacking.

Previous Research on KM and KMS Evaluation

Researchers (Grover & Davenport, 2001) have suggested a pragmatic framework for KM research based on the KM processes and the context in which the process is embedded. The KM processes can be divided into generation, codification, transfer, and realization. The elements of the embedded context include strategy, structure, people/culture, and technology. The framework can be applied for processes at individual, group, and organization levels. We adopt a similar classification for categorizing previous research on KM and KMS measures based on elements of evaluation (user, system, project, process, and organization).

Table 1. Selected studies on KMS users

Study	User	KMS	Sample
Constant, Sproull & Kiesler (1996)	Contributor factors on seeker	E-mail distribution list	Tandem Computers, 48 seekers and 263 contributors
Goodman & Darr (1998)	Contributor and seeker	Repository + electronic COP	Office equipment distributor, 1,500 respondents
Jarvenpaa & Staples (2000)	Contributor and seeker combined	All electronic media	1 university, 1,125 employees
Kuo, Young, Hsu, Lin, & Chiang (2003)	Contributor and seeker	Electronic COP	264 teachers in an online forum
Barreto & Heckman (2004)	Contributor and seeker	All electronic media	375 knowledge workers (322 surveys and 53 interviews)
Cummings (2004)	Contributor and seeker	All electronic media	182 work groups in a telecommunications firm
Clay, Dennis, & Ko (2005)	Contributor and seeker	Field sales KMS	Pharmaceutical firm, 1,013 sales representatives
Tiwana & Bush (2005)	Contributor and seeker	Electronic COP	4 COPs, 30 participants
Bock, Zmud, Kim, & Lee (2005)	Contributor	All electronic media	154 executives
Chay, Menkhoff, Loh, & Evers (2005)	Contributor	All electronic media	262 university staff, administrators and students
Kankanhalli, Tan, & Wei (2005a)	Contributor	Repository	150 knowledge workers
Wasko & Faraj (2005)	Contributor	Electronic COP	173 members of a national legal profession association
Han & Anantatmula (2006)	Contributor	All electronic media	235 knowledge workers
Wasko & Faraj (2000)	More emphasis on contributor	Electronic COP	3 Usenet groups, 342 participants
Zhang & Watts (2003)	Seeker	Electronic COP	145 participants in a travel forum
Gray & Meister (2004)	Seeker	All electronic media	313 knowledge workers
Levin & Cross (2004)	Seeker	All electronic media	127 knowledge workers
Kankanhalli, Tan, & Wei (2005b)	Seeker	Repository	160 knowledge workers
Desouza, Awazu, & Wan (2006)	Seeker	All electronic media	IT consulting organization, 175 employees

The previous research articles have been selected based on the following criteria. First, they are empirical articles that have proposed and tested measures for evaluating KMS users, KMS, project, KM process, or organizational outcomes. Second, they are chosen from reputed journals such as *MIS Quarterly*, *Management Science*, *Information Systems Research*, *ACM Transactions*, *IEEE Transactions*, *Journal of the American Society for Information Science and Technology*, *Journal of Management Information Systems*, *Organization Science*, *Administrative Science Quarterly*, *Journal of Strategic Information Systems*, *Decision Support Systems*, *Information and Management*, *Harvard Business Review*, *Sloan Management Review*, and *California Management Review*, as well as established conferences in Information Systems such as the International Conference on Information Systems, Hawaii International Conference on Systems Sciences, and Americas Conference on Information Systems. The articles span the period from 1998 to the present with the exception of the Constant, Sproull, and Kiesler (1996) article that is one of the first research articles in its area.

Table 2. Selected studies on KMS evaluation

Study	KMS	Performance Criteria	Sample
Ackerman (1998)	Answer Garden Knowledge Repository (FAQ) + Electronic COP (via e-mail)	• Usage—heavy, intermittent, tire-kicker • User evaluation in seeking answer • Expert evaluation of providing answer	2 university lab sites, 49 users (seeker), 7 experts (contributor)
Jennex, Olfman, Panthawi, & Park (1998)	Knowledge repository	• Individual job time, number of assignments, completeness of solutions, quality of solutions, complexity of assignment, client satisfaction • Organizational unit capability (problem correct) • Unplanned scrams (problem solve)	120 engineers in 50 nuclear power plants
Baek & Liebowitz (1999)	Knowledge repository	Contributor • Simplicity, richness, flexibility of creation • Ease of consistency checking, ease of knowledge change management Seeker • Ease of knowledge navigation and searching Both • Awareness, timeliness, fairness	2 multimedia design teams (3 members and 4 members)
Jennex & Olfman (2002)	Knowledge repository	• Integration • Adaptation • Goal attainment • Pattern Maintenance	83 engineers
Hendriks & Vriens (1999)	Knowledge based system (expert system)	Assessment of current knowledge • Establishment of strategic value of knowledge • Comparison of knowledge to competition • Establishment of required knowledge • Creation of new knowledge • Distribution of knowledge • Application of knowledge • Evaluation of knowledge	17 organizations (government, bank, insurance, manufacturing)

Table 2. continued

Study	KMS	Performance Criteria	Sample
Nissen (1999)	Koper Knowledge-based system	KM effects • Knowledge capture, organization, and formalization • Knowledge distribution and application • Analytical consistency and completeness • Knowledge integration	Large multi-site enterprise
Maier (2002)	Knowledge management system	DeLone and McLean IS success model-based criteria	73 organizations
Jennex (2005)	Knowledge management system	• Current KMS usage • Perceived benefit of KMS use	1 engineering organization, 125 users
Gonzalez, Giachetti, & Ramirez (2005)	Knowledge management system for help desk	Time of problem calls	4,965 help desk calls
Gottschalk (2000)	Data warehouse, executive IS, expert system, enterprisewide system, intranet	IT support for KM • Generating knowledge • Accessing knowledge • Transferring knowledge • Sharing knowledge • Codifying knowledge	73 law firms in Norway
Ruppel & Harrington (2001)	Intranet implementation	Level of implementation for knowledge sharing	44 organizations (different industries)
Freeze & Robles-Flores (2005)	Tool history management system	Usefulness, satisfaction with KMS, time per contract	169 technicians (users), 150 nontechnicians (nonusers)

KMS User Evaluation

The articles on user evaluation are tabulated according to the type of user (contributor or seeker), type of KMS, and the sample of the empirical study (see Table 1). The bulk of previous research at the user level has been studies to evaluate the motivation of users to contribute to or to seek knowledge from different types of KMS and in a few studies the consequent usage of KMS. Research has investigated both contributor and seeker motivations for using both repository and network model KMS. Most of the samples for the studies have been drawn from one organization or one online forum.

KMS Evaluation

The articles on KMS evaluation are tabulated according to the type of KMS, performance criteria suggested, and the sample of the empirical study (see Tables 2a and 2b). It can be seen that a variety of performance criteria have been proposed focusing on user, task, KM process, and organizational outcomes for different KMS. The samples in these studies have been drawn both from single organizations and multiple organizations.

Project Evaluation

Relatively fewer articles were found on KM measures related to project evaluation (see Table 3). These articles are tabulated based on the nature of the project, the performance criteria, and the sample of the study. The projects include software development, new product development, and process improvement projects. In the first study of the table, the performance criteria are in terms of the knowledge process, whereas in the other two studies, knowledge processes (sharing and creation) appear as mediators. All these studies are tested on single or multiple projects within a single organization.

KM Process and Organizational-Level Evaluation

Research on KM and KMS measures at the KM process and organizational levels often have been combined. Similar to the practice firm performance metrics, the research metrics at these levels also attempt to tease out the relationships between KM initiative, process, or capability, and firm performance, albeit with a theoretical emphasis. Literature in this area is tabulated based on the independent variables, performance criteria, and the sample (see Table 4). Effectiveness outcomes have been studied in single and multiple organizational settings.

Table 3. Selected studies on project evaluation

Study	Project	Performance Criteria	Sample
Mukherjee, Lapre, & Wassenhove (1998)	Total quality management project	Project performance, goal achievement, ability to specify impact, change in attention rules	62 projects in a Belgian multinational steel wire manufacturer
Verkasalo & Lappalainen (1998)	Hypertext annual plan project	Efficiency index for knowledge utilization • Process width = number of employees • Process delay = time taken to spread/distribute • Process effort = time to document, distribute, and perceive use (not collect and compile)	Nokia telecom factory
Hansen (1999)	New product development project	Project completion time (conception to market)	120 projects in a large electronics company

Table 4. Selected studies on process and organizational-level evaluation

Study	Impact of	Performance Criteria	Sample
Khalifa, Lam, & Lee (2001)	Overall KM initiative	KM effectiveness (organizational performance impacts)	185 KM practitioners from discussion forums
Becerra-Fernandez & Sabherwal (2001)	Knowledge internalization, externalization, combination, socialization + all KM tools use	KM satisfaction (availability, effectiveness of knowledge, KM at task, directorate, across organization, knowledge sharing)	Kennedy Space Center, 159 employees from 8 subunits
Lee & Choi (2003)	Knowledge creation (internalization, externalization, combination, socialization)	Organizational creativity Organizational performance • Degree of overall success • Market share • Growth rate profitability • Innovativeness	426 employees from 58 firms, manufacturing, service, and financial business
Gold, Malhotra, & Segars (2001)	KM process capability (knowledge acquisition, conversion, application, protection) Knowledge infrastructure capability	Organization effectiveness • Innovation and commercialization, coordination of unit • Anticipate and identify opportunities • Speed and adaptation to market • Avoid redundancy and streamline	323 executives, finance and manufacturing, large organizations
Lee, Lee, & Kang (2005)	Knowledge circulation process (creation, accumulation, sharing, utilization, internalization)	• Stock price • Price earnings ratio • R&D expenditure	101 firms listed in KOSDAQ market in Korea
Tanriverdi (2002)	IT knowledge relatedness	Market-based performance Tobin's Q	315 firms, manufacturing and service

Table 4. continued

Study	Impact of	Performance Criteria	Sample
Tanriverdi (2005)	IT Relatedness, KM Capability (knowledge creation, transfer, integration, leverage)	Market-based performance Tobin's Q	250 firms, manufacturing and service
Chen & Feng (2004)	KM adoption	Cost ratio	255 firms that adopted KM since 1999
McKeen, Zack, & Singh (2006)	KM Practices • Ability to locate and share existing knowledge • Ability to experiment and create new knowledge • Culture that encourages knowledge creation and sharing • Regard for the strategic value of knowledge and learning	Organizational performance • Extent of product and service innovation • Quality • Customer satisfaction and retention • Operating efficiency Financial performance • Return on assets or equity • Profitability	90 knowledge managers
Liao & Chuang (2006)	Technical KM resource, Social KM resource, Innovation speed, Innovation magnitude	Firm Performance • Market share gain • Sales growth • Profitability • Efficiency of operations • Quality of services	118 R&D managers

KM Evaluation in Specific Industries

In order for KM assessment to be effective, it is important to utilize measures that are in sync with business goals and that provide the right answers to the right questions (APQC, 2003; Tiwana, 2002). In other words, the chosen KM measure and corresponding performance indicators should reflect industry specifics and a firm's core competency. To this end, it is useful to distinguish between general, industry-specific, and firm-specific indicators. General indicators are those that are applicable to all contexts. Examples of such indicators include employee educational level and turnover. Industry-specific indicators are often those that relate to specific business processes (e.g., service expense per customer in service industries). Firm-specific indicators refer to those that require definition by the firm (e.g., employee value to the firm). It is also important to recognize that even when different industries or firms use the same set of indicators, certain indicators may be more important for one industry or firm compared to the other. In all cases, the selection of appropriate indicators is often a multi-criteria decision problem that requires resolution involving various stakeholders.

To illustrate the differences in KM measures across industries, selected industry-specific indicators for assessing KM in service, hotel, healthcare, and IT industries are listed for comparison (see Table 5).

Table 5. Comparison of selected indicators in different industries

Study	Industry	Industry-Specific Indicators
Engstrom, Westnes, & Westnes (2003)	Hotel	• Occupancy rate • Revenue per available room • F&B profit
Lim & Dallimore (2004)	Service (banking, telecom, tourism)	• Number of service award won • Utility of telemarketing • Number of new services provided • Workforce competence profile • Brand recognition index • Total training and education cost • Customer satisfaction index
Han & Han (2004)	Telecommunications	• Exploitation by competitor • Market overreaction • Government policy change
Leitner & Warden (2004)	R&D	• Total scientific staff • Total IT expenditure per employee • Publications in refereed journals • Number of patent applications • Income from licenses
Berler, Pavlopoulos, & Koutsouris (2005)	Healthcare	• Treatment cost • Mortality rate • Time in waiting list • Patient satisfaction • Medical device usage growth • Healthcare professional training rate
Wang & Chang (2005)	IT	• IT acceptance rate

Agenda for Future Research

From our literature review, we can infer certain gaps in research on KM and KMS evaluation in terms of unit or level of study. At the intersection of user and system level, most research tends to investigate motivations of users. There is a lack of research on usability of KMS and limited studies on usage of KMS. Both usability and usage studies, if well-designed, can provide a good indicator of user acceptance of KMS. For example, usability studies of both interactive and integrative KMS may be undertaken. Also, comparative studies of KMS usability may prove fruitful. Studies across multiple organizations or forums can add to existing studies.

At the system level, the majority of studies appear to focus on EKR, OMIS, knowledge-based systems, and overall KM technologies. There appears to be a lack of evaluation studies on electronic COP, since the majority of studies on COP appears to be anecdotal in nature. Therefore, future research can investigate suitable metrics for evaluating electronic COP, an integral part of the network model of KMS. Further, review studies can help to infer commonalities and differences among the measures for different forms of KMS.

There appears to be a relative paucity of KM evaluation studies at the group and team levels, except for a few virtual team studies (Alavi & Tiwana, 2002). Although there have been studies at the project level (see Table 3) that could be interpreted as group-level evaluations, these studies did not investigate group characteristics and team dynamics in relation to evaluation of KM. This area presents an opportunity for future research on team effectiveness in terms of KM. For example, studies of how effective KMS are in terms of facilitating group, team, and project KM may be useful. Additionally, a greater variety of projects can be studied to draw inferences about what measures are useful for particular types of projects. Alternatively, measures for a particular type of project can be compared across different organizations.

In relation to Grover and Davenport's (2001) framework, there appears to be a lack of studies focusing purposefully on evaluation of KM strategy and KM structure. Considering that both elements can be vital to the success of KM initiatives, research on these elements is required. Additionally, there is a gap between the micro-level assessment studies (user and system level) and the macro-level assessment studies (organization level). Possibly more research on team, project, and business unit level KM evaluation may serve to bridge this gap. Aggregation from user and system-level evaluation to team, project, and business unit-level evaluation and, subsequently, to organization-level KM evaluation could provide a worthwhile avenue for future research.

A preliminary comparison of the KM performance indicators (Table 5) in different industries revealed that there is more need for research on the applicability of KM metrics in different contexts (e.g., service vs. manufacturing industries, public vs. private organizations). In order to understand contextual variations, case studies may be conducted to examine the decision-making process by which organizations in different industries select and weigh their indicators. General criteria or methodologies for selection then may be generated through cross-case analysis in order to assist managers operating in various industries with diverse environmental dynamism, heterogeneity, and hostility to systematically and objectively apply KM measures and indicators. To this end, it also may be fruitful to incorporate ideas and methodologies (e.g., Analytic Hierarchy Process) from managerial decision-making literature (Han & Han, 2004).

Although limited by the fact that a complete review of literature cannot be claimed, this study throws light on the existing research on KMS and KM evaluation. It also serves to identify potential areas in which further evaluation research would be useful. Given that organizations are expending significant resources on implementing KM initiatives and KMS, more research on measures in these areas is warranted.

References

Ackerman, M. S. (1998). Augmenting organizational memory: A field study of answer garden. *ACM Transactions on Information Systems, 16*(3), 203-224.

Alavi, M., & Leidner, D. E. (1999). Knowledge management systems: Issues, challenges and benefits. *Communications of Association of Information Systems*, 1, 1-37.

Alavi, M., & Leidner, D. E. (2001). Review: Knowledge management and knowledge management systems: Conceptual foundations and research issues. *MIS Quarterly, 25*(1), 107-136.

Alavi, M., & Tiwana, A. (2002). Knowledge integration in virtual teams: The potential role of KMS. *Journal of the American Society for Information Science and Technology, 53*(12), 1029-1037.

APQC. (2006). *Process classification framework.* Retrieved Spetember 19, 2006, from http://www.apqc.org/portal/apqc/ksn/X.pdf?paf_gear_id=contentgearhome&paf_dm=full&pageselect=contentitem&docid=121388.

APQC. (2003). *Measure the impact of knowledge management.* Retrieved September 19, 2006, from http://www.apqc.org/portal/apqc/ksn/X.pdf?paf_gear_id=contentgearhome&paf_dm=full&pageselect=contentitem&docid=124934.

Baek, S., & Leibowitz, J. (1999). Designing a Web based knowledge repository in a virtual team and exploring its usefulness. In *Proceedings of the Fifth Americas Conference on Information Systems,* Milwaukee, Wisconsin.

Barreto, C., & Heckman, R. (2004). Understanding knowledge sharing motivators within knowledge management initiatives. In *Proceedings of the Tenth Americas Conference on Information Systems,* New York.

Becerra-Fernandez, I., & Sabherwal, R. (2001). Organizational knowledge management: A contingency perspective. *Journal of Management Information Systems, 18*(1), 23-55.

Berler, A., Pavlopoulos, S., & Koutsouris, D. (2005). Using key performance indicators as knowledge-management tools at a regional health-care authority level. *IEEE Transactions on Information Technology in Biomedicine, 9*(2), 184-192.

Bock, G. W., Zmud, R. W., Kim, Y. G., & Lee, J. N. (2005). Behavioral intention formation in knowledge sharing: Examining the roles of extrinsic motivators, social-psychological forces and organizational climate. *MIS Quarterly, 29*(1), 87-111.

Bontis, N. (2001). Assessing knowledge assets: A review of the models used to measure intellectual capital. *International Journal of Management Review, 3*(1), 41-60.

Boudreau, M., Gefen, D., & Straub, D. W. (2001). Validation in IS research: A state-of-the-art assessment. *MIS Quarterly, 25*(1), 1-24.

Brooking, A. (1996). *Intellectual capital: Core assets for the third millennium enterprise.* London: Thomson Business Press.

Brown, J. S., & Duguid, P. (1991). Organizational knowledge and communities of practice. *Organization Science, 2*(1), 40-57.

Chay, Y. W., Menkhoff, T., Loh, B., & Evers, H. D. (2005). Theorizing, measuring, and predicting knowledge sharing behavior in organizations—A social capital approach. In *Proceedings of 38th Annual Hawaii International Conference on System Sciences,* Hawaii.

Chen, E. T., & Feng, K. (2004). Knowledge management capability and firm performance: An empirical investigation. In *Proceedings of the Tenth Americas Conference on Information Systems,* New York.

Clay, P. F., Dennis, A. R., & Ko, D. G. (2005). Factors affecting the loyal use of knowledge management systems. In *Proceedings of the 38th Annual Hawaii International Conference on System Sciences.*

Constant, D., Sproull, L., & Kiesler, S. (1996). The kindness of strangers: The usefulness of electronic weak ties for technical advice. *Organization Science, 7*(2), 119-135.

Cook, T. D., & Campbell, D. T. (1979). *Quasi-experimentation: Design and analysis issues for field settings.* Boston: Houghton Mifflin.

Cummings, J. (2004). Work groups, structural diversity, and knowledge sharing in a global organization. *Management Science, 50*(3), 352-364.

Dept. of Navy. (2001). *Metrics guide for knowledge management initiatives.* Retrieved September 19, 2006, from http://openacademy.mindef.gov.sg/OpenAcademy/Central/HTML%20Folder/KM/bcp/downloads/KM_Metrics_Guide_Final_15AUG01.doc

Desouza, K. C., Awazu, Y., & Wan, Y. (2006). Factors governing the consumption of explicit knowledge. *Journal of the American Society for Information Science and Technology, 57*(1), 36-43.

Edvinsson, L., & Malone, M. S. (1997). *Intellectual capital: Realizing your company's true value by finding its hidden brainpower.* New York: Harper Business.

Engstrom, T. E. J., Westnes, P., & Westnes, S. F. (2003). Evaluating intellectual capital in the hotel industry. *Journal of Intellectual Capital, 4*(3), 287-333.

Freeze, R., & Robles-Flores, J. A. (2005). Knowledge management-centric help desk: Specification and performance evaluation. In *Proceedings of the Eleventh Americas Conference on Information Systems,* Omaha, Nebraska.

Glazer, R. (1998). Measuring the knower: Towards a theory of knowledge equity. *California Management Review, 40*(3), 175-194.

Gold, A. H., Malhotra, A., & Segars A. (2001). Knowledge management: An organizational capabilities perspective. *Journal of Management Information Systems, 18*(1), 185-214.

Gonzalez, L. M., Giachetti, R. E., & Ramirez G. (2005). Knowledge management-centric help desk: Specification and performance evaluation. *Decision Support Systems, 40*, 389-405.

Goodman, P. S., & Darr, E. D. (1998). Computer-aided systems and communities: Mechanisms for organizational learning in distributed environments. *MIS Quarterly, 22*(4), 417-440.

Gottschalk, P. (2000). Knowledge management in the professions: The case of IT support in law firms. In *Proceedings of the 39th Annual Hawaii International Conference on System Sciences.*

Gray, P. H., & Meister, D. (2004). Knowledge sourcing effectiveness. *Management Science, 50*(6), 821-834.

Grover, V., & Davenport, T. (2001). General perspectives on knowledge management: Fostering a research agenda. *Journal of Management Information Systems, 18*(1), 5-22.

Hall, B., Jaffe, A., & Trajtenberg, M. (2000). *Market value and patent citations: A first look* [working paper no. W7741]. Cambridge, MA: NBER.

Han, B. M., & Anantatmula, V. S. (2006). Knowledge management in IT organizations from employee's perspective. In *Proceedings of the 39th Annual Hawaii International Conference on System Sciences.*

Han, D., & Han, I. (2004). Prioritization and selection of intellectual capital measurement indicators using analytic hierarchy process for the mobile telecommunications industry. *Expert Systems with Applications, 26*, 519-527.

Hansen, M. T. (1999). The search-transfer problem: The role of weak ties in sharing knowledge across organization sub-units. *Administrative Science Quarterly, 44*(1), 82-111.

Hansen, M. T., Nohria, N., & Tierney, T. (1999). What's your strategy for managing knowledge? *Harvard Business Review, 77*(2), 106-116.

Hauser, J., & Clausing, L. (1988). The house of quality. *Harvard Business Review, 3*, 63-73.

Hendriks, P. H. J., & Vriens, D. J. (1999). Knowledge based systems and knowledge management. *Information and Management, 35*(2), 113-125.

Jarvenpaa, S. L., & Staples, D. S. (2000). The use of collaborative electronic media for information sharing: An exploratory study of determinants. *Journal of Strategic Information Systems, 9*(2-3), 129-154.

Jennex, M. E. (2005). The issue of system use in knowledge management systems. In *Proceedings of the 38th Annual Hawaii International Conference on System Sciences.*

Jennex, M., & Olfman, L. (2002). Organizational memory/knowledge effects on productivity: A longitudinal study. In *Proceedings of the 33rd Annual Hawaii International Conference on System Sciences.*

Jennex, M., Olfman, L., Panthawi, P., & Park, Y. (1998). An organizational memory information systems success model: An extension of DeLone and McLean's I/S success model. In *Proceedings of the 29th Annual Hawaii International Conference on System Sciences.*

Kankanhalli, A., Tan, B. C. Y., & Wei, K. K. (2005a). Contributing knowledge to electronic knowledge repositories: An empirical investigation. *MIS Quarterly, 29*(1), 113-143.

Kankanhalli, A., Tan, B. C. Y., & Wei, K. K. (2005b). Understanding seeking from electronic knowledge repositories: An empirical study. *Journal of the American Society for Information Science and Technology, 56*(11), 1156-1166.

Kaplan, R., & Norton, D. (1996). *Translating strategy into action: The balanced scorecard.* Boston: Harvard Business School Press.

Khalifa, M., Lam, R., & Lee, M. (2001). An integrative framework for knowledge management effectiveness. In *Proceedings of the Inter. Conf. on IS,* New Orleans, Louisiana.

KPMG. (2000). *Knowledge management research report 2000.*

Kuo, B. F. Y., Young, M., Hsu, M., Lin, C., & Chiang, P. (2003). A study of the cognition-action gap in knowledge management. In *Proceedings of the International Conference on Information Systems,* Seattle, Washington.

Lee, H., & Choi, B. (2003). Knowledge management enablers, processes, and organizational performance: An integrative view and empirical examination. *Journal of Management Information Systems, 20*(1), 179-228.

Lee, K. C., Lee, S., & Kang, I. W. (2005). KMPI: measuring knowledge management performance. *Information & Management, 42*(3), 469-482.

Leitner, K. H., & Warden, C. (2004). Managing and reporting knowledge-based resources and processes in research organizations: Specifics, lessons learned and perspectives. *Management Accounting Research, 15*(1), 22-51.

Levin, D. Z., & Cross, R., (2004). The strength of weak ties you can trust: The mediating role of trust in effective knowledge transfer. *Management Science, 50*(11), 1477-1490.

Liao, C., & Chuang, S. (2006). Exploring the role of knowledge management for enhancing firm's innovation and performance. In *Proceedings of the 39th Hawaii International Conference on System Sciences,* Hawaii.

Lim, L. L. K., & Dallimore, P. (2004). Intellectual capital: Management attitudes in service industries. *Journal of Intellectual Capital, 5*(1), 181-194.

Maier, R. (2002). *Knowledge management systems: Information and communication technologies for knowledge management.* Berlin: Springer.

McKeen, J. D., Zack, M. H., & Singh, S. (2006). Knowledge management and organizational performance: An exploratory survey. In *Proceedings of the 39th Hawaii International Conference on System Sciences.*

Mukherjee, A. S., Lapre, M. A., & Wassenhove, L. (1998). Knowledge driven quality improvement. *Management Science, 44*(11), S35-S49.

Nissen, M. E. (1999). Knowledge-based knowledge management in the re-engineering domain. *Decision Support Systems, 27*(1-2), 47-65.

Roos, J., Roos, G., Dragonetti, N. C., & Edvinsson, L. (1998). *Intellectual capital: Navigating in the new business landscape.* London: Macmillan.

Ruggles. (1998). The state of the notion: knowledge management in practice. *California Management Review, 40*(3), 80-89.

Ruppel, C. P., & Harrington, S. J. (2001). Sharing knowledge through intranets: A study of organizational culture and intranet implementation. *IEEE Transactions on Professional Communication, 44*(1), 37-50.

Schiemann, W. A., & Lingle, J. (1998). Seven greatest myths of measurement. *IEEE Engineering Management Review, 26*(1), 114-116.

Stein, E. W., & Zwass, V. (1995). Actualizing organizational memory with information systems. *Information Systems Research, 6*(2), 85-117.

Stewart, T. (1997). *Intellectual capital: The new wealth of organizations*. New York: Doubleday.

Sveiby, K. E. (1997). *The new organizational wealth: Managing and measuring knowledge-based assets*. San Francisco: Barrett-Kohler.

Tanriverdi, H. (2002). Does IT knowledge relatedness differentiate performance of multi-business firms? *International Conference on Information Systems*, Barcelona, Spain.

Tanriverdi, H. (2005). Information technology relatedness knowledge management capability and performance of multibusiness firms. *MIS Quarterly, 29*(2), 311-334.

Tiwana, A. (2000). *The knowledge management toolkit: Practical techniques for building a knowledge management system*. Upper Saddle River, NJ: Prentice Hall.

Tiwana, A. (2002). *The knowledge management toolkit: Orchestrating IT, strategy, and knowledge platforms*. Upper Saddle River, NJ: Prentice Hall.

Tiwana, A., & Bush, A., (2005). Continuance in expertise-sharing networks: A social perspective. *IEEE Transactions on Engineering Management, 52*(1), 85-101.

Verkasalo, M., & Lappalainen, P. (1998). A method of measuring the efficiency of the knowledge utilization process. *IEEE Transactions on Engineering Management, 45*(4), 414-423.

Wang, W., & Chang, C. (2005). Intellectual capital and performance in causal models. *Journal of Intellectual Capital, 6*(2), 222-236.

Wasko, M. M., & Faraj, S. (2000). It is what one does: Why people participate and help others in electronic communities of practice. *The Journal of Strategic Information Systems, 9*(2-3), 155-173.

Wasko, M. M., & Faraj, S. (2005). Why should I share? Examining social capital and knowledge contribution in electronic networks of practice. *MIS Quarterly, 29*(1), 35-57.

Wei, C., Hu, P. J., & Chen, H. (2002). Design and evaluation of a knowledge management system. *IEEE Software, 19*(3), 56-59.

Zach, M.H. (1999). Managing codified knowledge. *Sloan Management Review 40*(4), 45-58.

Zhang, W., & Watts, S. (2003). Knowledge adoption in online communities of practice. In *Proceedings of the International Conference on Information Systems,* Seattle, Washington.

Endnote

1 These two models alternately have been labeled as integrative and interactive architectures, respectively (Zack, 1999).

Chapter XI

Knowledge Management Success Factors and Models

Murray E. Jennex, San Diego State University, USA

Lorne Olfman, Claremont Graduate University, USA

Abstract

This chapter surveys knowledge management (KM) and knowledge management system (KMS) success factors and models. It also provides a framework for assessing KM and KMS success models. The framework uses three criteria: how well the model fits actual KMS success factors, the degree to which the model has a theoretical foundation, and if the model can be used for both types of KMS. The framework then is applied to four KMS success models found in the literature and is determined to be a useful framework for assessing KMS success models.

Introduction

Knowledge management systems (KMS) are systems designed to manage organizational knowledge. Alavi and Leidner (2001) clarify KMS as IT-based systems developed to support/enhance the processes of knowledge creation, storage/retrieval, transfer, and application. Additionally, KMS support knowledge management through the creation of network-based organizational memory (OM) and support for virtual project teams and organizations and Communities of Practice. A final goal of KMS is to support knowledge/OM creation.

There are several taxonomies of KMS from Zack's (1999) integrative and interactive KMS to KMS classified based on knowledge life cycle (Alavi & Leidner, 2001), KM spectrum (Hahn & Subramani, 2000), KM Architecture (Borghoff & Pareschi, 1998), and so forth. However, this chapter classifies KMS by the context captured and the users that are targeted, resulting in two approaches to building KMS: the process/task approach and the infrastructure/generic approach. The process/task approach focuses on the use of knowledge/OM by participants in a process, task, or project in order to improve the effectiveness of that process, task, or project. This approach identifies the information and knowledge needs of the process, where they are located, and who needs them. This approach requires KMS to capture minimal context, because users are assumed to understand the milieu of the knowledge that is captured and used.

The infrastructure/generic approach focuses on building a system to capture and distribute knowledge/OM for use throughout the organization. Concern is with capturing context to explain the captured knowledge and the technical details needed to provide good mnemonic functions associated with the identification, retrieval, and use of knowledge/OM. The approach focuses on network capacity, database structure and organization, and knowledge/information classification.

Both approaches may be used to create complete KMS. The process/task approach supports specific work activities, while the infrastructure/generic approach integrates organizational knowledge into a single system that can be leveraged over the total organization instead of just a process or project. Morrison and Weiser (1996) support the dual approach concept by suggesting that an organizationwide KMS be designed to combine an organization's various task/process-based KMS into a single environment and integrated system.

Once a KMS is implemented, whichever type it is, its success or effectiveness needs to be determined. Turban and Aronson (2001) list three reasons for measuring the success of a knowledge management system (KMS):

- To provide a basis for company valuation
- To stimulate management to focus on what is important
- To justify investments in KM activities

All are good reasons from an organizational perspective. Additionally, from the perspective of KM academics and practitioners, the measurement of KMS effectiveness or success is crucial to understanding how these systems should be built and implemented.

To meet this need, several KMS success/effectiveness models have been proposed. It is the purpose of this chapter to propose a framework for assessing the usefulness of these models. In order to do this, the chapter describes an evaluation model based on comparing the KMS success model to KMS success factors, determining the degree to which the model has a theoretical foundation and determining if the model can be applied to both approaches to building a KMS.

The chapter first will define the assessment framework. Then five KM/KMS success/effectiveness models will be described, followed by an analysis with respect to how well the models match the assessment framework and a conclusion on the usefulness of the framework. KM/KMS success/effectiveness will not be defined, because we found that each model defines success/effectiveness as part of the model.

Methodology

The proposed assessment framework consists of three main questions: How well does the KMS success model meet KM/KMS success criteria? What is the degree of the model's theoretical foundation? Can it be applied to both approaches to building a KMS? Stinchcombe (1968) says to test theories by determining how well they reflect observed data and that the more observations that can be compared, the better. The proposed framework does this by comparing the KMS success models to a set of KMS success criteria. The set of KMS success criteria was determined through a literature survey. Several studies were found that reported issues affecting the success of a KMS. The studies used in this chapter utilize a variety of methods, including surveys, case studies, Delphi studies, and experimentation. A total of 78 projects or organizations were investigated using case studies. Three surveys were administered, and one Delphi study and experiment were performed.

The second criterion is the theoretical foundation of the KMS success model. This criterion is based on being able to generalize the model. It is proposed that a model that is based on accepted theory or other widely supported models will be more generalizable. The theoretical foundation is determined by reviewing the publication in which the model is presented. A judgment is made as to the appropriateness of the theoretical foundation.

The third criterion is for the KMS success model to be applicable to both KMS approaches. This criterion is determined by judging the focus of the model to determine if it is specific to either the task/process approach or the generic/infrastructure approach.

KM/KMS Success Factors

A successful KMS should perform well the functions of knowledge creation, storage/retrieval, transfer, and application. However, other factors can influence KMS success. This section creates a KMS success factor framework by reviewing research related to identifying KMS success factors. Additionally, findings from studies looking at knowledge management (KM) and organizational memory (OM) success also are included. KM is included as a Churchman (1979) view of a KMS, which can be defined to include the KM initiative driving the implementation of a KMS (also the counter view is valid, as looking at KM also can include looking at the KMS). OM is included, as Jennex and Olfman (2002) found that KM and OM are essentially the same with the difference being the players. End users tend to do KM in which KM is concerned with the identification and capture of key knowledge. Information systems (IS) personnel tend to be concerned with OM in which OM is the storage, search, retrieval, manipulation, and presentation of knowledge. KMS and organizational memory systems (OMS) are the systems built to support KM and OM and are essentially systems designed to manage organizational knowledge. As stated previously, Alavi and Leidner (2001) clarify KMS as IT-based systems developed to support/enhance the processes of knowledge creation, storage/retrieval, transfer, and application. Additionally, a KMS supports knowledge management through the creation of network-based OM, and support for virtual project teams and organizations and communities of practice. A final goal of a KMS is to support knowledge/OM creation. Stein and Zwass (1995) define the OMS as the processes and IT components necessary to capture, store, and bring to bear knowledge created in the past on decisions currently being made. Jennex and Olfman (2002), using these definitions of KMS and OMS and a Churchman (1979) view of systems, combined the KMS and OMS into a single system.

A success factor framework is constructed by reviewing the literature by author. This is done so that the context resulting in the generation of the success factor can be presented. The identified success factors then are analyzed for similar concepts and combined into composite success factors. The composite success factors are ranked based on the number of authors mentioning the factor. Basing the ranking on the number of authors mentioning the success factor is problematic but is done as it implies greater consensus on the existence of the success factor (i.e., the more often a success factor is mentioned, the greater the consensus that it is a success factor and the greater the likelihood it is important). Table 1, presented at the end of this discussion, provides the ranked list of composite success factors.

Mandviwalla, Eulgem, Mould, and Rao (1998) summarized the state of the research and described several strategy issues affecting the design of a KMS. These include the focus of the KMS (who are the users), the quantity of knowledge to be captured and in what formats; who filters what is captured, and what reliance and/or limitations are placed on the use of individual memories. Additional technical issues affecting KMS design include knowledge storage/repository considerations, how information and knowledge is organized so that it can be searched and linked to appropriate events and use, and processes for integrating the various repositories and for re-integrating information and knowledge extracted from specific events. Some management issues include how long the knowledge is useful, access locations as users rarely access the KMS from a single location (leads to network needs and security concerns), and the work activities and processes that utilize the KMS.

Ackerman (1994) studied six organizations that had implemented his Answer Garden system. Answer Garden is a system designed to grow organizational memory in the context of help-desk situations. Only one organization had a successful implementation, because expectations of the capabilities of the system exceeded the actual capabilities. Ackerman and Mandel (1996) found that a smaller task-based system was more effective on the sub-organization level because of its narrower expectations. They refer to this narrower system as "memory in the small."

Jennex and Olfman (2000) studied three KM projects to identify design recommendations for building a successful KMS. These recommendations include the following:

- Develop a good technical infrastructure by using a common network structure, adding KM skills to the technology support skill set, using high-end PCs, integrated databases, and standardizing hardware and software across the organization.
- Incorporate the KMS into everyday processes and IS by automating knowledge capture.
- Have a enterprisewide knowledge structure.
- Have Senior Management support.
- Allocate maintenance resources for OMS.
- Train users on use and content of the OMS.
- Create and implement a KM Strategy/Process for identifying/maintaining the knowledge base.
- Expand system models/life cycles to include the KMS and assess system/process changes for impact on the KMS.
- Design security into the KMS.
- Build motivation and commitment by incorporating KMS usage into personnel evaluation processes, implementing KMS use/satisfaction metrics, and identifying organizational culture concerns that could inhibit KMS usage.

Additionally, Jennex and Olfman (2002) performed a longitudinal study of KM on one of these organizations and found that new members of an organization do not use the computerized KMS due to a lack of context for understanding the knowledge and the KMS. They found that these users needed pointers to knowledge more than codified knowledge.

Jennex, Olfman, and Addo (2003) investigated the need for having an organizational KM strategy to ensure that knowledge benefits gained from projects are captured for use in the organization by surveying Year 2000 (Y2K) project leaders. They found that benefits from Y2K projects were not being captured because the parent organizations did not have a KM strategy/process. Their conclusion was that KM in projects can exist and can assist projects in utilizing knowledge during the project.

Davenport, DeLong, and Beers (1998) studied 31 projects in 24 companies. Eighteen projects were determined to be successful, five were considered failures, and eight were too new to be rated. Eight factors were identified that were common in successful KM projects. These factors are as follows:

- Senior management support
- Clearly communicated KMS purpose/goals
- Linkages to economic performance
- Multiple channels for knowledge transfer

- Motivational incentives for KM users
- A knowledge-friendly culture
- A solid technical and organizational infrastructure
- A standard, flexible knowledge structure

Malhotra and Galletta (2003) identified the critical importance of user commitment and motivation through a survey study of users of a KMS being implemented in a healthcare organization. They found that using incentives did not guarantee a successful KMS. They created an instrument for measuring user commitment and motivation that is similar to Thompson, Higgins, and Howell's (1991) Perceived Benefit model but based on self-determination theory that uses the Perceived Locus of Causality.

Ginsberg and Kambil (1999) explored issues in the design and implementation of an effective KMS by building a KMS based on issues identified in the literature and then experimentally implementing the KMS in a field setting. They found knowledge representation, storage, search, retrieval, visualization, and quality control to be key technical issues and incentives to share and use knowledge to be the key organizational issues.

Alavi and Leidner (1999) surveyed executive participants in an executive development program with respect to what was needed for a successful KMS. They found organizational and cultural issues associated with user motivation to share and use knowledge to be the most significant. They also found it important to measure the benefits of the KMS and to have an integrated and integrative technology architecture that supports database, communication, and search and retrieval functions.

Holsapple and Joshi (2000) investigated factors that influenced the management of knowledge in organizations through the use of a Delphi panel consisting of 31 recognized KM researchers and practitioners. They found leadership and top management commitment/support to be crucial. Resource influences such as having sufficient financial support, skill level of employees, and identified knowledge sources are also important.

Koskinen (2001) investigated tacit knowledge as a promoter of success in technology firms by studying 10 small technology firms. Key to the success of a KMS was the ability to identify, capture, and transfer critical tacit knowledge. A significant finding was that new members take a long time to learn critical tacit knowledge, and a good KMS facilitates the transference of this tacit knowledge to new members.

Barna (2003) studied six KM projects with various levels of success (three were successful, two failed, and one was an initial failure that turned into a success) and identified two groups of factors important to a successful KMS. The main managerial success factor is creating and promoting a culture of knowledge sharing within the organization by articulating a corporate KM vision, rewarding employees for knowledge sharing, creating communities of practice, and creating a best-practices repository. Other managerial success factors include obtaining senior management support, creating a learning organization, providing KMS training, and precisely defining KMS project objectives.

Design/construction success factors include approaching the problem as an organizational problem and not a technical one; creating a standard knowledge submission process,; methodologies and processes for codification, documentation and storage of knowledge, and processes for capturing and converting individual tacit knowledge into organizational knowledge; and creating relevant and easily accessible knowledge-sharing databases and knowledge maps.

Cross and Baird (2000) propose that KM would not improve business performance simply by using technology to capture and share the lessons of experience. It was postulated that in order for KM to improve business performance, it had to increase organizational learning through the creation of organizational memory. To investigate this, 22 projects were examined. The conclusion was that improving organizational learning improved the likelihood of KM success. Factors that improved organizational learning include the following:

- Supporting personal relationships between experts and knowledge users
- Providing incentives to motivate users to learn from experience and to use the KMS
- Providing distributed databases to store knowledge and pointers to knowledge
- Providing work processes for users to convert personal experience into organizational learning
- Providing direction to what knowledge the organization needs to capture and learn from

Sage and Rouse (1999) reflected on the history of innovation and technology and identified the following issues:

- Modeling processes to identify knowledge needs and sources
- KMS strategy for the identification of knowledge to capture and use and who will use it
- Provide incentives and motivation to use the KMS
- Infrastructure for capturing, searching, retrieving, and displaying knowledge
- An understood enterprise knowledge structure
- Clear goals for the KMS
- Measuring and evaluating the effectiveness of the KMS

Yu, Kim, and Kim (2004) explored the linkage of organizational culture to knowledge management success. They found that KM drivers such as a learning culture, knowledge sharing intention, KMS quality, rewards, and KM team activity significantly affected KM performance. These conclusions were reached through a survey of 66 Korean firms.

Chan and Chau (2005) determined lessons learned from a failed case of KM in a Hong Kong organization. Key findings were the need for continued top management support and involvement, a knowledge-sharing culture, integrated infrastructure and enterprise data, appropriate incentives, and an appropriate technology strategy that focuses on use of IT and knowledge, sources of knowledge, and the expected benefits of knowledge use.

Lam and Chua (2005) in their study of KM abandonment in four KM projects identified critical success factors (CSFs) for KM from the literature. Their CSFs include the following:

- A clear KM vision and strategy (Maier & Remus, 2003; Von Krogh, 1998)
- Alignment of KM strategy to business goals (Malone, 2002)
- A learning culture (Goh, 2002; McDermott & O'Dell, 2001; van Zolingen, Streumer, & Stooker, 2001)
- Incentives for knowledge creation and reuse (Lynne, 2001)
- A specific community that provides a context in which KM can flourish (Dixon, 2000; Wenger, McDermott, & Snyder, 2002)
- Continuous top management support (Storey & Barnett, 2000)
- Employee empowerment (Liebowitz & Beckman, 1998; Stenmark, 2003)
- A positive attitude to knowledge sharing (Bock & Kim, 2002)
- A flexible organization structure (Forcadell & Guadamillas, 2002)
- Usable and up-to-date KM systems (Davenport & Prusak, 1999)
- Knowledge governance structure for maintaining quality of knowledge content (Dilnutt, 2002)

These studies provide several success factors. As previously discussed, in order to analyze the factors, they have been reviewed and paraphrased into a set of ranked composite success factors in which the ranking is based on the number of sources citing them. Table 1 lists the final set of success factors in their rank order. Additionally, success factors SF1 through SF4 are considered the key success factors, as they were mentioned by at least half of the success factor studies.

Knowledge Management Success Models

Bots and de Bruijn: Knowledge Value Chain

Bots and de Bruijn (2002) assessed KM and determined that the best way to judge good KM was through a knowledge value chain. In this evaluation process, KM is assessed for effectiveness at each step of the knowledge process and is good if each of the indicated activities is performed well with the ultimate factor being if the KMS enhances competitiveness. Figure 1 illustrates the KM value chain. The model was developed by viewing and contrasting KM through an analytical (technical) perspective and an actor (user) perspective. These perspectives are conflicting, and KM assessment occurs by determining how well the KMS meets each perspective at each step.

Table 1. KMS success factor summary

ID	Success Factor	Source
SF1	A Knowledge Strategy that identifies users, sources, processes, storage strategy, knowledge, and links to knowledge for the KMS	Barna (2003), Chan and Chau (2005), Dixon (2000), Ginsberg and Kambil (1999), Holsapple and Joshi (2000), Jennex, Olfman, and Addo (2003), Koskinen (2001), Maier and Remus (2003), Mandviwalla et al. (1998), Sage and Rouse (1999), Von Krogh (1998), Wenger et al. (2002), Yu et al. (2004)
SF2	Motivation and Commitment of users, including incentives and training	Alavi and Leidner (1999), Barna (2002), Chan and Chau (2005), Cross and Baird (2000), Davenport et al. (1998), Ginsberg and Kambil (1999), Jennex and Olfman (2000), Liebowitz and Beckman (1998), Lynne (2001), Malhotra and Galletta (2003), Stenmark (2003), Yu et al. (2004)
SF3	Integrated Technical Infrastructure, including networks, databases/ repositories, computers, software, KMS experts	Alavi and Leidner (1999), Barna (2003), Cross and Baird (2000), Chan and Chau (2005), Davenport and Prusak (1999), Davenport et al. (1998), Ginsberg and Kambil (1999), Jennex and Olfman (2000), Mandviwalla et al. (1998), Sage and Rouse (1999), Yu et al. (2004)
SF4	An organizational culture and structure that supports learning and the sharing and use of knowledge	Alavi and Leidner (1999), Barna (2003), Bock and Kim (2002), Chan and Chau (2005), Davenport et al. (1998), Forcadell and Guadamillas (2002), Jennex and Olfman (2000), Sage and Rouse (1999), Yu et al. (2004)
SF5	A common enterprisewide knowledge structure that is clearly articulated and easily understood	Barna (2003), Chan and Chau (2005), Cross and Baird (2000), Davenport et al. (1998), Ginsberg and Kambil (1999), Jennex and Olfman (2000), Mandviwalla et al. (1998), Sage and Rouse (1999)
SF6	Senior Management support, including allocation of resources, leadership, and providing training	Barna (2003), Chan and Chau (2005), Davenport et al. (1998), Dilnutt (2002), Holsapple and Joshi (2000), Jennex and Olfman (2000), Storey and Barnett (2000), Yu et al. (2004)
SF7	Learning Organization	Barna (2003), Cross and Baird (2000), Goh (2002), McDermott and O'Dell (2001), Sage and Rouse (1999), Yu et al. (2004), van Zolingen et al. (2001)
SF8	There is a clear goal and purpose for the KMS	Ackerman (1994), Barna (2003), Chan and Chau (2005), Cross and Baird (2000), Davenport et al. (1998), Malone (2002)
SF9	Measures are established to assess the impacts of the KMS and the use of knowledge as well as to verify that the right knowledge is being captured	Alavi and Leidner (1999), Chan and Chau (2005), Davenport et al. (1998), Jennex and Olfman (2000), Sage and Rouse (1999)
SF10	The search, retrieval, and visualization functions of the KMS support easy knowledge use	Alavi and Leidner (1999), Ginsberg and Kambil (1999), Mandviwalla et al. (1998)
SF11	Work processes are designed that incorporate knowledge capture and use	Barna (2003), Cross and Baird (2000), Jennex and Olfman (2000)
SF12	Security/protection of knowledge	Jennex and Olfman (2000), Sage and Rouse (1999)

Figure 1. Bots and de Bruijn (2002) KM value chain

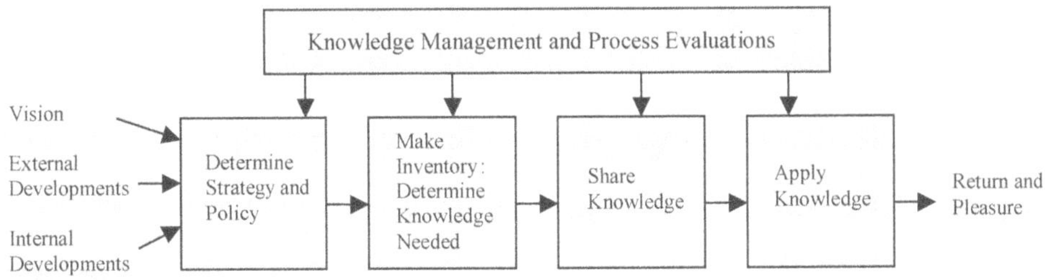

Massey, Montoya-Weiss, and Driscoll KM Success Model

Massey, Montoya-Weiss, and O'Driscoll (2002) present a process-based KM success model derived from their Nortel case study. The case study suggested that KM cannot be applied generically and that a process approach to KM will help an organization to understand how it can apply KM in order to improve organizational performance. The model is based on the framework proposed by Holsapple and Joshi (2001) and reflects that KM success is based on understanding a process-oriented KM strategy and its effects on the organization, its knowledge users, and how they use knowledge. It recognizes that KM is an organizational change process and that KM success cannot separate itself from the organizational change success with the result that the KM success essentially is defined as improving organizational or process performance. The model is presented in Figure 2. Key components of the model are as follows:

KM strategy: Defines the processes using knowledge and what that knowledge is, the sources, users, and form of the knowledge, and the technology infrastructure for storing the knowledge.

Key managerial influences: Defines management support through leadership, allocation and management of project resources, and oversight of the KMS through coordination and control of resources and the application of metrics for assessing KMS success.

Key resource influences: These are the financial resources and knowledge sources needed to build the KMS.

Key environmental influences: Describe the external forces that drive the organization to exploit its knowledge in order to maintain its competitive position.

Figure 2. Massey et al. (2002) KM success model

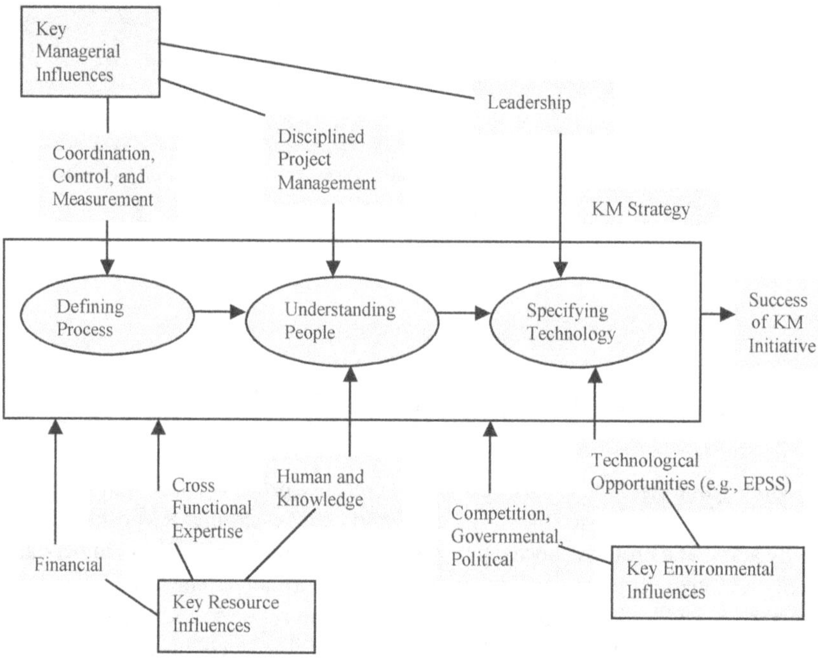

Lindsey KM Effectiveness Model

Lindsey (2002) proposes a conceptual KM effectiveness model based on combining organizational capability perspective theory (Gold, Malhotra, & Segars, 2001) and contingency perspective theory (Becerra-Fernandez & Sabherwal, 2001). The model defines KM effectiveness in terms of two main constructs: knowledge infrastructure capability and knowledge process capability, with the knowledge process capability construct being influenced by a Knowledge Task. Knowledge infrastructure capability represents social capital, the relationships between knowledge sources and users, and is operationalized by technology (the network itself), structure (the relationship), and culture (the context in which the knowledge is created and used). Knowledge process capability represents the integration of KM processes into the organization and is operationalized by acquisition (the capturing of knowledge), conversion (making captured knowledge available), application (degree to which knowledge is useful), and protection (security of the knowledge). Tasks are activities performed by organizational units and indicate the type and domain of the knowledge being used. Tasks ensure that the right knowledge is being captured and used. KM success is measured as satisfaction with the KMS and is considered a weak definition of success. It is proposed that research be conducted into KMS effectiveness to find ties into organizational

Figure 3. Lindsey KM effectiveness model (Lindsey, 2002)

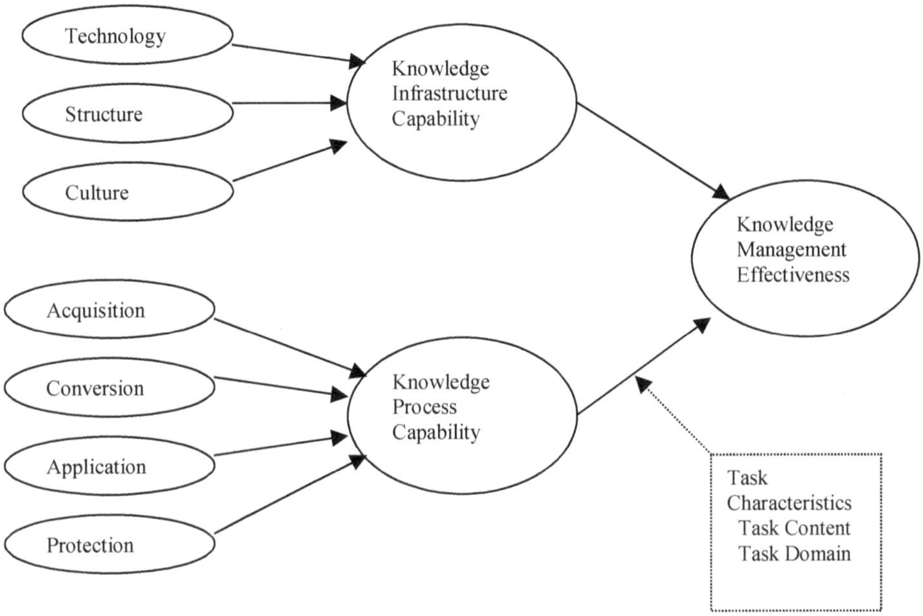

effectiveness. Kaplan and Norton's (1992) Balanced Scorecard may be useful in establishing measures for KMS effectiveness. Figure 3 illustrates the Lindsey model.

KMS Success Models Based on the DeLone and McLean IS Success Model

Jennex and Olfman (2006) present a KM/KMS Success Model that is based on the respecified DeLone and McLean (2003) IS success model. Figure 4 shows the KM/KMS Success Model. This model was derived from a longitudinal case study, a quantitative study across an industry, and action research applying the model in the field. The model evaluates success as an improvement in organizational effectiveness based on use of and impacts from KM. Descriptions of the dimensions of the model follow:

System Quality: Defines how well the KMS performs the functions of knowledge creation, storage/retrieval, transfer, and application; how much of the OM is codified and included in the computerized portion of the OM, and how the KMS is supported by the infrastructure.

Figure 4. Jennex and Olfman (2006) KM success model

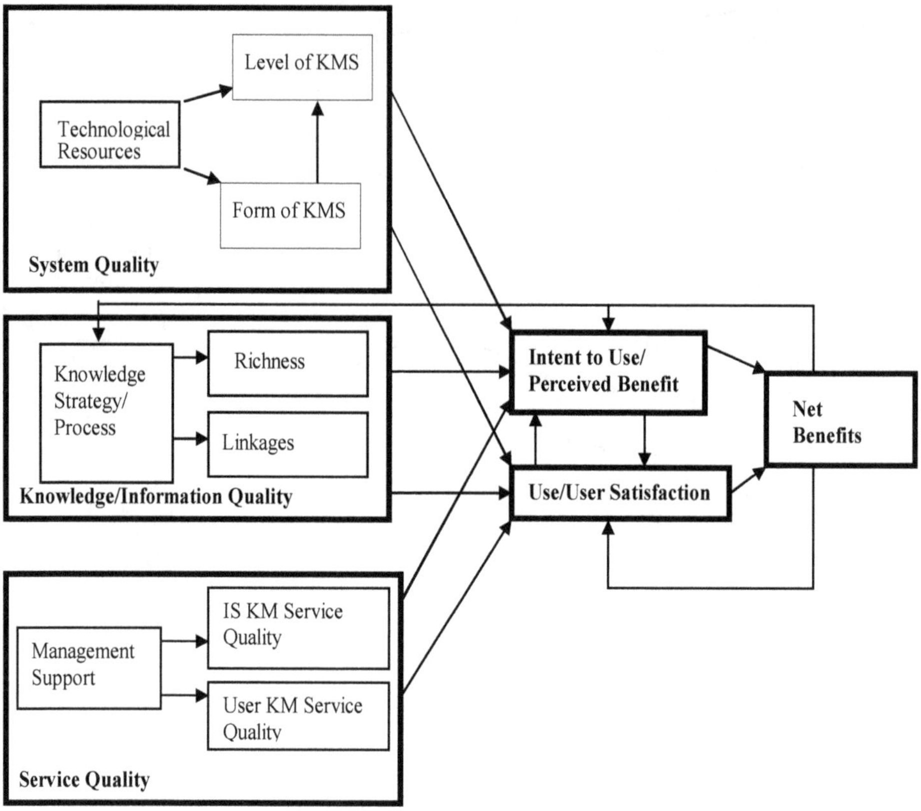

Knowledge/Information Quality: Ensures that the right knowledge/OM with sufficient context is captured and available for the right users at the right time.

Service Quality: Ensures that the organization has adequate service support from management, user organizations, and the IS organization.

Use/User Satisfaction: Indicates actual levels of KM use as well as the satisfaction of KM users. Actual use is most applicable as a success measure when the use of a system is required. User satisfaction is a construct that measures satisfaction with the KM by users. It is considered a good complementary measure of KM use when use of KM is required, and effectiveness of use depends on users being satisfied with the KMS.

Perceived Benefit: Measures perceptions of the benefits and impacts of the KMS by users and is based on Thompson et al.'s (1991) perceived benefit model. It is good for predicting continued KM use when use of KM is voluntary, and the amount and/or effectiveness of KM use depends on meeting current and future user needs.

Figure 5. Maier (2002) KMS success model

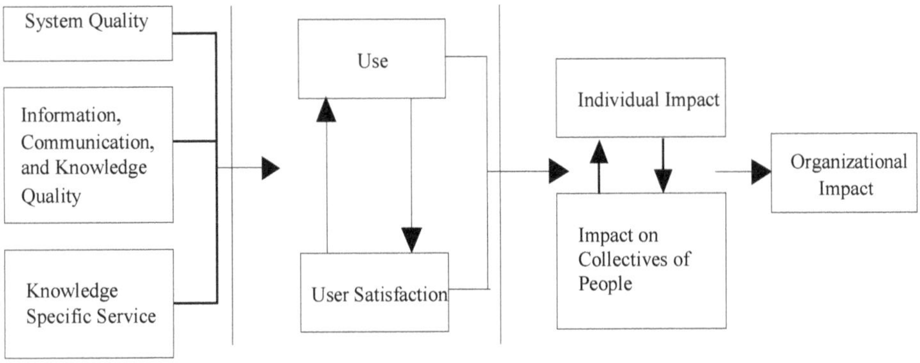

Net Impact: An individual's use of KM will produce an impact on that person's performance in the workplace. Each individual impact, in turn, will have an effect on the performance of the whole organization. Organizational impacts are typically not the summation of individual impacts, so the association between individual and organizational impacts is often difficult to draw, which is why this construct combines all impacts into a single construct. This model recognizes that the use of knowledge/OM may have good or bad benefits and allows for feedback from these benefits to drive the organization to either use more knowledge/OM or to forget specific knowledge/OM.

Maier (2002) also proposes a KMS success model based on the DeLone and McLean (1992) IS success model. This model is similar to the Jennex and Olfman (2006) model. Breakdown of the dimensions into constructs is not provided, but specific measures for each dimension are identified. This model is illustrated in Figure 5 and uses the following dimensions:

System Quality: Taken directly from DeLone and McLean (1992) and refers to overall quality of the hardware and software.

Information, Communication, and Knowledge Quality: Refers to the quality of the stored data, information, and knowledge and to the quality of knowledge flow methods.

Knowledge-Specific Service: Refers to how well subject matter experts and KMS managers support the KMS.

System Use/User Satisfaction; Taken directly from DeLone and McLean (1992) and refers to actual KMS use and the satisfaction users have with that use.

Individual Impact: Taken directly from DeLone and McLean (1992) and refers to the impacts KMS use has on the individual's effectiveness.

Impact on Collectives of Peoples: Refers directly to the improved effectiveness within teams, work groups, and/or communities that comes from using the KMS.

Organizational Impacts: Taken directly from DeLone and McLean (1992) and refers to improved overall organizational effectiveness as a result of KMS use.

Application of the Framework

To illustrate the use of the framework, the KMS Success models first are analyzed by comparing them to the identified set of success factors and determining how well the models reflect the set of success factors. Table 2 summarizes this comparison. Assessing responsiveness to the top four success criteria finds that the Value Chain, Maier, and Lindsey models are not as good at reflecting the observed data as the Massey et al. and Jennex and Olfman models. Also, the only difference between the Massey et al. and Jennex and Olfman models is SF5, Culture. Given that this would be the next most important success factor, it is determined that the Jennex and Olfman model most closely fits the observed data, as reflected by the success factors model. It should be noted that further derivation of the Maier model dimensions may improve its fit to the KMS success factors.

Looking at the theoretical foundation for the KMS success models finds that all four have some theoretical foundation. The Value Chain model uses the commonly used value chain approach. The Massey et al. model relies on the Holsapple and Joshi (2001) framework. The Lindsey model utilizes organizational capability perspective theory and contingency perspective theory. The Jennex Olfman and Maier models utilize the widely accepted DeLone and McLean IS success model. Assessing the ability to generalize from the theory, it can be determined that the value chain, Jennex and Olfman, and Maier models are utilizing theory that is utilized more widely for assessing effectiveness. However, the Massey et al. and Lindsey models' theoretical foundation may be proven to be more widely applicable after being applied and studied in a variety of organizations and applications.

Assessing the KM success models for applicability to both approaches for building a KMS, it can be determined that the Jennex and Olfman model has no characteristics that would limit its applicability to either KMS approach, while the Massey et al., value chain, Maier, and Lindsey models could be interpreted as being specific to an approach. The Value Chain model typically is applied to organizational systems in order to determine strategic processes, focusing this model on generic/infrastructure uses of a KMS. The Massey et al., Maier, and Lindsey models specifically incorporate task-specific components that may make it difficult to focus the models on assessing organizational effectiveness. However, it can be concluded that all four models could be applied to both KMS approaches if the user is aware of the differences between the approaches and the limitations of the models.

In summary, the proposed framework provides a user with a measuring stick for selecting a KMS success model. Users wanting a model based on widely accepted success models and that fits the observed data (as expressed in the KMS success factors) would rank the five models in order of preference as Jennex and Olfman, Massey et al., Value Chain, Lindsey,

Table 2. KM success models vs. KM success factors

Success Factor ID	Value Chain	Massey et al.	Lindsey	Jennex and Olfman	Maier
SF1	Strategy stage	KM Strategy	Task and Acquisition constructs	KM Strategy/ Process Construct	Information, Communication, and Knowledge quality
SF2	Weak—Apply knowledge stage	Key Management Influences	No clear tie	Perceived Benefit and User Service Quality Constructs	No clear tie
SF3	No clear tie—Share knowledge stage	KM Strategy	Technology construct—Networks	Technical Resources and Service Quality Constructs	System Quality and Knowledge-Specific Service Quality
SF4	No clear tie	No clear tie	Culture construct	Perceived Benefit Construct	No clear tie
SF5	No clear tie	KM Strategy	Structure and Conversion constructs	Form Construct	Information, Communication, and Knowledge quality
SF6	Implied—No clear tie	Key Management Influences	No clear tie	Perceived Benefit Construct	No clear tie
SF7	No clear tie	No clear tie	No clear tie	Management Support Construct	No Clear Tie
SF8	Strategy stage	KM Strategy	Task construct	KM Strategy/ Process Construct	Information, Communication, and Knowledge quality
SF9	Return stage	Key Management and Environmental Influences	Task construct	Net Impacts Construct	Impact dimensions, Information, Communication, and Knowledge quality
SF10	Share knowledge and apply knowledge stages	KM Strategy	Conversion and Task constructs	Level Construct	System Quality
SF11	Apply knowledge stage	KM Strategy	Application construct	Perceived Benefit Construct	No clear tie
SF12	No clear tie	No clear tie	Protection Construct	No clear tie	No clear tie

and Maier. Users wanting a model specifically for assessing a project/task KMS may opt for the Massey et al., Maier, or Lindsey models. Users focusing on generic/infrastructure KMS may opt for the value chain model. Users implementing both types of KMS and wanting a single KMS effectiveness model may opt for the Jennex and Olfman model.

Conclusion

The proposed framework for assessing KMS success models appears to be useful. It allows users to validate that the KMS success model they are using reflects observed factors that have been found to affect KMS success. The use of the KMS Success Factors to assess this fit is very powerful and is the major contribution of this chapter. The KMS success factors were identified from a large number of studies, projects, and KMSs providing a broad view of KMS success.

The use of the other two criteria of the framework is less powerful but still important. It is important to determine that a KMS success model has a theoretical foundation, since otherwise, it simply could be a reflection of a single data point's success criteria and may not be applicable to the KMS to be assessed. Additionally, it is also important to ensure that the KMS success model being used applies to the approach of the KMS under consideration. It is inappropriate to apply an organizational effectiveness model to a task/process KMS and vice versa.

References

Ackerman, M. (1994). Definitional and contextual issues in organizational and group memories. In *Proceedings of the Twenty-Seventh Annual Hawaii International Conference on System Sciences.* IEEE Computer Society Press, 191-200.

Ackerman, M., & Mandel, E. (1996). Memory in the small: An application to provide task-based organizational memory for a scientific community. In *Proceedings of the Twenty-Ninth Annual Hawaii International Conference on System Sciences, IEEE Computer Society Press* (pp. 323-332).

Alavi, M., & Leidner, D. E. (1999). Knowledge management systems: Emerging views and practices from the field.In *Proceedings of the 32nd Hawaii International Conference on System Sciences.* IEEE Computer Society.

Alavi, M., & Leidner, D. E. (2001). Review: Knowledge management and knowledge management systems: Conceptual foundations and research issues. *MIS Quarterly, 25*(1), 107-136.

Barna, Z. (2003). *Knowledge management: A critical e-business strategic factor* [unpublished masters thesis]. San Diego: San Diego State University.

Becerra-Fernandez, I., & Sabherwal, R. (2001). Organizational knowledge management: A contingency perspective. *Journal of Management Information Systems, 18*(1), 23-55.

Bock , G. W., & Kim, Y. (2002). Breaking the myths of rewards: An exploratory study of attitudes about knowledge sharing. *Information Resources Management Journal, 15*(2), 14-21.

Borghoff, U. M., & Pareschi, R. (1998). *Information technology for knowledge management.* Berlin: Springer-Verlag.

Bots, P. W. G., & de Bruijn, H. (2002). Effective knowledge management in professional organizations: Going by the rules. In *Proceedings of the 35th Hawaii International Conference on System Sciences.* IEEE Computer Society Press.

Chan, I., & Chau, P. Y. K. (2005). Getting knowledge management right: Lessons from failure. *International Journal of Knowledge Management, 1*(3), 40-54.

Churchman, C. W. (1979). *The systems approach.* New York: Dell Publishing.

Cross, R., & Baird, L. (2000). Technology is not enough: Improving performance by building organizational memory. *Sloan Management Review, 41*(3), 41-54.

Davenport, T. H., DeLong, D. W., & Beers, M. C. (1998). Successful knowledge management projects. *Sloan Management Review, 39*(2), 43-57.

Davenport, T. H., & Prusak, L. (1999). *Working knowledge.* Boston: Harvard Business School Press.

DeLone, W. H., & McLean, E. R. (1992). Information systems success: The quest for the dependent variable. *Information Systems Research, 3*, 60-95.

DeLone, W. H., & McLean, E. R. (2003). The DeLone and McLean model of information systems success: A ten-year update. *Journal of Management Information Systems, 19*(4), 9-30.

Dilnutt, R. (2002). Knowledge management in practice, three contemporary case studies. *International Journal of Accounting Information Systems,* 3(2), 75-81.

Dixon, N. (2000). *Common knowledge: How companies thrive by sharing what they know.* Boston: Harvard Business School Press.

Forcadell, F. J., & Guadamillas, F. (2002). A case study on the implementation of a knowledge management strategy oriented to innovation. *Knowledge and Process Management, 9*(3), 162-171.

Ginsberg, M., & Kambil, A. (1999). Annotate: A Web-based knowledge management support system for document collections. In *Proceedings of the 32nd Hawaii International Conference on System Sciences.* IEEE Computer Society Press.

Goh, S. C. (2002). Managing effective knowledge transfer: An integrative framework and some practice implications. *Journal of Knowledge Management, 6*(1), 23-30.

Gold, A. H., Malhotra, A., & Segars, A. H. (2001). Knowledge management: An organizational capabilities perspective. *Journal of Management Information Systems, 18*(1), 185-214.

Hahn, J., & Subramani, M. R. (2000). A framework of knowledge management systems: Issues and challenges for theory and practice. In *Proceedings of the Twenty-first International Conference on Information Systems, Association for Information Systems,* 302-312.

Holsapple, C. W., & Joshi, K. D. (2000). An Investigation of factors that influence the management of knowledge in organizations. *Journal of Strategic Information Systems, 9*, 235-261.

Holsapple, C. W., & Joshi, K. D. (2001). Knowledge management: A three-fold framework. *The Information Society, 18*(1), 47-64.

Jennex, M. E., & Olfman, L. (2000). *Development recommendations for knowledge management/organizational memory systems.* Information Systems Development Conference.

Jennex, M. E., & Olfman, L. (2002). Organizational memory/knowledge effects on productivity, a longitudinal study. In *Proceedings of the 35th Hawaii International Conference on System Sciences, HICSS35.* IEEE Computer Society.

Jennex, M. E., & Olfman, L. (2006). A model of knowledge management success. *International Journal of Knowledge Management, 2*(3), 51-68.

Jennex, M. E., Olfman, L., & Addo, T. B. A. (2003). The need for an organizational knowledge management strategy. In *Proceedings of the 36th Hawaii International Conference on System Sciences, HICSS36.* IEEE Computer Society.

Kaplan, R., & Norton, D. (1992). The balanced scorecard—Measures that drive performance. *Harvard Business Review, 70*(1), 71-79.

Koskinen, K. U. (2001). Tacit knowledge as a promoter of success in technology firms. In *Proceedings of the 34th Hawaii International Conference on System Sciences.* IEEE Computer Society.

Lam, W., & Chua, A. (2005). Knowledge management project abandonment: An explanatory examination of root causes. *Communications of the Association for Information Systems, 16*, 723-743

Liebowitz, J., & Beckman, T. (1998). *Knowledge organizations: What every manager should know.* Boca Raton, FL: CRC Press.

Lindsey, K. (2002). Measuring knowledge management effectiveness: A task-contingent organizational capabilities perspective. In *Proceedings of the Eighth Americas Conference on Information Systems,* 2085-2090.

Lynne, M. (2001). Toward a theory of knowledge reuse: Types of knowledge reuse situations and factors in reuse success. *Journal of Management Information Systems, 18*(1), 57-93.

Maier, R. (2002). *Knowledge management systems: Information and communication technologies for knowledge management.* Berlin: Springer-Verlag.

Maier, R., & Remus, U. (2003). Implementing process-oriented knowledge management strategies. *Journal of Knowledge Management, 7*(4), 62-74.

Malhotra, Y., & Galletta, D. (2003). Role of commitment and motivation as antecedents of knowledge management systems implementation. In *Proceedings of the 36th Hawaii International Conference on System Sciences.* IEEE Computer Society.

Malone, D. (2002). Knowledge management: A model for organizational learning. *International Journal of Accounting Information Systems, 3*(2), 111-123.

Mandviwalla, M., Eulgem, S., Mould, C., & Rao, S. V. (1998). Organizational memory systems design [unpublished working paper for the Task Force on Organizational Memory]. In *Proceedings of the 31st Annual Hawaii International Conference on System Sciences.*

Massey, A. P., Montoya-Weiss, M. M., & O'Driscoll, T. M. (2002). Knowledge management in pursuit of performance: Insights from Nortel Networks. *MIS Quarterly, 26*(3), 269-289.

McDermott, R., & O'Dell, C. (2001). Overcoming cultural barriers to sharing knowledge. *Journal of Knowledge Management, 5*(1), 76-85.

Morrison, J., & Weiser, M. (1996). A research framework for empirical studies in organizational memory. In *Proceedings of the Twenty-Ninth Annual Hawaii International Conference on System Sciences.* IEEE Computer Society Press.

Sage, A. P., & Rouse, W. B. (1999). Information systems frontiers in knowledge management. *Information Systems Frontiers, 1*(3), 205-219.

Stein, E.W. & Zwass, V. (1995). Actualizing organizational memory with information systems. *Information Systems Research, 6*(2), 85-117.

Stenmark, D. (2003). Knowledge creation and the Web: Factors indicating why some intranets succeed where others fail. *Knowledge and Process Management, 10*(3), 207-216.

Stinchcombe, A. L. (1987). *Constructing social theories*. Chicago: The University of Chicago Press.

Storey, J., & Barnett, E. (2000). Knowledge management initiatives: Learning from failure. *Journal of Knowledge Management, 4*(2), 145-156.

Thompson, R. L., Higgins, C. A., & Howell, J. M. (1991). Personal computing: Toward a conceptual model of utilization. *MIS Quarterly, 15*(1), 125-143.

Turban, E., & Aronson, J. E. (2001). *Decision support systems and intelligent systems* (sixth ed.). Prentice Hall.

van Zolingen, S. J., Streumer, J. N., & Stooker, M. (2001). Problems in knowledge management: A case-study of a knowledge-intensive company. *International Journal of Training and Development, 5*(3), 168-184.

Von Krogh, G. (1998). Care in knowledge creation. *California Management Review, 40*(3), 133-153.

Wenger, E. C., McDermott, R., & Snyder, W. M. (2002). *Cultivating communities of practice.* Boston, MA: Harvard Business School Press.

Yu, S-H, Kim, Y-G, & Kim, M-Y. (2004). Linking organizational knowledge management drivers to knowledge management performance: An exploratory study. In *Proceed-*

ings of the 37th Hawaii International Conference on System Sciences, HICSS36. IEEE Computer Society.

Zack, M. H. (1999). Managing codified knowledge. *Sloan Management Review, 40*(4), 45-58.

Chapter XII

Knowledge Management Success: Empirical Assessment of a Theoretical Model

Shih-Chen Liu, Chihlee Institute of Technology, Taiwan

Lorne Olfman, Claremont Graduate University, USA

Terry Ryan, Claremont Graduate University, USA

Abstract

This chapter presents the empirical testing of a theoretical model of knowledge management system (KMS) success. The Jennex and Olfman model of knowledge management success was developed to reflect the DeLone and McLean model of information systems success in the knowledge management context. A structural equation model representing the Jennex and Olfman theoretical model is developed. Using data from a prior study aimed at knowledge management system use and individual learning, this model is tested. The overall fit of the model to the data is fair, although some interpretation of the estimated model parameters is problematic. The results of the model test provide limited support for the Jennex and Olfman theoretical model but indicate the value of continued investigation and refinement of it. Suggestions for future research are provided.

Introduction

Involvement with a knowledge management system (KMS) generally leads to the desire to determine how successful it is. Practically, the measurement of KMS success (or effectiveness) can be valuable in a number of ways, including the justification of knowledge management (KM) investments (Turban & Aronson, 2001). Academically, the conceptualization of information system (IS) effectiveness is one of the most important research domains in the IS discipline (ISWorld, 2004a). A valid specific model of KMS success would have value for KM researchers in much the same way that a valid general model of IS success would have for the IS field.

The DeLone and McLean (D&M) (1992, 2002, 2003) model of IS success is currently the most widely accepted conceptualization of IS effectiveness among researchers (ISWorld, 2004b). The D&M model comprises six theoretical dimensions: information quality, system quality, service quality, intention to use/use, user satisfaction, and net benefits (DeLone & McLean, 2003). Each of these dimensions constitutes a well-trodden research path in its own right, as indicated by the separate pages devoted to each on the ISWorld Web site (ISWorld, 2004a). Figure 1 illustrates the model.

The DeLone and McLean model is a general framework for understanding IS effectiveness and must be adapted to specific contexts. For example, DeLone and McLean (2003) provide an adaptation of the most recent iteration of their model to e-commerce. Jennex et al. have adapted the D&M model to the KM context (Jennex & Olfman, 2002, 2004; Jennex, Olfman, & Addo, 2003; Jennex, Olfman, Pituma, & Park, 1998). This adaptation, which can be labeled the Jennex and Olfman (J&O) model, can claim both empirical and theoretical

Figure 1. DeLone and McLean (2003) IS success model

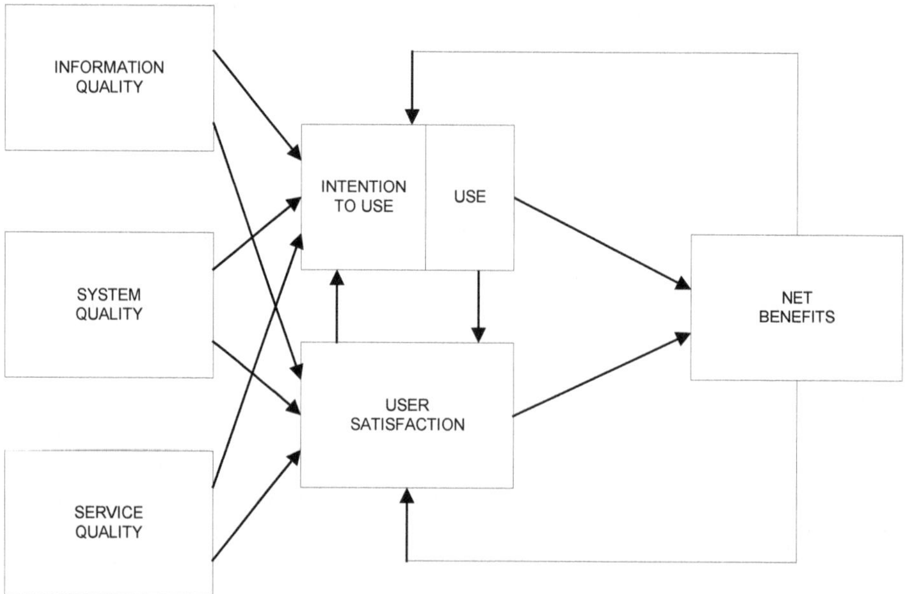

Figure 2. Jennex and Olfman (2004) IS success model

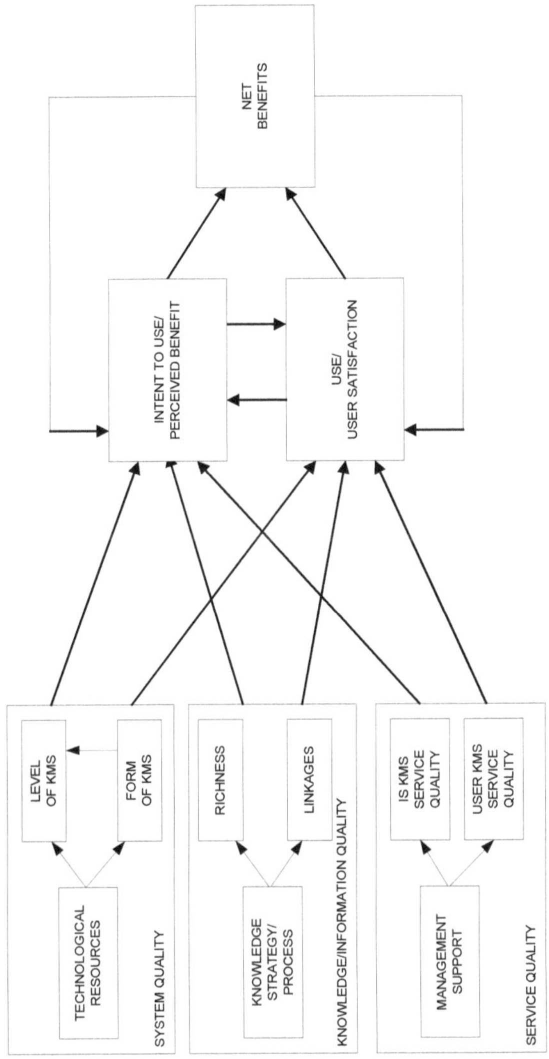

justification. The earliest version of the model (Jennex et al., 1998) was informed empirically by an ethnography concerning KMS use in an engineering setting and theoretically by the 1992 D&M model, along with thinking at that time about KM and organization memory, such as Stein and Zwass (1995). A revision of the model was informed empirically by a longitudinal study of engineering use of a KMS over a five-year period and theoretically by the 2002 revised D&M model, along with thinking at that time about KM, such as Alavi and Leidner (2001). The latest version of the J&O model reflects the reasoning given for the latest version of the D&M model (DeLone & McLean, 2003), along with the maturation of thinking of researchers in the KM field. Figure 2 depicts the J&O model in its current incarnation (Jennex & Olfman, 2004).

Although the J&O model was developed to reflect system success in a KM context, as is true for any theoretical model, its value as an explanation is open to empirical test. This research constitutes such a test; that is, it aims to assess how well the J&O model describes KMS success in the world. More specifically, the chapter reports the testing of a structural equation modeling (SEM) model conforming to the J&O theoretical model with survey data collected from KMS users. This test provides an evaluation of the adequacy of the J&O model in its present form, along with suggestions of improvements that might be made to it.

Background

Relationship Between the D&M Model and the J&O MO

The J&O model is an adaptation of the KM context of the well-accepted D&M model of IS success. The J&O model conceptualizes the basic dimensions of success in much the same ways that the D&M model does, but the ideas involved in the J&O model are targeted more to the KM setting than are the concepts constituting the D&M model. The J&O model consists of the same number of dimensions with the same fundamental relationships among them as the D&M model; the differences between the two models lie in the subdimensions proposed by Jennex, Olfman, and their colleagues to map the D&M dimensions to the KM setting. In the following paragraphs, the mapping is explained between each D&M dimension and its corresponding J&O dimension.

The D&M system quality dimension is conceptualized in the J&O model as involving three subdimensions. The first of these subdimensions is technological resources, which involves the capability of an organization to develop, operate, and maintain a KMS (Jennex & Olfman, 2004). This construct captures ideas about the networks, databases, and other hardware involved in the KMS, as well as the experience and expertise behind the KMS initiative and the usage competence of typical KMS users. The second system quality subdimension is Form of KMS, which has to do with the extent to which the knowledge andknowledge management processes are computerized and integrated (Jennex & Olfman, 2004). This concept reflects the amount of knowledge that is accessible through the KMS interface as well as the extent of automation and integration of the interface and the activities of knowledge creation, storage, retrieval, transfer, and application. The third system quality subdimension is Level of KMS. This is defined as the ability of the KMS to bring knowledge to bear upon current activities (Jennex & Olfman, 2004); it is centered on the nature and implementation of the KMS's search and retrieval functions. These subdimensions jointly cover the aspects of a KMS that theory and empirical observation point to as most critical in understanding what system quality is in KM settings.

The D&M dimension information quality is relabeled in the J&O model as knowledge/information quality. A high value for this dimension occurs whenever the right knowledge with sufficient context is captured and available for the right users at the right time (Jennex & Olfman, 2004). The dimension involves three subdimensions. The first of these, knowledge strategy/process, captures three ideas: the processes used for identifying the knowledge that can be captured and reused (and the users who can capture and reuse it); the formality of the processes, including how much planning occurs; and the format and content of the knowledge to be captured. This subdimension has evolved to reflect ideas of personalization and codification (Hansen, Nohria, & Tierney, 1999); it recognizes that evolution occurs in how knowledge is captured and reused. The second subdimension involved in knowledge/information quality is richness. This notion "reflects the accuracy and timeliness of the stored knowledge as well as having sufficient knowledge context to make the knowledge useful" (Jennex & Olfman, 2004). The third subdimension for this dimension, Linkages, is intended to reflect the knowledge and topic maps and/or listings of expertise available to the organization (Jennex & Olfman, 2004).

The D&M dimension service quality is defined in the J&O model as those aspects of a KMS that ensure the KMS has adequate support for users to use the KMS effectively (Jennex & Olfman, 2004). The dimension comprises three subdimensions. The first of these, Management Support, has to do with the allocation of adequate resources, encouragement and direction, and adequacy of control. The second service quality subdimension, User KM service quality, involves support from the user organization in how to use the KMS, how to capture knowledge as part of the work, and how to use the KMS in the normal course of business processes. The third of these subdimensions, IS KM service quality, centers on support from the IS organization in KMS tools, maintenance of the knowledge base, maps of databases, and reliability and availability of the KMS.

The D&M dimension intention to use/use in the J&O model becomes intent to use/perceived benefit. This dimension measures perceptions of the benefits of the KMS by the users (Jennex & Olfman, 2004). It reflects intention to use in that it concerns prediction of future usage behavior; it does not reflect use, which Jennex, Olfman, and their colleagues view as a different matter—in the J&O model, use is aligned with user satisfaction (see the following). The reflection of intention to use in the J&O model is extended in theoretical terms by incorporating perceived benefit, a concept originally advanced by Triandis (1980) and adapted to the IS context by Thompson, Higgins, and Howell (1991). This extension of the dimension allows it to reflect social and job-related characteristics of KMS user expectations that otherwise would not be captured (Jennex & Olfman, 2004).

The D&M dimension user satisfaction maps to use/user satisfaction in the J&O model. The J&O dimension combines use and user satisfaction because Jennex, Olfman, and their colleagues see the two concepts as complementary notions in the KM setting. In their view, when system use is optional, how much the system is used serves as a good indicator of success, and user satisfaction can be considered a complementary indicator. User satisfaction becomes a more useful indicator of success when system use is not optional. Beyond this, in situations in which a KMS is needed only occasionally (in situations where the absolute amount of usage is unimportant), employing use as a measure would underestimate KMS success; satisfaction provides a better indicator in that case.

The final D&M dimension, net benefits, corresponds to a J&O model dimension of the same name. The conceptualizations of this dimension are essentially the same in the two models.

Relationship between the J&O Model and the SEM Model

The SEM model tested in this study corresponds to the J&O model in most respects. Figure 3 depicts the dimensions in the SEM model and the scales—corresponding to subdimensions—used in this study. Figure 4 depicts the structural aspects of the SEM model.

The J&O model and the SEM model differ in two important ways. The first involves the elimination of feedback paths in the SEM model to allow its estimation as a recursive model. The second involves the limitation of certain theoretical content in the SEM model's dimensions to map them to the data available in this study.

According to Kline (1998), the statistical demands for SEM analysis are greatly simplified for recursive models—those in which all causal effects are unidirectional and all disturbances are mutually independent. The likelihood of a problem in the analysis of a nonrecursive

Figure 3. Indicators of SEM model dimensions

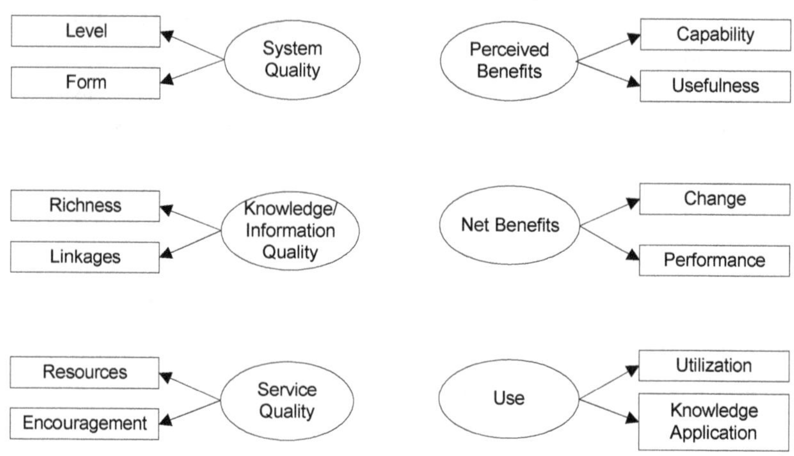

Figure 4. SEM structural model

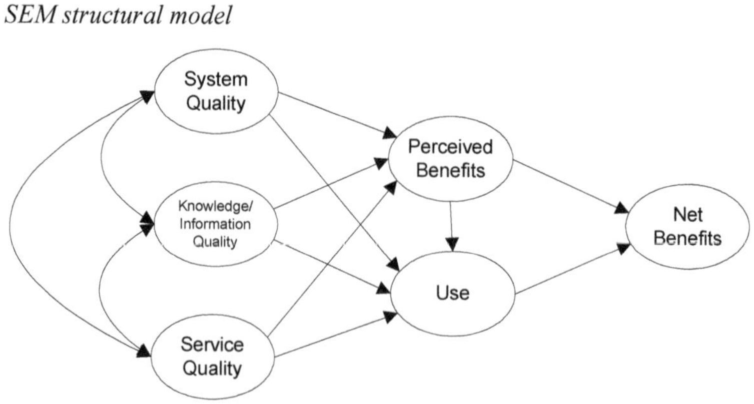

model is much greater than for a recursive model (Kline, 1998, p. 107). Formulating the SEM model as a recursive one (with one-way causal effects among endogenous variables but without disturbance correlations) guaranteed it would be identified (Kline, 1998).[1]

To convert the J&O model to a recursive form, three feedback paths were dropped: (1) from net benefits to intent to use/perceived benefit; (2) from net benefits to use/user satisfaction; and (3) from use/user satisfaction to intent to use/perceived benefit. The first two of these paths were viewed as being more appropriate for inclusion in a longitudinal study, which this study was not intended to be. The third path was viewed as less important theoretically than the path from intent to use/perceived benefit to use/user satisfaction. It was felt that perceptions of possible benefit influence system use more strongly than the other way around. As compromises to allow the testing of a recursive form of the SEM model, it was felt that these path deletions were reasonable.

The dimensions of the SEM model are limited in terms of how much of the conceptual content of the J&O model's dimensions they carry. The primary reason for this limitation is that the data used to test the SEM model were collected in an earlier study aimed at assessing individual learning in KMS situations (Liu, 2003). The data from Liu's study reflect most of the theoretical content of the J&O model's dimensions, but not all of it.[2] Where some theoretical content was not reflected in the indicators that Liu selected or created for

Table 1. Correspondence of dimensional theoretical content between J&O and SEM models

J&O Model Dimension	**J&O Model Subdimensions**	**SEM Model Dimension**	**SEM Model Subdimensions**
System Quality	Form	System Quality	Form
	Level		Level
	Technological Resources		<Missing>[3]
Knowledge/Information Quality	Linkages	Knowledge/ Information Quality	Linkages
	Richness		Richness
	Knowledge Strategy/ Process		<Missing>[5]
Service Quality	Management Support	Service Quality	Encouragement
	User KM Service Quality		Resources
	IS KM Service Quality		<Missing>[5]
Intent to Use/Perceived Benefit	Capability	Intent to Use/ Perceived Benefit	Capability
	Usefulness		Usefulness
Use/User Satisfaction	Utilization	Use/User Satisfaction	Utilization
	Knowledge Application		Knowledge Application
Net Benefits	Change	Net Benefits	Change
	Performance		Performance

Figure 5. Proposed SEM model (full)

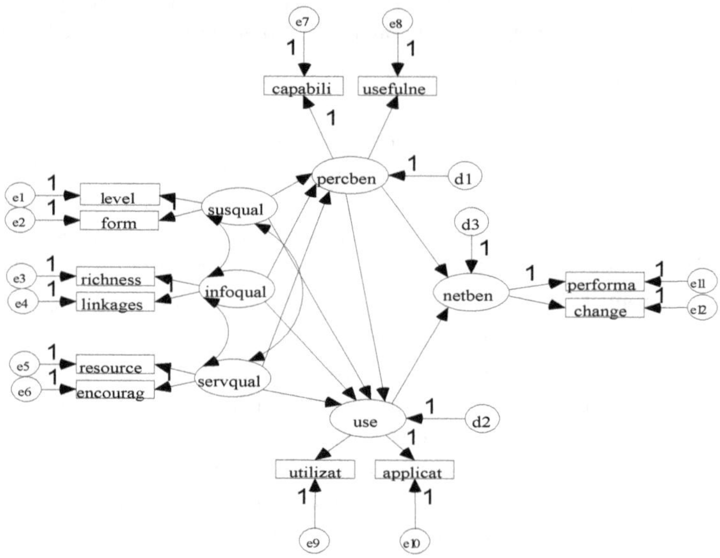

her study, at least two reasons were active. First, Liu did not feel such content to be relevant in understanding individual learning. Second, Liu had reference to an earlier version of the J&O model (Jennex & Olfman, 2002). Nonetheless, the indicators Liu used show enough correspondence to the theoretical content of the current (2004) J&O model to allow an SEM model reflecting it to be tested here. Table 1 summarizes the theoretical dimensions of the J&O model reflected in the SEM model. Note that 12 of the 15 subdimensions included in the J&O model are mapped to the SEM model.

It is prudent to be concerned that three of the 15 subdimensions of the J&O model (technical resources, knowledge strategy/process, and IS KM service quality) are not represented in the SEM model. On the other hand, the SEM model only employs the J&O model's subdimensions as indicators of its dimensions. Keeping in mind that any indicator reflects only imperfectly the theoretical ideas it represents, it was decided that the SEM model that could be specified with the available indicators was acceptable as a representation of the J&O model. Figure 5 depicts the modified SEM model in its full form.

Connecting the SEM Model with Type of System and Stakeholder

Seddon, Staples, Patnayakuni, and Bowtell (1999) assert that how one assesses information systems success should reflect the type of system and the system's stakeholders. They present a taxonomy of IS effectiveness measures organized by six types of systems and five types of stakeholders. For this study, a type of IT application (KMS) is considered, as it is used to benefit individual stakeholders (distinct KMS users).[4] These two focuses lead to a concentration on a benefit that any KMS might provide to any individual user. For purposes of this study, this benefit is individual learning, an outcome of KMS use that leads to individual better-offness (Seddon et al., 1999, p. 7). Individual learning is unquestionably important as a KMS outcome. Argyris and Schön (1996) argue "that individuals are the only subjects of learning" (p. 188), asserting that organizations learn only through the experiences and actions of individuals. While outcomes of KMS use other than individual learning might be considered, it appears to be one acceptable point for anchoring the Net Benefits dimension in terms of outcomes that would matter to individual stakeholders. The focus on individual learning, along with the emphasis on use by individual users, allows reconciliation of this study's investigation of the J&O model with Seddon et al.'s (1999) advice about assessing IS effectiveness.

Method

Liu (2003) gathered the data used to assess the SEM model tested in this study through a study of individual learning in a KM setting. Liu designed an online survey, using an early version of the J&O model (Jennex & Olfman, 2002) as a general guide. The survey included 54 items and was developed using, with some modification, the three-stage instrument development process proposed by Moore and Benbasat (1991). First, an initial version of the survey instrument was developed based on theory-grounded in operationalization of the constructs. Additionally, demographical items were included in the survey to capture information about gender, age, length of time with the organization and in current position, years using KMS, industry employed, job title and function, and the highest education attained. Published forms of items were used whenever possible, relying on work by Jennex and Olfman (2002), Dewitz (1996), Doll and Torkzadeh (1988), Bahra (2001), Gold, Malhotra, and Segars (2001), Thompson et al. (1991), King and Ko (2001), and Davis (1989), and constructing new items only when necessary. Second, based on solicited input from people with expertise in KMS and instrument development, the instrument was restructured and reworded to make it focused, brief, and clear (Alreck & Settle, 1995). Third, the instrument underwent a pilot study utilizing 56 KMS users from various firms to pretest the revised questionnaire, resulting in the final revision of the instrument.

This study uses data items from Liu's (2003) survey assembled into subdimension scales to serve as indicators for the SEM model depicted in Figure 3. The Appendix section details the items as they were worded in the survey and assembled into subdimension scales for this study. Respondents rated each item on a rating scale from *strongly agree* (5) to *strongly disagree* (1), although they had the option of rating any item as *not applicable*. Analysis of items and subdimension scales was done with SPSS; estimation of the SEM model was done with Amos, a package for SEM analysis.[5]

Individuals who, through a business firm or other organization, used a KMS for acquisition, organization, storage, or dissemination of knowledge, were invited to participate. This sampling procedure was purposive in nature; it was oriented toward obtaining as many survey responses as possible rather than sampling from a particular sampling frame.

Table 2. Scale descriptive statistics, reliability, and mv multicollinearity estimates (N=354)

Name	Mean	Std. Dev.	Skewness	Kurtosis	Alpha	R^2 to Test MV Multicollinearity
Level	2.14	0.58	1.06	1.67	0.76	0.31
Form	2.38	0.77	0.95	0.94	0.84	0.47
Richness	2.00	0.70	1.13	1.51	0.89	0.67
Linkages	2.33	0.79	1.16	1.79	0.79	0.56
Resources	2.30	0.78	0.97	1.77	0.52	0.71
Encouragement	2.33	0.84	1.00	1.38	0.82	0.63
Capability	1.67	0.60	0.87	0.83	0.79	0.47
Usefulness	1.63	0.61	1.34	2.11	0.73	0.45
Change	2.25	0.73	0.82	2.33	0.83	0.48
Performance	2.19	0.69	0.38	1.07	0.87	0.57
Utilization	1.87	0.89	1.63	3.07	0.85	0.77
Knowledge Application	2.21	0.70	0.137	0.50	0.75	0.73

Table 3. Scale correlations (2-tailed significance, N=354)

	Level	Form	Richness	Linkages	Resources	Encouragement
Level	1 .000	.681 .000	.699 .000	.680 .000	.513 .000	.418 .000
Form	.681 .000	1 .	.625 .000	.643 .000	.402 .000	.265 .000
Richness	.699 .000	.625 .000	1 .	.695 .000	.495 .000	.424 .000
Linkages	.680 .000	.643 .000	.695 .000	1 .	.520 .000	.426 .000
Resources	.513 .000	.402 .000	.495 .000	.520 .000	1 .	.576 .000
Encouragement	.418 .000	.265 .000	.424 .000	.426 .000	.576 .000	1 .
Capability	.491 .000	.361 .000	.617 .000	.448 .000	.385 .000	.282 .000
Usefulness	.543 .000	.397 .000	.556 .000	.506 .000	.374 .000	.474 .000
Change	.541 .000	.335 .000	.428 .000	.419 .000	.369 .000	.416 .000
Performance	.480 .000	.300 .000	.313 .000	.415 .000	.360 .000	.388 .000
Utilization	.296 .000	.138 .009	.359 .000	.180 .001	.272 .000	.360 .000
Knowledge App.	.292 .000	.151 .004	.170 .001	.204 .000	.220 .000	.275 .000

Table 3. continued

	Capability	Usefulness	Change	Performance	Utilization	Knowledge App.
Level	.491 .000	.543 .000	.541 .000	.480 .000	.296 .000	.292 .000
Form	.361 .000	.397 .000	.335 .000	.300 .000	.138 .009	.151 .004
Richness	.617 .000	.556 .000	.428 .000	.313 .000	.359 .000	.170 .001
Linkages	.448 .000	.506 .000	.419 .000	.415 .000	.180 .001	.204 .000
Resources	.385 .000	.374 .000	.369 .000	.360 .000	.272 .000	.220 .000
Encouragement	.282 .000	.474 .000	.416 .000	.388 .000	.360 .000	.275 .000
Capability	1 .	.522 .000	.351 .000	.356 .000	.378 .000	.136 .010
Usefulness	.522 .000	1 .	.575 .000	.614 .000	.388 .000	.356 .000
Change	.356 .000	.614 .000	1 .	.823 .000	.344 .000	.600 .000
Performance	.351 .000	.575 .000	.823 .000	1 .	.297 .000	.642 .000
Utilization	.378 000	.388 .000	.344 .000	.297 .000	1 .	.326 .000
Knowledge App.	.136 .010	.356 .000	.600 .000	.642 .000	.326 .000	1 .

Results

Three hundred sixty-nine people provided responses to Liu's (2003) survey. Nine cases were dropped due to noncompletion of the survey or nonuse of a KMS, leaving a total of 360 respondents. Most of these (52.8%) were from engineering or manufacturing organizations, most (61.9%) were male, and most (71.2%) were between 30 and 49 years of age. This analysis dropped an additional six cases due to one of more missing scale values, leaving a sample size of 354.

Scale scores were calculated as the averages of relevant item scores to serve as measured variables in the SEM model. Table 2 provides descriptive statistics for the scale scores as well as reliability coefficients for each scale and R^2 estimates for the regression of each scale score on the set of all the others (as a basis for judging multivariate multicollinearity). Table 3 provides a correlation matrix for the scale scores.

In order to avoid problems in SEM analysis, one must check the data for normality, outliers, and multicollinearity (Kline, 1998). The distributions of scale scores are roughly symmetrical, and estimates of skewness and kurtosis are not too large. There are no outliers, with no scale score as much as three standard deviations from its mean. There are no extremely large bivariate correlations, and none of the R^2 values for regressions of scale scores on the sets of all other scale scores exceeds 0.90, indicating no multivariate multicollinearity problems. The data, at least in these terms, seem to be adequate to conduct SEM analysis.

Byrne (2001) describes the core parameters of the SEM model (those that must be estimated typically) as including the regression coefficients for measurements and structure, the variances for errors and disturbances, and the factor variances and covariances. Based on these rules for counting parameters, the proposed model requires that 36 parameters be estimated. With 12 observed variables, there are 78 available data points. This implies that the SEM model is overidentified, having 43 degrees of freedom above what would have been a just identified model.[6]

As indicated by Nidumolu and Knotts (1998), sample size significantly influences statistical conclusion validity. Sample size requirements for SEM models are related to model complexity, but no definitive relationship exists between sample size and model complexity (Kline, 1998). One standard dictates that the sample size must be 50 observations more than eight times the number of variables (Garson, 2004); by this rule, the minimum sample size for this study would be 194 respondents. Another standard says that there should be 15 cases for every indicator (Stevens, 1996 [reported by Garson, 2004]); given this model has 12 indicators, the implication is that at least 180 respondents would be needed. Yet another standard advises that there should be 10 cases per parameter estimate (Kline, 1998), which means a sample size of no less than 360 would be required. Irrespective of the guideline followed, the achieved sample size, 354, can be considered adequate.

Evaluating the Proposed Model

Evaluation of an SEM model considers both the estimates of individual parameters and the overall fit of the model to the data (Byrne, 2001). According to Byrne, there are three aspects of individual parameters to consider: (1) all should be reasonable—no correlations larger than 1, no negative variances, and positive definite matrices of correlations and covariances); (2) estimates should be significant, having critical ratios greater than or equal to 1.96; and (3) standard errors should not be too large or too small, although no clear standards are available for what too large or too small would be. Table 4 presents values for individual parameter estimates and related statistics. In these terms, the proposed model can be considered to produce fairly reasonable individual parameters. The biggest problem noted with individual parameters is the occurrence of some low values for critical ratios, particularly for two of the structural regression coefficients. The estimates for the paths from service quality to perceived benefits and from system quality to perceived benefits have critical ratios of 0.601 and -0.875, respectively. This indicates that the values for these parameters cannot be distinguished with confidence from 0.

Three other regression parameter estimates have low critical ratios as well, but not so low as the ones just mentioned and probably within the range of acceptability.

Besides individual parameters, the overall fit of the model must be examined. It is common in reports of SEM analysis to present a variety of statistics that reflect different aspects of overall model fit.

Table 4. Individual parameter estimates and related statistics for proposed SEM model

			Estimate	S.E.	C.R.	P
percben	<---	servqual	.050	.083	.601	.548
percben	<---	infoqual	.698	.304	2.294	.022
percben	<---	sysqual	-.250	.286	-.875	.382
use	<---	servqual	.400	.226	1.770	.077
use	<---	infoqual	-2.296	1.353	-1.697	.090
use	<---	sysqual	1.631	1.037	1.573	.116
use	<---	percben	1.720	.735	2.340	.019
netben	<---	percben	.482	.130	3.696	***
netben	<---	use	.893	.138	6.459	***
form	<---	sysqual	1.000			
level	<---	sysqual	.872	.052	16.759	***
linkages	<---	infoqual	1.000			
richness	<---	infoqual	.926	.050	18.447	***
encourag	<---	servqual	1.000			
resource	<---	servqual	1.017	.091	11.172	***
applicat	<---	use	1.000			
utilizat	<---	use	.662	.100	6.632	***
capabili	<---	percben	1.000			
usefulne	<---	percben	1.256	.104	12.039	***
performa	<---	netben	1.000			
change	<---	netben	1.030	.044	23.433	***

Kline (1998) describes a variety of indicators of overall model fit. He asserts that a minimum set of these indicators should be reported, including "the X^2 statistic and its degrees of freedom and significance level; an index that describes the overall proportion of explained variance; an index that adjusts the proportion of explained variance for model complexity; and an index based on the standardized residuals" (p. 130).

Kline (1998) cautions that researchers should bear in mind the limitations of fit indexes: (1) they are indicative of average fit; (2) they are not indicative of theoretical meaning; and (3) they are not indicative of a model's predictive power. Table 5 presents overall model fit indexes for the proposed SEM model.

In order for a model to have a fair level of fit to data, according to Kline (1998), "Low and non-significant values of the X^2 index are desired" (p. 128). Because the X^2 index is sensitive to sample size, researchers sometimes employ X^2/df. A significant X^2 value means "an unconstrained model fits the covariance/correlation matrix as well as the given model" (Garson, 2004); a nonsignificant value suggests that the fit of the data to the model is adequate. The X^2 statistic calculated for the proposed model is significant (CMIN = 253.3, df = 42, p=.000), which suggests that the fit of the model is not entirely adequate. On the other hand, according to Garson (2004), "many researchers who use SEM believe that with a reasonable sample size (ex. > 200) and good approximate fit as indicated by other fit tests (ex., NNFI, CFI, RMSEA, ...), the significance of the chi-square test may be discounted" (p. 11).

The GFI (goodness of fit index) reflects the degree to which the observed covariances are explained by the covariances implied by the proposed model. The standard for GFI values to indicate a good fit is values greater than or equal to .90 (Garson, 2004). The GFI value achieved for the proposed model is .894. Although this is below the conventional cutoff value, GFI values are biased downward at times, such as when the number of degrees of freedom is large relative to the sample size and when the number of parameters is not large. Garson (2004) reports Steiger's suggestion to use an adjusted GFI to account for GFI's downward bias. The adjusted GFI for this study, calculated with the formula Garson provides, is .980.

The CFI (comparative fit index) contrasts the fit of the proposed model with that of a model that assumes no correlation among the latent variables (Garson, 2004). Values above .90 indicate a good fit of the model to the data. The value of CFI for this study is .913.

Table 5. Overall model fit indexes for the proposed SEM model

Model	NPAR	CMIN	DF	P	CMIN/DF
Default model	36	253.327	42	.000	6.032
Saturated model	78	.000	0		
Independence model	12	2508.159	66	.000	38.002

GFI	CFI	IFI	NFI	NNFI (TLI)	RMR	RMSEA
.894	.913	.914	.899	.864	.041	.119

The IFI (incremental fit index) "should be equal to or greater than .90 to accept the model" (Garson, 2004). The IFI value obtained in this research is .914.

The NNFI (non-normed fit index) is also known as the TLI (tucker-lewis index). It expresses, in a manner adjusted for model complexity, how much the proposed model improves fit, compared with a null model—one having random variables. Garson (2004) reports several guidelines for judging goodness of fit using the NNFI, with the most lenient being values greater than or equal to .80, and the most strict being values greater than or equal to .95. The value of NNFI achieved for the proposed model is .864.

The RMR (root mean square residual) is an index that indicates good fits with small values—the closer to 0, the better. According to Kline (1998), "in a well-fitting model this value will be small, say, .05 or less" (p. 82). This index represents the average of residual differences between the variances and covariances observed and those hypothesized. In this study, RMR had a value of .041.

The RMSEA (root mean square error of approximation) takes into account the error of approximation in the population. RMSEA values over .10 indicate poor fit (Byrne, 2001). The value achieved for the proposed model is .119.

Across the set of indicators, the proposed model shows some evidence of having an acceptable fit to the data (in terms of the Adjusted GFI, CFI, IFI, NFI, NNFI/TLI, and RMR), and it shows some evidence of unacceptable fit (in terms of the X^2 and RMSEA).

Discussion

The J&O model of KM success received fair support from the results presented previously. Although the data used were collected in an earlier study with different research aims, were concerned with the intersection of KMS use and individual learning (Liu, 2003), and hence, weren't explicitly intended to serve for testing the J&O model, the fit of the proposed SEM model to the data can be characterized as adequate, if not particularly good.

To the extent that the J&O model is more credible as a whole in light of these findings, some implications of the research merit additional attention. First, the relationships involving perceived benefit, use, and net benefits in the J&O model can be treated as more plausible. The regression coefficients corresponding to these relationships were significant and substantial. These findings support the theoretical relationships, flowing through the J&O model from the D&M model, that higher levels of perceived benefit associated with a KMS leads to higher levels of use—users make use of the system when they perceive benefits from doing so.

Second, the covariance relationships involving system quality, knowledge/information quality, and service quality were confirmed as well. The coefficients calculated for these relationships in the model were all sizeable but not too large. This finding supports the ideas from the J&O (and D&M) model that the three KMS (IS) quality factors are interrelated but distinct qualities.

Third, the relationships involving the effects of system quality, knowledge/information quality, and service quality on perceived benefits and use were not confirmed consistently. The calculated coefficients—six in all—showed a decidedly mixed pattern of significance: two of the calculated coefficients should not be viewed as significant, three should be taken as marginally significant, and one should be considered significant. The significant coefficient for the influence of knowledge/information quality on perceived benefit had a value of .698 (p = .02). This estimate confirms the notion that an increase in the amount of knowledge that a KMS provides leads to an increase in the amount by which individuals view the KMS as providing benefit. As such, this is good news for the J&O model. The marginally significant estimates provide news of a more mixed nature. Two of these—from system quality to use (1.631, p = .12) and from service quality to use (.400, p = .08)—provide the suggestion of support to the J&O model, but the other—from knowledge/information quality to use (-2.296, p = .09)—is in the opposite direction suggested by the J&O model. The nonsignificant estimates—from system quality to perceived benefits (-.250, p = .382) and from service quality to perceived benefits (.050, p = .548)—are not supportive of the J&O model.

What to make of these estimates as a group is somewhat puzzling. While together they do not overwhelmingly support the J&O model, neither do they disconfirm it. Rather, one should conclude from these findings that there is now enough empirical support for the J&O model to justify additional efforts to confirm and refine it. The following section contains suggestions for how such research might be done effectively.

To provide a convincing test of the J&O model, better data will be needed. In order for the data employed in this study to have been completely acceptable, several changes would have been needed. Most important of these changes would have been the inclusion of omitted scales. The data Liu (2003) collected did not include items that could serve to represent several subdimensions of the J&O model, including technological resources, knowledge strategy/process, IS KM service quality, and user KM service quality. While other data from the Liu study allowed a partial coverage of the conceptual content of the system quality, knowledge/information quality, and service quality dimensions, it is likely that the theoretical underrepresentation of the J&O model in the data used made the test conducted here less precise than it might have been. Future research attempting to assess the J&O model should be sure to represent all subdimensions.

A second change in the data that likely would have improved the fit to the proposed model would have involved additional refinement of the scales employed. Since the Liu (2003) data were not collected explicitly to represent the subdimensions of the J&O model, they do not provide as many items for each subdimension as would be desirable, nor do they obviously represent the constructs related to these subdimensions in any certain fashion. Future research would benefit from instrument development and validation efforts targeted explicitly to the testing of the J&O model's conceptualizations of dimensions and subdimensions of KMS success.

A third change in the data that would have improved the fit to the proposed model would have involved a different sampling strategy. The Liu (2003) sampling strategy, which amounted to snowball sampling (Atkinson & Flint, 2001), did not assure that all respondents had interacted with similar KMSs. Neither did it employ random selection of participants from a well-defined sampling frame. If future research can identify a group of potential respondents that employs information systems that are similar in their adherence to some definition of a KMS, then random selection of individuals from this group probably would improve the research's chances of reducing the level of extraneous variance. This should allow better estimates to be derived. Future research should strive to attain a random sample of users of a known type of KMS.

A fourth change in the data that would have improved fit would have been to recruit a larger sample size for the study. The sample Liu (2003) collected (N = 354), although not small, was certainly no larger than what the analysis minimally required. If a future study could attract a much larger group of respondents, the chances of calculating better estimates would improve. It also would make it possible to retest the model in the form it was tested here and then to test respecifications of it that might be suggested by such retests. One of the virtues of an SEM approach to research is that, given sufficient sample size, researchers can identify opportunities for model improvement with one subsample and then attempt to confirm such improvements with another subsample. The current study had insufficient data to take on this task, but replications might have an adequate number of respondents to do so. Future research should strive to attract enough respondents to allow the testing of multiple versions of the model.

Despite the need for future research to be conducted somewhat differently in order to foster progress in confirming and modifying the J&O model of knowledge management success, the current research provides some support for the model, certainly enough to prompt continued investigation. Additional work to develop this model will result, it is hoped, in an improved version that will provide researchers and practitioners with a sound explanation of success in knowledge management.

References

Alavi, M., & Leidner, D. E. (2001). Review: Knowledge management and knowledge management systems: Conceptual foundations and research issues. *MIS Quarterly, 25*(1), 107-136.

Alreck, P. L., & Settle, R. B. (1995). *The survey research handbook: Guidelines and strategies for conducting a survey* (2nd ed.). New York: McGraw-Hill.

Argyris, C., & Schön, D.A. (1996). Organizational learning II—Theory, method, and practice. Reading, MA: Addison-Wesley.

Atkinson, R., & Flint, J. (2001). Accessing hidden and hard-to-reach populations: Snowball research strategies. *Social Research Update, 33*. Retrieved September 30, 2004, from http://www.soc.surrey.ac.uk/sru/SRU33.html

Bahra, N. (2001). *Competitive knowledge management*. New York: Palgrave.

Byrne, B. M. (2001). *Structural equation modeling with AMOS: Basic concepts, applications, and programming*. Mahwah, NJ: Lawrence Erlbaum Associates.

Davis, F. D. (1989). Perceived usefulness, perceived ease of use, and user acceptance of information technology. *MIS Quarterly, 13*(3), 319-339.

DeLone, W. H., & McLean, E. R. (1992). Information system success: The quest for the dependent variable. *Information Systems Research, 3*(1), 60-95.

DeLone, W. H., & McLean, E. R. (2002). Information systems success revisited. In *Proceedings of the Thirty-Fifth Annual Hawaii International Conference on System Sciences*. IEEE Computer Society.

DeLone, W. H., & McLean, E. R. (2003). The DeLone and McLean model of information systems success: A ten-year update. *Journal of Management Information Systems, 19*(4), 9-30.

Dewitz, S. D. (1996). *Systems analysis and design and the transition to objects*. New York: McGraw-Hill.

Doll, W. J., & Torkzadeh, G. (1988). The measurement of end-user computing satisfaction. *MIS Quarterly, 12*(2), 259-274.

Garson, G. D. (2004). Quantitative research in public administration. Retrieved October 1, 2004, from http://www2.chass.ncsu.edu/garson/pa765

Gold, A. H., Malhotra, A., & Segars, A. H. (2001). Knowledge management: An organizational capabilities perspective. *Journal of Management Information Systems, 18*(1), 185-214.

Hansen, M. T., Nohria, N., & Tierney, T. (1999). What's your strategy for managing knowledge? *Harvard Business Review, 77*(2), 106-116.

ISWorld. (2004a). *Information system effectiveness: IS effectiveness.* Retrieved September 29, 2004, from http://business.clemson.edu/ISE/html/is_effectiveness.html

ISWorld. (2004b). *Information system effectiveness: Home.* Retrieved September 29, 2004, from http://business.clemson.edu/ISE

Jennex, M. E., & Olfman, L. (2002). Organizational memory/knowledge effects on productivity, a longitudinal study. In *Proceedings of the Thirty-Fifth Annual Hawaii International Conference on System Sciences.* IEEE Computer Society.

Jennex, M. E., & Olfman, L. (2004). Modeling knowledge management success. *Proceedings of the Conference on Information Science and Technology Management, CISTM.*

Jennex, M. E., Olfman, L., & Addo, T. B. A. (2003). The need for an organizational knowledge management strategy. In *Proceedings of the Thirty-Sixth Annual Hawaii International Conference on System Sciences.* IEEE Computer Society.

Jennex, M. E., Olfman, L., Pituma, P., & Park, Y. T. (1998). An organizational memory information systems success model: An extension of DeLone and McLean's IS success model. In *Proceedings of the Thirty-First Annual Hawaii International Conference on System Sciences.* IEEE Computer Society.

King, W. R., & Ko, D. G. (2001). Evaluating knowledge management and the learning organization. *Communications of the AIS, 5*(14), 1-26.

Kline, R. B. (1998). *Principles and practice of structural equation modeling*. New York: Guilford.

Liu, S.-C. (2003). *A study of the factors the facilitate use of knowledge management systems and the impact of use on individual learning* [doctoral dissertation]. Claremont, CA: Claremont Graduate University.

Moore, G. C., & Benbasat, I. (1991). Development of an instrument to measure the perceptions of adopting an IT innovation. *Information Systems Research, 2*(3), 192-222.

Nidumolu, S. R., & Knotts, G. W. (1998). The effects of customizability and reusability on perceived process and competitive performance of software firms. *MIS Quarterly, 22*(2), 105-137.

Seddon, P. B., Staples, S., Patnayakuni, R., & Bowtell, M. (1999). Dimensions of information systems success. *Communications of the AIS, 2*(20).

Stein, E. W., & Zwass, V. (1995). Actualizing organizational memory with information systems. *Information Systems Research, 6*(2), 85-117.

Thompson, R. L., Higgins, C. A., & Howell, J. M. (1991). Personal computing: Toward a conceptual model of utilization. *MIS Quarterly, 15*(1), 125-143.

Triandis, H. C. (1980). *Beliefs, attitudes, and values*. University of Nebraska Press.

Turban, E., & Aronson, J. E. (2001). *Decision support systems and intelligent systems* (6th ed.). Prentice Hall.

Endnotes

1 If a model is not identified, it is not theoretically possible to calculate unique estimates of its parameters.

2 The data collected in the Liu (2003) model are discussed below when the survey that collected them is described.

3 No data included in Liu (2003).

4 Seddon et al. (1999) found that the combination of type of IT application and individual stakeholder was the second most common of the 30 possible combinations in their taxonomy in terms of its appearance in an analysis they performed of 186 studies in three journals over a nine-year period.

5 Information about Amos can be found at Assessment Systems Corporation (http://www.assess.com/frmSoftCat.htm).

6 To be able to test an SEM model, it must be overidentified.

Appendix

Model Dimensions, Scales, and Items Used in Study

System Quality Level

1. **Completeness of Search.** Your KS allows you to do both information and people searches.
2. **Effectiveness—Knowledge Base.** Whenever you search the KS knowledge base and/or yellow pages, the retrieved knowledge is always what you need.

3. **Effectiveness—Linkage.** Whenever you search the KS knowledge base and/or yellow pages, the returned linkage always directs you to the right person.
4. **Speed of Retrieval.** Whenever you search the KS knowledge base and/or yellow pages, the retrieved results normally display quickly.
5. **Ease of Search.** Your KS search function is easy to use.
6. **Reliability.** Your KS is not subject to frequent problems and crashes.

Form

1. **Computerization.** Your KS allows you to find most of the organizational information/knowledge online.
2. **Integration.** Whenever you search the KS, you don't need to try different ways to locate the needed information.
3. **Integration.** Whenever you search the KS, you don't need to try different ways to locate the right person.
4. **Integration.** Whenever you search the KS, you don't need to access more than one system to locate the needed information.
5. **Integration.** Whenever you search the KS, you don't need to access more than one system to locate the right person.

Information Quality Richness

1. **Relevance.** Your KS provides information/knowledge that is exactly what you need.
2. **Understandability.** Your KS provides information/knowledge that uses recognized vocabulary rather than highly specialized terminology.
3. **Adequacy.** Your KS provides information/knowledge that is adequate for you to complete tasks.
4. **Contextuality.** Your KS provides contextual information/knowledge so that you truly can understand what is being accessed.
5. **Contextuality.** Your KS provides contextual information/knowledge so that you can easily apply it to your work.
6. **Currency.** Your KS provides up-to-date information/knowledge.

Linkages

1. **Completeness of Linkage.** The knowledge portal of your KS links you to a complete collection of documents and data.

2. **Accuracy of Linkage.** The yellow pages of your KS guide you to connect to the people with the know-how that you are seeking.
3. **Currency of Linkage.** Your organization keeps updating its knowledge portal so that you have access to current documents and data.
4. **Currency of Linkage.** Your organization keeps updating its yellow pages so that you can locate newly hired or acquired expertise without a problem.

Service Quality Resources

1. **Technical Support.** Whenever you have difficulties with your KS, there is a specific person (or group) available to help you.
2. **Allow Sufficient Time for Dialogue.** You have sufficient time to dialogue online with your co-workers about important problems and solutions.

Encouragement

1. **Encouragement From Peers.** You are encouraged to do online exploration and experimentation by your peers.
2. **Encouragement From Supervisor.** You are encouraged to do online exploration and experimentation by your supervisor.
3. **Endorse Knowledge Sharing.** Your organization actively endorses knowledge sharing.
4. **Encourage Online Discussion.** Your organization encourages online discussion of new ideas and working methods.

Perceived Benefits Capability

1. **Self-Efficacy.** You can use your KS without needing someone's help.
2. **Cognitive Capability.** You find it easy to understand the information/knowledge you found in the knowledge base.
3. **Cognitive Capability.** You find it easy to use the information/knowledge you found in the knowledge base.

Usefulness

1. **Willingness to Search.** When job-related problems occur, you are willing to do an online search of your KS for solutions.

2. **Tendency to Analyze.** You analyze and interpret what is brought to your attention in your KS.
3. **Perceived Usefulness.** You find your KS useful in your job.

Use Utilization

1. **Distribution.** Your KS helps your daily work by distributing customized knowledge to you.
2. **Distribution.** Your KS helps your daily work by distributing customized knowledge to others.

Knowledge Application

1. **Decision Making and Problem Solving.** You use knowledge from the KS to perform decision-making and problem-solving tasks.
2. **Questioning Rules and Routines.** You use knowledge from the KS to question existing rules and routines.
3. **Exploring Alternatives.** You use knowledge from the KS to search for and explore alternatives.

Net Benefits Change

1. **Cognitive Change.** Your KS helps you to detect work-related problems.
2. **Cognitive Change.** Your KS enlightens you to new ways of thinking.
3. **Behavioral Change.** Your KS changes the way you do things in a way beneficial to the organization's overall interest.

Performance

1. **Better Decisions.** Your KS improves the decisions you make.
2. **Fewer Mistakes.** Your KS helps you to make fewer mistakes.
3. **Better Experience Transfer/Knowledge Reuse.** Your KS allows better experience transfer and knowledge reuse.
4. **Reduce Duplicate Work.** Your KS reduces duplicate work.
5. **Better Cycle Time.** Your KS allows you faster cycle time to problem resolution.

Chapter XIII

Knowledge Management Information Technology User Acceptance:

Assessing the Applicability of the Technology Acceptance Model

William Money, The George Washington University, USA

Arch Turner, The George Washington University, USA

Abstract

This chapter presents the results of a study investigating the applicability of Davis' technology acceptance model (TAM) to user acceptance of a knowledge management system (KMS) in a modern organizational environment. The objective of the study was to expand empirical research of two important and complex research questions: (1) What are the important factors, conditions, and mechanisms that affect people's acceptance and usage of collaborative and interdependent KMS in the modern organizational environment? and (2) How applicable is the TAM and the substantial body of information technology (IT) research around this model to user acceptance and usage of a KMS in a modern organizational environment in which collaboration, knowledge sharing, and role-based system usage is necessary in order for the organization to function competitively? The study provided preliminary evi-

dence suggesting that previous TAM research may serve as a foundation for research of KMS user acceptance. Relationships among primary TAM constructs found in this study were in substantive agreement with those of previous research. These findings are relevant and significant because they suggest that the considerable body of previous TAM-related IT research may be applied usefully to the knowledge management (KM) domain in which interdependent social processes that require knowledge creation, storage and retrieval, transfer, and application are required for effective organizational functioning.

Introduction

Although business investment in IT has declined somewhat in recent years, firms around the world still spend more than $2 trillion a year on IT (Carr, 2003). It is also estimated that IT investment comprises approximately 50% of U.S. business capital investment, making it the top capital investment area for American businesses (Carr, 2003). With these continuing enormous business resource investments, understanding and creating conditions under which IT will be accepted and used in the organization remains a high priority within the IT research community (Venkatesh & Davis, 2000). Understanding why individuals accept or reject IT systems has proved to be one of the most challenging issues in information systems research (Doll, Hendrickson, & Xiandong, 1998). User acceptance of IT—a phenomenon that is not yet well-understood—and usage are widely considered to be crucial factors in the ultimate determination of information system success, since information systems that are not used are of little value (Mathieson, Peacock, & Chin, 2001). Nevertheless, as will be discussed later, system usage alone may not be entirely representative of KMS organizational benefits.

A preponderance of research and accumulated knowledge of the factors affecting IT acceptance has as its foundation the technology acceptance model (TAM). TAM was conceived originally by Fred Davis in 1986 and is an intentions-based model derived from the Theory of Reasoned Action (TRA) tailored to meet the needs of information technology research (Davis, Bagozzi, & Warshaw, 1989). Since its inception, TAM has enjoyed growing acceptance and has proved to be a reasonably accurate predictor of both users' intentions to use IT and of IT usage (Ma & Liu, 2004). Evidence of the research community's growing acceptance of TAM is reflected in the fact that the Institute for Scientific Information Social Science Citation Index recently (January 2006) listed more than 1,150 journal citations of the initial TAM research papers published by Davis (1989) (628 citations) and Davis et al. (1989) (531 citations).

A second related topic of considerable interest in the business world is the multifaceted concept widely referred to as knowledge management (KM). KM can be defined broadly as the set of systematic and disciplined actions that an organization can take to obtain the greatest value from the knowledge available to it (Marwick, 2001) and/or to efforts aimed at "identifying and leveraging the collective knowledge in an organization to help the organization compete" (Alavi & Leidner, 2001, p. 113). KM rapidly is becoming a critical integral business function as organizations increasingly realize that their competitiveness in the intensely competitive global marketplace hinges on effective management of intellectual resources (Davenport & Grover, 2001). Increased interest and investment in KM can be attributed to the growing recognition that one of a firm's most unique and inimitable resources is the intellectual capabilities of its workers.

Reflecting this interest, recent literature is replete with research of a wide range of important issues related to the question of how organizations can best capitalize on their knowledge resources, develop processes to support KM, and broadly integrate KMS into organizational functioning. A cursory sampling of key KM issues reported recently include KM and new organizational structures (Malhotra, 2000), assessments of KM organizational capability prerequisites (Gold, Malhotra, & Segars, 2001), KM strategies and taxonomies (Earl, 2001; Zack, 1999), the relative importance of various knowledge types (Lam, 2000), general discussions of KM benefits and challenges (Alavi & Leidner, 1999), the mapping of organizational knowledge (Vaill, 1999), the integration of information technology to enhance organizational KM (Bourdreau & Couillard, 1999), and the development of a KM research agenda (Grover & Davenport, 2001). Grant (1996) and Spender (1996) elevate the KM agenda further through their discussions of a knowledge-based theory of the firm.

An important topic in the evolving scholarly KM discourse is empirical research of factors, conditions, and mechanisms affecting individual acceptance and usage of IT implemented as a KMS. In their survey of KMS conceptual foundations and research issues, Alavi and Leidner (2001) observed that while "the majority of knowledge management initiatives involve at least in part, if not to a significant degree, information technology … little research exists in the design, use, or success of systems to support knowledge management" (p. 115). Since then, measurable progress has been made. Ericsson and Avdic (2003) provided insightful ideas and directions relative to KM enablers and facilitating conditions. Jennex and Olfman (2002, 2003) have performed empirical KMS research based on the Information System Success model of Delone and Mc Lean (1992). Their research incorporates the intention to use construct to help to predict voluntary KMS usage and revealed that the perceived benefit model was useful for predicting continued use of a KMS in an engineering organization. Jennex and Olfman (2004) assessed KMS success factors and proposed a theory-based KMS Success Model that was shown to be a useful framework for assessing KMS success models.

Nevertheless, there remains a need for empirical field research into factors affecting KMS acceptance and usage. Evidence of this need can be found in Legris, Ingham, and Collerette (2002), who synthesized findings of 22 scholarly IT acceptance studies. None of the studies considered in this meta-analysis addressed KMS acceptance. Furthermore, the majority was situated in educational settings that bear little resemblance to the modern organizational environment. Indeed, a survey of IT acceptance literature indicates that most findings are drawn from analysis of individual usage IT and, for the most part, are situated in settings not representative of the modern organizational environment. The importance of the distinction between the IT acceptance research that has been done to date and that which is asserted to be needed around KMSs is reflected by Alavi and Leidner (2001) who observed:

> *[K]nowledge management consists of a dynamic and continuous set of processes and practices embedded in individuals, as well as in groups and physical struc-*

> *tures. At any point in time and in any part of a given organization, individuals and groups may be engaged in several different aspects and processes of knowledge management. Thus, knowledge management is not a discrete, independent, and monolithic organizational phenomenon.* (p. 123)

There is a consensus among organizational scholars and practitioners regarding the importance to organizations of leveraging their unique human intellectual resources. Additionally, in spite of a substantial research effort focused on IT acceptance and usage, numerous questions remain. As noted by Legris et al. (2002) and Kankanhalli, Tan, and Wei (2001), IT implementation in general, and KMS implementation in particular, continue to experience difficult challenges in the organizational environment. Due to the widely acknowledged social nature of successful organizational KM, KMS acceptance and usage represents a more complex phenomenon than individual acceptance and use of IT. It may involve a mix of voluntary and mandatory behaviors, integrated organization processes, and complex social forces and relationships. KMS acceptance research can be seen as the intersection of these two important areas and warrants increased attention.

Knowledge Management Systems

KM is widely acknowledged to encompass a diverse mix of complex and dynamic components. Furthermore, scholar and practitioner admonitions to resist the temptation to research exclusively technology matters at the expense of complex KM social and behavioral issues seem ubiquitous (Alavi & Leidner, 2001). Nevertheless, modern IT is unquestionably critical to current organizational KM. The ability of modern IT to synchronously or asynchronously span previously insurmountable organizational, time, and geographic barriers is a critical enabler that must be viewed as a catalyst for increased KM interest. Thus, it is not surprising that modern information technology has been the center of gravity for most enterprise knowledge management initiatives (Alavi & Leidner, 1999; Grover & Davenport, 2001). As Taylor (2003) and Kankanhalli et al. (2001) observe collectively, modern IT can be viewed as a virtually necessary, albeit not sufficient, component of successful organizational KM.

Information-related technologies that support knowledge management include collaboration and community of interest/practice support technologies, structured and unstructured data indexing, categorization, taxonomy-producing tools, common databases, data warehousing technology, search and retrieval, and document management tools, to name just a few.

As with any IT implementation, the success of a KMS begins inevitably with individual acceptance. This research attempts to expand our understanding of the linkages between two important IT research topics: user IT acceptance and organizational KM. Davis' TAM is used as a framework to investigate the implementation of a KMS within an organizational unit of a large private consulting and technical services firm. TAM was selected for this research due to its broad and seemingly growing adoption among IT researchers, the well-established reliability and validity of its constructs, and the realization that the model had not been applied to KMS acceptance.

The firm studied has a global presence and is involved in a broad range of high technology product and service business areas. The specific organizational unit studied provides highly technical command and control, communications, computer, intelligence, surveillance, reconnaissance project management, and related consulting services to predominantly U.S. Department of Defense clients. The vast majority of the members of the organizational unit studied is well-educated professionals. They work in a highly competitive business environment, and the work they perform can be described accurately as knowledge work. The KMS studied is a Web-based document repository and management tool intended primarily to support three organizational objectives: (1) improvement of internal software development processes to achieve Software Engineering Institute Level 2 accreditation; (2) enhanced diffusion of internal research and development (IR&D) products throughout the organization by providing employees better access to IR&D products; and (3) enhanced business process and employee professional development by providing convenient electronic access to current and past project information and documentation.

Research Questions

Given the broad reach of TAM IT acceptance research and the potential benefit of improved understanding of KMS acceptance, this research focused on questions critical to KM acceptance using TAM. It was intended to develop preliminary answers to the following two basic questions:

1. What are the important factors, conditions, and mechanisms that affect people's acceptance and usage of a KMS in an organizational environment in which collaboration, knowledge sharing, and role-based system usage is necessary in order for the organization to compete?
2. How well does TAM substantiate the predicted relationships and mechanisms relative to user acceptance and usage of a KMS in a modern organizational environment in which collaboration, knowledge sharing, and role-based system usage is necessary in order for the organization to compete?

Theoretical Background and Discussion of the Technology Acceptance Model

Davis developed the TAM to provide an explanation of the determinants of computer acceptance that is general, capable of explaining user behavior across a broad range of end-user computing technologies and user populations, while also being both parsimonious and theoretically justified (Davis, 1989). TAM cuts a wide theoretical swath that includes the adoption of innovations, the cost-benefit paradigm, expectancy theory, and self-efficacy

Figure 1. Technology acceptance model (Davis, 1989)

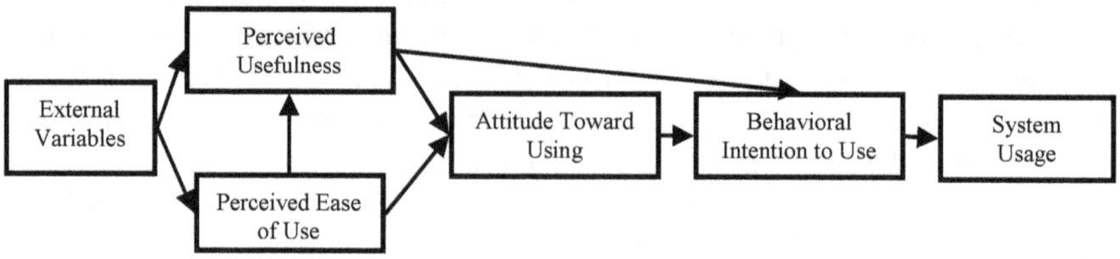

Figure 2. Theory of reasoned action (Fishbein & Ajzen, 1975)

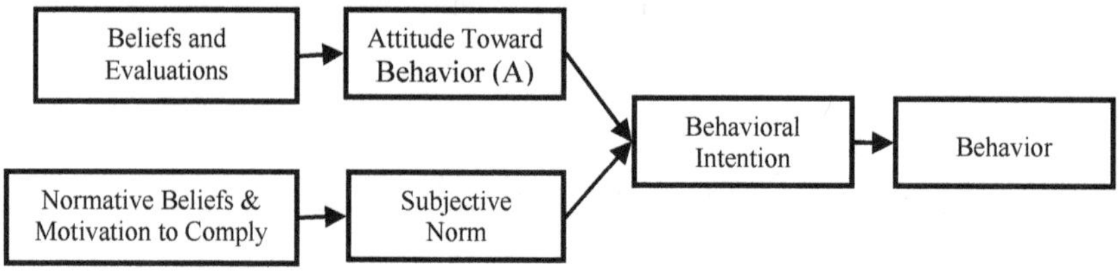

theory (Davis, 1989). Davis' original technology acceptance model is shown in Figure 1.

TAM is a derivative of the theory of reasoned action (TRA) model developed by Fishbein and Ajzen (1975). TRA focuses on situation-specific combinations of personal beliefs and attitudes and the effects of the beliefs of others close to the individual (Szajna, 1996). The fundamental premise of TRA is that individuals will adopt a specific behavior if they perceive it will lead to positive outcomes (Compeau & Higgins, 1995). TAM is a TRA derivative tailored to the study of a broader range of user behavior in the context of IT acceptance (Davis, 1989). TAM includes two primary belief constructs hypothesized by Davis to affect a potential user's attitude and intention to use an information technology: perceived usefulness (the degree to which a person believes that using a particular system would enhance his or her job performance), and perceived ease of use (the degree to which a person believes that using a particular system would be free of effort) (Davis et. al., 1989). Perceived usefulness and perceived ease of use and other TAM constructs relevant to this research will be discussed next briefly.

Perceived Usefulness. In developing TAM, Davis theorized that an individual's perception of usefulness would influence intention to use the technology primarily through the creation of a positive attitude. This was consistent with TAM's theoretical precursor, the TRA, which held that attitude (an individual's positive or negative feelings [evaluative affect] about performing a target behavior) mediated the effects of beliefs (and subjective norm) on behavioral intention.

In a departure from the TRA, a direct effect of perceived usefulness on intention to use was also included in Davis' original TAM. Davis rationalized this effect by theorizing that users may decide to accept and use a technology, regardless of their affective attitude toward it, if they have a sufficiently strong perception that it will contribute positively to their job performance (Davis et al., 1989).

Given the importance of perceived usefulness to the formation of a positive intention to use an IT, subsequent research has been directed to developing an understanding of the antecedents of usefulness. Venkatesh and Davis (2000) examined job relevance, output quality, and result demonstrability as antecedents to perceived usefulness. All three constructs were related positively to perceptions of information system usefulness.

Perceived Ease of Use. TAM's second principal belief construct, perceived ease of use, reflects an individual's assessment of the ease of use and ease of learning of a given system (Gefen & Straub, 2000). Davis (1989) and Davis et al. (1989) argued for the inclusion of ease of use as a separate belief construct based largely on the concept of self-efficacy (an individual's judgment of his or her ability to organize and execute tasks necessary to perform a behavior). They also cited factor analyses showing that usefulness and ease of use are statistically distinct constructs.

Venkatesh (2000) noted that a "vast body of research in behavioral decision making and IS demonstrate that individuals attempt to minimize efforts in their behaviors, thus supporting a relationship between perceived ease of use and usage behavior, albeit through intention as suggested by TAM," and that "other theoretical perspectives studying user acceptance have also employed similar constructs" (p. 344).

The relationship between perceived ease of use and perceived usefulness has proved complex. There is a lack of consensus in the literature regarding how perceived ease of use affects attitudes and/or intentions to use IT. In the original TAM, Davis hypothesized perceived ease of use to affect both attitude and usefulness directly. As already noted, Davis et al. (1989) suggested elimination of the attitude construct, postulating a direct relationship between perceived ease of use and intention to use rather than an indirect one through attitude. Finding that when the effects of usefulness are controlled, the effect of ease of use all but vanishes, Davis (1989), suggested that ease of use might be an antecedent to usefulness rather than a direct determinant of intention and usage. Subsequent research hypothesized dual TAM constructs: one for scenarios in which potential users were briefly introduced to a system and another for scenarios in which users had acquired hands-on experience with the system (Szajna, 1996). According to this line of thinking, perceived ease of use would have a different effect in the two scenarios.

In brief introduction scenarios, ease of use was thought to have a direct effect on user intention to use that was anchored by the individual's computer self-efficacy. In scenarios in which users accumulated considerable experience with the system, it was hypothesized that perceived ease of use affected user intentions and usage only indirectly through perceived usefulness. This suggested that after hands-on experience, users create their own perception of a system's ease of use and consciously or subconsciously consider it a dimension of system usefulness. This stream of research also suggested that only after direct hands-on experience did a system's objective ease of use (as measured by comparing expert/novice task achievement times) become a factor in an individual's perception of system ease of use (Venkatesh & Davis, 1996).

Szajna (1996) found that perceived ease of use influenced intentions only through usefulness and suggested that a single version of the TAM with this causal path was sufficient. Like others, including Davis, Szajna interpreted these findings to imply that unless individuals perceive a technology to be useful, its perceived ease-of-use characteristics are not critical. However, once an individual perceives a technology to be useful, then increased perceived ease of use contributes to its usefulness.

Gefen and Straub (2000) contributed another interesting perspective to the perceived ease-of-use discourse: that the effect of perceived ease of use on user acceptance would be task-dependent. They noted that most research had failed to address the nature of the task to be performed, focusing only on use or intent to use. Their research, performed in the context of e-commerce technology, investigated the hypothesis that when a task was extrinsic to IT (e.g., buying from an e-commerce site), ease of use was not a determinant of adoption. In contrast, when the task performed was intrinsic to IT (e.g., gathering information), the individual's ease-of-use perception would affect his or her decision to adopt. Their findings supported this hypothesis, providing a potentially new interpretation of the effect of ease of use on IT acceptance that merits further investigation.

Most TAM research has substantiated Davis et al. (1989), who concluded that perceived usefulness is a major determinant of individuals' intentions to use computers and that perceived ease of use is a significant secondary determinant of usage intentions. This logic is rationalized with the argument that users will tolerate ease-of-use shortcomings if they perceive the system is useful in their job. Conversely, users will not accept and use a system that is not useful, regardless of how easy it is to use.

Davis (1993) found evidence that perceived usefulness, perceived ease of use, and attitude fully mediated the effects of system design features on intention to use and usage. This research also found that perceived usefulness influenced attitude more than four times as much as perceived ease of use and that perceived usefulness was 50% more influential than perceived ease of use in explaining an individual's intention to use an information technology.

Attitude. In his original TAM, Davis theorized that an individual's perceptions of usefulness and ease of use would influence intention to use the technology primarily through the creation of a positive attitude.

Subsequent research suggests that the role of attitude as a mediator of the effects of perceived usefulness on intention to use is less clear. Davis et al. (1989) found that attitude was, at best, a partial mediator of the effect of perceived usefulness on intention to use and that it added little causal explanatory power. Davis (1993) found that the direct effect of perceived usefulness on intention to use was more than twice the influence of attitude on usage. Together, these findings led to the suggestion that attitude be eliminated from TAM to create an even more parsimonious model reflecting a direct influence of usefulness and ease-of-use perceptions on behavioral intention to use.

Behavioral Intention. Behavioral intention, a measure of the strength of one's intention to perform a specified behavior (Davis et al., 1989) is a construct borrowed from the domain of social psychology. Behavioral intention has been an important construct in most previous TAM research. The significance of behavioral intention derives from the theoretical perspective that intentions—as determined by a combination of attitudes and subjective norms—are the best predictor of an individual's behavior (Jackson, Chow, & Leitch, 1997). Davis et al. (1989) validated the notion that behavioral intention to use IT is a reasonably reliable predictor of use. Venkatesh, Morris, Davis, and Davis (2003) noted, "The role of intention as a predictor of behavior (e.g., usage) is critical and has been well established in IS and the reference disciplines" (p. 427). Evidence substantiating the hypothesized positive relationship between intentions and behavior is found in a meta-analysis of 86 TRA studies conducted by Shephard, Hartwick, and Warshaw (1988) that found a mean correlation of .54 between intentions and actions.

As Straub, Limayem, and Karahanna-Evaristo (1995) observe, the purpose in measuring intention is to predict future behavior. Thus, in research scenarios associated with the introduction of an IT, the TAM intention to use construct is particularly critical in order to predict future acceptance and usage.

Saga and Zmud (1994) viewed intention to use an IT as one of three dimensions or indicators of individual IT acceptance. Their conceptualization of IT acceptance postulated a causal chain beginning with the formation of a favorable attitude followed by an intention to use and, finally, by the behavior of usage (Saga & Zmud 1994).

External Variables. Davis defined this construct to include system design characteristics, user characteristics (e.g., cognitive style and other personality variables), and task characteristics (nature of the development or implementation process, political influences, and organizational structure) that might affect attitude, intentions to use, and/or usage. It is a central tenet of TAM that the effects of these external factors on attitude, intentions, or usage are mediated by the individual's perceived usefulness and perceived ease-of-use beliefs.

System Usage. System usage is a construct of some controversy in IT acceptance research. With relatively few exceptions (Straub et al., 1995; Szajna & Mackay, 1995; Taylor & Todd, 1995; Venkatesh & Phillips, 2002), most TAM-related research has measured the effect of the perceived usefulness and perceived ease of use constructs on intention to use (Jackson et al., 1997) and/or self-reported usage (Davis et al., 1989) as opposed to objective (actual) usage data. This widespread practice is justified by social psychology research that has "found that subjective and objective measures of neutral activities (e.g., computer use) are highly consistent (Straub et al., 1995, p. 1332). Nevertheless, there remain significant questions regarding its effect on research findings and a number of researchers who value actual usage over self-reported usage.

The research of Taylor and Todd (1995), using actual usage data, supported previous TAM research based on self-reported usage, concluding that an assessment of the usefulness of TAM in predicting usage behavior requires exploration of the model using objective actual usage data. However, Straub et al. (1995) found that self-reported usage and actual usage constructs were not strongly related. Their findings also suggested the fundamental TAM constructs, perceived usefulness and perceived ease of use, were more strongly related to self-report usage data than to actual usage data.

Straub et al. (1995) also posed thought-provoking questions concerning the relevance of IT usage measurement in general. If one accepts the widespread assumption that system usage is the primary variable through which IT affects performance (Straub et al., 1995), it seems legitimate to question which of several alternative usage dimensions best captures the desired effect; frequency of use, duration of use per time period, type of use, diversity of use are just a few candidates. One even might question the legitimacy of usage as the real variable of interest when attempting ultimately to understand and/or measure IT benefits. It could be argued that usage is but a surrogate for what researchers really need to capture (i.e., a measure of the benefit or utility derived from IT).

Szajna (1996) found statistically significant differences in the effects of user intentions on self-reported and actual usage data (stronger relationship for self-reported than actual) and low correlation (.26 at $< .05$ significance) between the two. These results led her to conclude that substitution of self-reported usage for actual usage should be discouraged.

Igbaria, Guimaraes, and Davis (1995) cited prior indications that users may overreport usage when self-reporting and called for additional research into the potential differences in the relationships between the TAM constructs and self-report usage data and actual usage data.

In summary, while there remain some inconsistent findings and beliefs surrounding the stream of TAM-centered research, a significant and growing body of work has tended to confirm the model's dominance as an IT acceptance research tool. As human knowledge increasingly has been identified as the modern organization's most valuable resource and as a foundation of competitive advantage, there is widespread interest and investment in IT KMS systems (Kankanhalli et al., 2001). However, most IT acceptance research to date, including studies focused on TAM, has studied individual acceptance of individual-use IT and has been conducted largely in educational environments (Fichman, 1992). Individual acceptance and usage of multi-user IT implementations intended to support KM objectives in an organizational scenario represent a more complex phenomenon. Empirical research is required to determine if past TAM research can help to inform the understanding of factors affecting individual acceptance and use of IT implemented to support organizational KM objectives.

Research Model and Research Hypotheses

The research model is presented in Figure 3. It supports the specific objective of this research to assess the relationships among TAM's two primary belief constructs—perceived usefulness and perceived ease of use and users' intentions to use, and their usage of the target knowledge management system. The research model is similar to Davis' original TAM (Figure 1). In accordance with the findings of Davis et al. (1989), the attitude construct is not included. The external variables constructs also are not included in the research model, since there is no intention here to investigate antecedents to perceived usefulness and ease of use.

Figure 3. Research model

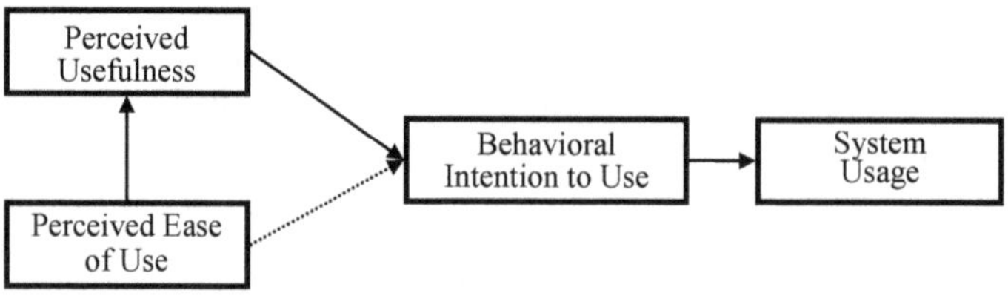

The research hypotheses investigated in this study are as follows:

H1: Perceived usefulness of the knowledge management system will exhibit a significant positive direct relationship with behavioral intention to use the system.

H2: Perceived ease of use will have both a direct and indirect effect on behavioral intention to use the system. The indirect effect (mediated by perceived usefulness) will be stronger than the direct effect.

H3: Behavioral intention to use the system will exhibit a significant positive relationship with system usage.

Methodology

A 14-item survey comprised of tailored measurement scales designed to measure each of the four constructs was used in this research. Measurement scales for each research model construct were drawn from previous IT acceptance research.

Perceived usefulness and perceived ease of use were measured using four-item measurement scales consistently demonstrating excellent psychometric qualities in previous research (Venkatesh & Davis, 1996). A three-item scale measuring behavioral intention to use the target system was adapted from Venkatesh and Davis (2000). Subsequent research has shown it to have both high reliability and excellent psychometric qualities.

Three items were used to measure user self-reported system usage. The first self-report usage item was adapted from Davis, Bagozzi, and Warshaw (1992). The third item also was drawn from the work of Davis et al. (1992) and has been used widely in information technology acceptance research.

Employees in two major Northeastern U.S. metropolitan areas with system access were identified by the organization as survey subjects. Each was provided password access to the survey that was hosted on a university server. Employee participation was optional but encouraged through correspondence from management that authorized employees to charge time spent completing the anonymous survey. Access to completed surveys was limited strictly to the researchers.

Results

Fifty-one employees, approximately 20% of the identified survey subjects, responded. Sixteen responses had to be excluded: three due to significant incompleteness and 13 due to respondent comments indicating no system awareness and/or no experience using it. (This finding within the survey target group was an unhappy surprise to management, although, given the broad organizational scope of the KMS studied, this might have been a predictable finding.). Demographic information collected with the survey confirmed the 35 usable responses were submitted by a cross section of organization personnel that included division managers, project managers, technical specialists, configuration control technicians, and administrative personnel.

Data Analysis

Data analyses included reliability and validity analyses, correlation analysis, and simple and multiple regression. All statistical analyses were performed using statistical analysis system (SAS) Version 8.0.

Cronbach Alpha measurement scale reliability coefficients calculated for each construct are shown in Table 1. The reliability of all measurement scales was comfortably above the recommended minimum level of .70 for social science research (Hatcher, 1994) and the accepted desirable level of .80 for social science research. The overall weighted reliability of the survey instrument was .938.

Table 1. Cronbach alpha reliability coefficients

Construct (Items)	Cronbach Alpha
Perceived Usefulness (4)	.978
Perceive Ease of Use (4)	.938
Behavioral Intention to Use (3)	.925
Frequency of Use (3)	.896
Overall (14)	.938

Table 2. Perceived usefulness and perceived ease of use validity factor analysis results

Item	Factor 1 Loading	Factor 2 Loading
Using the KMS improves my job performance.	.82	.14
Using the KMS increases my productivity.	.89	.05
Using the KMS enhances my effectiveness on the job.	.82	.16
I find the KMS to be useful in my job.	.85	.10
My interaction with KMS is clear.	.32	.59
Interaction with the KMS does not require a lot of mental effort.	.02	.87
I find the KMS easy to use.	.12	.86
I find it easy to get the KMS to do what I want it to do.	.14	.79

Table 3. Behavioral intention construct factor analysis results

Item	Factor Loading
Assuming I had access to the KMS and its use was voluntary, I would intend to use it to search for information while creating work products.	.96
Assuming I had access to the KMS and its use was voluntary, I would intend to use it to obtain, retrieve, and output stored products.	.81
Assuming I had access to the KMS and its use was voluntary, I would intend to use it to research topics relevant to my current work product.	.92

The perceived usefulness and perceived ease-of-use instruments used in this study have been validated extensively through prior research. Nevertheless, factor analysis was performed to assess their validity here. The eight questionnaire items comprising these two measures were subjected to exploratory factor analysis using squared multiple correlations as prior estimates of communality. Questionnaire items associated respectively with the two constructs loaded heavily (all but one well in excess of .80) on two different factors, thereby replicating the findings of Davis that the two beliefs comprise distinct constructs. Results of the factor analysis appear in Table 2.

The three behavioral intentions to use construct items also were subjected to factor analysis. Results presented in Table 3 indicate they all loaded heavily on a single factor, providing evidence of construct validity.

Correlation analysis results appear in Table 4 and include the observed correlations and associated p-values (probability of observed correlation value under the null hypothesis of zero correlation) and in the context of the research model in Figure 4. Data analysis results are discussed next in the context of the research hypotheses.

H1 postulated a significant positive relationship between perceived usefulness and behavioral intention to use the KMS. Evidence supporting confirmation of this hypothesis can be found in the positive correlation (.716, p value <.0001) and regression analysis showing that perceived usefulness explained 49.6% of the variation in behavioral intention to use in the sample.

H2 hypothesized positive direct and indirect relationships between perceived ease of use and behavioral intention to use the system. It was further hypothesized that the indirect effect (mediated by perceived usefulness) would be greater. This hypothesis was examined using multiple regression techniques. The ordinary least squares regression methodology was deemed most appropriate for this study due to the research model's simple factor structure and the fact that the TAM model structure has been researched extensively.

To establish perceived usefulness as a mediator of the relationship between perceived ease of use and behavioral intention to use, it is necessary to demonstrate the existence of four conditions: (1) a significant bivariate relationship between perceived ease of use and perceived usefulness; (2) a significant bivariate relationship between perceived ease of use and behavioral intention to use; (3) a significant relationship between perceived usefulness and intention to use when perceived ease of use is controlled for; and (4) a reduced or diminished relationship between perceived ease of use and behavioral intention to use when the effects of perceived usefulness are controlled for (Baron & Kenney, 1986).

Table 4. Correlation matrix

	Perceived Usefulness	Perceived Ease of Use	Behavioral Intention To Use	System Usage
Perceived Usefulness	1	.795 <.0001	.716 <.0001	.573 .0003
Perceived Ease of Use	.795 <.0001	1	.672 <.0001	.463 .005
Behavioral Intention To Use	.716 <.0001	.672 <.0001	1	.454 .006
System Usage	.573 .0003	.463 .005	.454 .006	1

Regression analysis results relevant to Hypothesis 2 are presented in Table 4. The first two data rows of the table illustrate necessary statistically significant bivariate relationships between perceived ease of use and perceived usefulness and perceived ease of use and behavioral intention to use. Results of the multiple linear regression of behavioral intention to use on perceived usefulness and perceived ease of use appear in the third row. These results reveal the hypothesized multivariate relationship in the survey data. Consistent with required mediation conditions three and four, the relationship between perceived usefulness and behavioral intention to use is statistically significant when the effects of ease of use are controlled for (p value of .018). Additionally, as required, the relationship between perceived ease of use and behavioral intention to use is diminished (smaller regression coefficient (β) estimate and statistically nonsignificant) when the effects of usefulness are controlled for in the multiple regression.

These results comply with the mediation criteria of Baron and Kenny (1986) and support confirmation of Hypothesis 2 that the effects of perceived ease of use on behavioral intention to use are mediated by perceived usefulness. They also show that when the effects of perceived usefulness are controlled for in the multiple regression, perceived ease of use contributes very little unique explanatory value with regard to intention to use. Perceived usefulness uniquely accounted for more than seven times as much variation in intention to use (10.8%), as did perceived ease of use (1.49%). However, these two beliefs combined to uniquely account for only 12.3% of the variation in intention to use. Thus, almost 39% of the variation in intention to use was explained by the two constructs jointly (this is consistent with the high correlation [.795] between theses two constructs, shown in Table 4).

Hypothesis 3 hypothesized a positive and significant relationship between behavioral intention to use the system and system usage. Table 4 presents evidence of confirmation of this hypothesis in the positive correlation (.454, p-value .006) between these constructs. Regression analysis indicates behavioral intention explains 20.6% of the variation in usage in the sample. The intention-behavior correlation is below the .53 average intentions-behavior correlation reported by Sheppard et al. (1988) in their meta-analysis. However, it

Table 5. Mediation analysis regression results (hypothesis 2)

Condition	β Ease of Use (p-value)	β Usefulness (p-value)	R^2
1. Bivariate: Usefulness = *f* (Ease of Use)	.783 (<.0001)	N/A	.632
2. Bivariate: Intention to Use = *f* (Ease of Use)	.600 (<.0001)	N/A	.451
3. Multivariate: Intention to Use = *f* (Ease of Use & Usefulness)	.249 (.167)	.448 (.018)	.541

is higher than the .35 intentions-behavior correlation reported by Davis et al. (1989) for intentions-behavior measurements taken at different times. While some TAM research has yielded weaker intention-behavior correlations (Szajna & Mackay, 1995), many have found evidence of a stronger tie between these variables.

Lower than anticipated intentions-behavior correlation and the relatively small percentage of usage variance explained by behavioral intention (20.6%) motivated further investigation of the role of the intentions construct in this research. Earlier TAM research has found behavioral intention to be the strongest predictor of usage behavior (Davis et al., 1989; Taylor & Todd, 1995). For example, Davis et al. (1989), referring to perceived usefulness and perceived ease of use, reported that "consistent with the theories," intentions "fully mediated the effects of these other variables on usage" (Davis et al., 1989, pp. 992–993).

Similarly, Taylor and Todd (1995) noted that "BI [behavioral intention] has long been recognized as an important mediator in the relationships between behavior and other factors such as attitude, subjective and perceived behavioral control" (Taylor & Todd, 1995, p. 165). They found removal of behavioral intention from their TAM model resulted in substantial "drop in predictive power," which is "consistent with Fishbein and Ajzen's (1975) identification of intentions as an important mediating variable (Taylor & Todd, 1995, p. 165) and concluded that "behavioral intention is the primary direct determinant of behavior" (Taylor & Todd, 1995, p. 165).

In this research, multiple regression of system usage on perceived usefulness and perceived ease of use (i.e., removal of behavioral intention from the model) resulted in an almost 60% increase in system usage variance explanatory power (i.e., R^2 increased from .206 to .329). This finding is in conflict with much previous research. After removal of the behavioral intention construct, usefulness and ease-of-use perceptions explained substantially more of the variance in system usage. Additional mediation analysis of the relationship among behavioral intentions, perceived usefulness, and usage confirmed conclusively that intentions did not mediate the effect of perceived usefulness on usage and that perceived usefulness was a stronger mediator of the effect of perceived ease of use on usage than on behavioral intentions.

Discussion

The only findings of this research not consistent with previous TAM IT acceptance research concern the role of behavioral intention as a mediator of the effects of usefulness and ease-of-use beliefs on system usage. Perceived usefulness and perceived ease of use combine to explain 51.1% of the variation in behavioral intention to use the KMS. This is consistent with previous TAM research. The individual relative effects of the two belief constructs are also consistent with previous findings. Both beliefs exhibit significant bivariate relationships with intention to use. However, when the effects of perceived usefulness and perceived ease of use are isolated in multiple regression, it can be seen that the effect of perceived ease of use on behavioral intention to use the KMS actually derives from potential users' perceptions of its usefulness (i.e., perceived usefulness mediates the relationship between perceived ease of use and intention to use the system). This is generally consistent with most previous TAM research addressing individual IT use, although an even smaller direct effect was anticipated.

The correlation between intention to use the KMS and self-reported system usage found in this study is somewhat lower than typically has been observed in previous TAM research. Perhaps even more noteworthy are findings that behavioral intentions exhibited less system usage variance explanatory power than the usefulness and ease-of-use beliefs and did not exhibit mediation of the effects of these two beliefs on system usage.

Table 6. Summary of research hypothesis findings

Hypothesis	Comments
1. Perceived usefulness will exhibit a significant positive relationship with behavioral intention to use.	• Confirmed, consistent with previous research
2. Perceived ease of use will have both a direct and indirect (mediated by perceived usefulness) relationship with behavioral intention to use.	• Strong mediation effect of perceived usefulness confirmed, consistent with most prior research • No unique perceived ease-of-use effect when perceived usefulness controlled for
3. Behavioral intention to use will have a significant positive relationship with system usage.	• Confirmed • Intention-behavior relationship not as strong as many earlier studies

Figure 4. Research model and observed correlations (p-values)

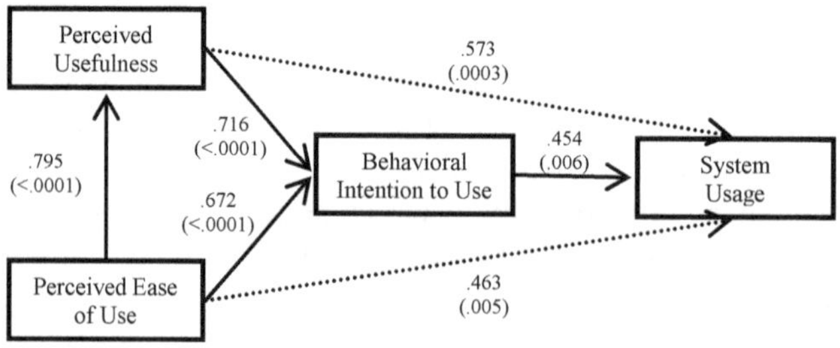

These potentially inconsistent findings may be attributable to the nature of the study scenario. The research data were collected long after the introduction of the target KMS. As Straub et al. (1995) observed, the value of measuring intention is to predict future behavior. Therefore, in research scenarios characterized by a brief introduction of an IT intended to help predict future acceptance and usage, the behavioral intention construct is critical. In this research, the intent was not to predict usage but to interpret experience with the target KMS through the lens of the TAM. Research by Davis et al. (1989) indicated that the strength of the link between intentions and behavior is correlated positively with user experience and correlated negatively with the elapsed time between the measurement of intentions and behavior (Venkatesh & Davis, 1996). In this case, respondents had extended the period in which to form their beliefs regarding the usefulness and ease of use of the target KMS. We believe that measurement of intentions significantly post facto is a plausible explanation for the ambiguous results surrounding the relationship between the intention and usage constructs. In retrospect, it might have been more appropriate to follow the reasoning of Straub et al. (1995) by adopting a research model that did not include the behavioral intention model.

Other explanations of the unexpected intentions-usage relationship centering on whether or not frequency or accumulated time of system usage is really an important metric with regard to KMS usage and/or whether or not potential users' intentions to use was affected adversely by the quality of the materials accessible in the KMS or its output quality is also potentially plausible explanations (Kankanhalli et al., 2001).

Conclusion

The results of this research provide preliminary evidence that previous IT acceptance research based on TAM can serve as a basis for critically needed empirical research of KMS user acceptance. Relationships among primary TAM constructs found in this research largely are consistent with those typical in previous TAM research, with the exception of the correlation between intention to use the KMS and self-reported system usage. This correlation was found to be lower than what typically has been observed in previous TAM research. This potentially inconsistent finding may be attributable to the data collection long after the introduction of the target KMS, to questions reading the metrics with regard to KMS usage, and to whether or not potential users' intentions to use was affected adversely by the quality of the materials accessible in the KMS.

Perceived usefulness and perceived ease of use combined to explain 51% of behavioral intention to use the system. This result is consistent with a significant body of previous TAM research in which these two constructs typically have been found to explain 40% to 60% of the intention to use/usage variance. Significant positive relationships among perceived usefulness, perceived ease of use, intention to use, and the strong mediating effect of perceived usefulness on the effect of perceived ease of use on intention to use were consistent with previous TAM research. Behavioral intention, typically found to be the best predictor of usage in IT acceptance research, explained only 20.6% of system usage variance. An IT acceptance model with only perceived usefulness and perceived ease of use as usage predictors accounted for 32.9% of system usage variance.

Limitations

This research bears inevitable generalizability limitations of any study of one information system in one organizational environment. Additional investigation of KMS implemented in a representative range of modern organizational settings is essential to increase understanding of those factors, conditions, and mechanisms critical to their success. A second key shortcoming of this research is its limited sample size, both in absolute terms and relative to the population of potential organizational users of the KMS. With approximately 12% of the population providing a usable response to the voluntary survey, one cannot summarily ignore the possibility of sample bias. However, the broad range of responses received across all measured constructs mitigates this concern somewhat. Another potential shortcoming of this research is its reliance upon self-reported usage. As noted previously, previous researchers have raised credible and largely unanswered questions regarding the fidelity of self-report usage data and their relationship to key TAM constructs when compared to actual usage data. Finally, it must be recognized that this study covers only a single point in time, whereas KMS adoption and usage issues are likely to evolve over longer time cycles. Given the complexity of KMS usage and adoption, longitudinal studies could provide valuable research insights not otherwise available.

Future Research

Additional study of diverse KMS in a range of modern organizational settings is necessary to support the accumulation of knowledge and development of sound theory regarding the factors, conditions, and mechanisms critical to KM success. Given the complex and diverse nature of KMS and the approaches used to implement them (i.e., task/process-oriented vs. infrastructure/generic), a spectrum of situation-specific models/constructs may be required.

When possible, future researchers should attempt to collect and analyze both self-report and objective actual usage data. This will contribute to resolution of lingering questions regarding this important construct and its effect on previous TAM research. In addition and as already suggested, further investigation of the relationship between intention to use and KMS usage is warranted.

Furthermore, although this research suggests that previous TAM-based IT user acceptance research can serve as a basis for future investigation of KMS user acceptance, it seems likely that other factors associated with the complex sociocultural organizational implications of KMS use/acceptance must be explored. For example, Agarwal (2000) conceptualizes individual acceptance of IT as being influenced by a complex set of factors, including beliefs and attitudes, managerial interventions, situational influences, social influences, and individual differences.

To increase the explanatory power of the model relative to KMS, it will be necessary to incorporate additional theory-based constructs to the TAM. Given the social nature of KM, interesting candidates include items that measure important organizational culture and/or subjective norm influences. The research of Kraut, Rice, Cool, and Fish (1998), Fulk, Schmitz, and Steinfield (1990), and Fulk (1993), although focused on interactive communications technology, highlights the potential of social influence to affect technology acceptance and usage in the organizational environment.

Antecedents to the current TAM beliefs need to be researched in order to understand the components of perceived usefulness and perceived ease of use relative to KMS. For example, what are the dimensions or constituents of usefulness in the context of a KMS IT technology? This would be consistent with current trends in general IT user acceptance research, such as those reported by Venkatesh et al. (2003), Venkatesh and Davis (2000), and Venkatesh (2000).

Finally, longitudinal studies of organizational and user patterns that comprise the totality of a KMS over its development, implementation, and adoption should be pursued. Longitudinal studies provide a unique opportunity to investigate how the relative influence of various IT acceptance/usage factors change over time (Agarwal & Prasad, 1997) and to observe IT usage patterns and benefits that emerge only over time and are unlikely to surface in cross-sectional studies (Devaraj & Kohli, 2003). Longitudinal research is essential in order to develop a better understanding of how the organization and KMS undergo mutual interactive adaptation over time.

Given the critical importance of knowledge to the modern organization and the pivotal role of IT in organizational knowledge-leveraging initiatives, further investigation of user KMS acceptance factors would seem to be a critical research priority.

References

Agarwal, R. (2000). Individual acceptance of information technologies. In R. Zmud (Ed). *Framing the domains of IT management: Projecting the future through the past*, (pp. 85-104). Pinnaflex Press.

Agarwal, R., & Prasad, J. (1997). The role of innovation characteristics and percieved voluntariness in the acceptance of information technologies. *Decision Sciences, 28*(3), 557-582.

Alavi, M., & Leidner, D. (1999). Knowledge management systems, issues, challenges, and benefits. *Communications of the Associations of Information Management Systems, 1*(7), 1-37.

Alavi, M., & Leidner, D. (2001). Review: Knowledge management and knowledge management systems: Conceptual foundations and research issues. *MIS Quarterly, 25*(1), 107-136.

Baron, R., & Kenney, D. (1986). The moderator-mediator variable distinction in social psychological research: Conceptual, strategic, and statistical considerations. *Journal of Personality and Psychology, 51*(6), 1173-1182.

Bourdreau, A., & Couillard, G. (1999). Systems integration and knowledge management. *Information Systems Management, 16*(4), 24-32.

Carr, N. (2003). IT doesn't matter. *Harvard Business Review, 81*(5), 41.

Compeau, D., & Higgins, C. (1995). Computer self-efficacy: Development of a measure and initial test. *MIS Quarterly, 19*(2), 189-211.

Davenport, T., & Grover, V. (2001). General perspectives on knowledge management: Fostering a research agenda. *Journal of Management Information Systems, 18*(1), 5-22.

Davis, F. (1989). Perceived usefulness, perceived ease of use, and user acceptance of information technology. *MIS Quarterly,* 13, 319-339.

Davis, F. (1993). User acceptance of information technology: System characteristics, user perceptions and behavioral impacts. *International Journal of Man-Machine Studies, 38,* 475-487.

Davis, F., Bagozzi, R., & Warshaw, P. (1989). User acceptance of computer technology: A comparison of two theoretical models. *Management Science, 35*(8), 982-1003.

Davis, F., Bagozzi, R., & Warshaw, P. (1992). Extrinsic and intrinsic motivation to use computers in the workplace. *Journal of Applied Psychology, 22*(14), 1111-1132.

DeLone, W., & McLean, E. (1992). Information systems success: The quest for the dependent variable. *Information Systems Research, 3*, 60-95.

Devaraj, S., & Kohli, R. (2003). Performance impacts of information technology: Is actual usage the missing link? *Management Science, 49*(3), 273-289.

Doll, W., Hendrickson, A., & Xiandong, D. (1998). Using Davis's perceived usefulness and ease-of-use instruments for decision-making: A confirmatory and multi-group invariance analysis. *Decision Sciences, 29*(4), 839-869.

Earl, M. (2001). Knowledge management strategies: Toward a taxonomy. *Journal of Management Information Systems, 18*(1), 214-233.

Ericsson, F., & Avdic, A. (2003). Knowledge management system acceptance. In E. Coakes (Ed.), *Knowledge and management: Current issues and challenges* (pp. 39-51). Hershey, PA: IRM Press.

Fichman, R. (1992). Information technology diffusion: A review of empirical research. Cambridge, MA: MIT Sloan School of Management.

Fishbein, M., & Ajzen, I. (1975). *Belief, attitude, intention and behavior: An introduction to theory and research.* Reading, MA: Addison-Wesley.

Fulk, J. (1993). Social construction of communication technology. *Academy of Management Journal, 36*(5), 921-950.

Fulk, J., Schmitz, J., & Steinfield, C. (1990). A social influence model of technology use. In J. Fulk, & C. Steinfield (Eds.), *Organizations and communication technology,* pp. 117-142. Newbury Park, CA: SAGE Publishing.

Gefen, D., & Straub, D. (2000). The relative importance of perceived ease of use in IS adoption: A study of e-commerce adoption. *Journal of the Association of Information Systems, 1*(8), 1-28.

Gold, A., Malhotra, A., & Segars, A. (2001). Knowledge management: An organizational capabilities perspective. *Journal of Management Information Systems, 18*(1), 185-214.

Grant, R. (1996). Toward a knowledge based theory of the firm. *Strategic Management Journal, 17*(Winter Special Issue), 109-122.

Hatcher, L. (1994). *A step-by-step approach to using the SAS® system for factor analysis and structural equation modeling*. Cary, NC: SAS® Institute.

Igbaria, M., Guimaraes, T., & Davis, G. (1995). Testing the determinants of microcomputer usage via a structural equation model. *Journal of Management Information Systems, 11*(4), 87-114.

Jackson, C., Chow, S., & Leitch, R. (1997). Toward an understanding of the behavioral intention to use an information system. *Decision Sciences, 28*(2), 357-389.

Jennex, M., & Olfman, L. (2002). Organizational memory/knowledge effects on productivity, a longitudinal study. In *Proceedings of the 35th Annual Hawaii International Conference on Systems Sciences.* IEEE Computer Society.

Jennex, M., & Olfman, L. (2003). A knowledge management success model: An extension of Delone and Mc Lean's IS success model. In *Proceedings of the Ninth Americas Conference on Information Systems* (pp. 2529-2539).

Jennex, M., & Olfman, L. (2004). Assessing knowledge management success/effectiveness models. In *Proceedings of the 37th Hawaii International Conference on Systems Sciences.* IEEE Computer Society.

Kankanhalli, A., Tan, B., & Wei, K. (2001). Seeking knowledge in electronic knowledge repositories: An exploratory study. In *Proceedings of the Twenty Second International Conference on Information Systems* (pp. 123-133).

Kraut, R., Rice, R., Cool, C., & Fish, R. (1998). Varieties of social influence: The role of utility and norms in the success of a new communication medium. *Organization Science, 9*(4), 437-453.

Lam, A. (2000). Tacit knowledge: Organizational learning and societal institutions: An integrated framework. *Organization Studies, 21*(3), 187-209.

Legris, P., Ingham, J., & Collerette, P. (2002). Why do people use information technology? A critical review of the technology acceptance model. *Information & Management, 40,* 191-204.

Ma, Q., & Liu, L. (2004). The technology acceptance model: A meta-analysis of empirical findings. *Journal of Organizational and End User Computing 16*(1), 59-72.

Malhotra, Y. (2000). Knowledge management and new organization forms. *Information Resources Management Journal, 13*(1), 5-14.

Marwick, A. (2001). Knowledge management technology. *IBM Systems Journal, 40*(4), 814-830.

Mathieson, K., Peacock, E., & Chin, W. (2001). Extending the technology acceptance model: The influence of perceived user resources. *The DATABASE for Advances in Information Systems*, *32*(3), 86-112.

Saga, V. & Zmud, R. (1994). The nature and determinants of IT acceptance, routinization, and infusion. In L Levine (ed). *Diffusion transfer, and implementation of information technology*, (pp. 67-86). North Holland: Amsterdam.

Sheppard, B. H., Hartwick, J., & Warshaw, P. R. (1988). The theory of reasoned action: A meta-analysis of past research with recommendations for modifications and future research. *Journal of Consumer Behavior, 15*(3), 325-343.

Spender, J. C. (1996). Making knowledge the basis of a dynamic theory of the firm. *Strategic Management Journal, Special Edition, 17,* 45-72.

Straub, D., Limayem, M., & Karahanna-Evaristo, E. (1995). Measuring system usage: Implications for IS theory testing. *Management Science, 41*(8), 1328-1342.

Szajna, B. (1996). Empirical evaluation of the revised technology acceptance model. *Management Science, 42*(1), 85-92.

Szajna, B., & Mackay, M. (1995). Predictors of learning performance in a computer-user training environment: A path-analytic study. *International Journal of Human-Computer Interaction, 7*(2), 167-186.

Taylor, S., & Todd, P. (1995). Understanding information technology usage: A test of competing models. *Information Systems Research, 6*(2), 144-176.

Taylor, W. (2003). *Computer mediated knowledge sharing and individual user differences: An exploratory study* [working paper].

Vaill, E. (1999). Knowledge mapping: Getting started with knowledge management. *Information Systems Research, 16*(4), 16-23.

Venkatesh, V. (2000). Determinants of perceived ease of use: Integrating control, and emotion into the technology acceptance model. *Information Systems Research, 11*(4), 342-365.

Venkatesh, V., & Davis, F. (1996). A model of the antecedents of perceived ease of use: Development and test. *Decision Science, 27*(3), 451-481.

Venkatesh, V., & Davis, F. (2000). A theoretical extension of the technology acceptance model: Four longitudinal studies. *Management Science, 46*(2), 186-204.

Venkatesh, V., & Johnson, P. (2002). Telecommuting technology implementations: a within- and between- subjects longitudinal field study. *Personnel Psychology, 55*(3), 661-687.

Venkatesh, V., Morris, M., Davis, G., & Davis. F. (2003). User acceptance of information technology: Toward a unified view. *MIS Quarterly, 27*(3), 425-478.

Zack, M. (1999). Developing a knowledge strategy. *California Management Review, 41*(3), 125-146.

Section IV

Knowledge in Organizations

Chapter XIV

The Role of Context and Its Explication for Fostering Knowledge Transparency in Modern Organizations

Stefan Smolnik, European Business School (ebs), Germany

Stefan Kremer, The Information Management Group (IMG AG), Switzerland

Lutz Kolbe, University of St. Gallen, Switzerland

Abstract

In order for a company to be oriented consistently toward its customers and their processes, it needs to customize its intracorporate processes and systems. The solution seems to be customer process-oriented portals that integrate companies' systems and provide transparent access to information objects stored in these systems. A key problem in this regard is finding relevant information objects in systems that not only are growing but also also are being disseminated. An additional challenge is making knowledge available at the right time and at the right place. A company's competitive advantage is rooted in this knowledge advantage as well as in the capability to transform this superior knowledge into market-driven business processes. The research questions addressed in this chapter are how the value of information objects is affected by the context in which it is considered and how associated contexts can be uncovered for given situations. We introduce a continuum of context explication comprised of the relationships among data, information objects, knowledge, and their contexts according to their degree and ease of context explication. The extremes of the continuum, therefore, would be data with no context to explicate and knowledge with rich, person-specific context. We conclude that discovering implicit meanings and expressing those meanings explicitly increase information objects' potential values. In addition, we evaluate the full-text search, attribute-based search, and topic maps as approaches for knowledge discovery through customer process-oriented portals as well as providing patterns that indicate when to apply which approach. Two small case studies are presented of knowledge discovery through such portals. We conclude with suggestions for future research, based on our final deductions with respect to the study.

Introduction and Overview

Challenge

The use of information technology has given many organizations access to vast internal and external information repositories. Intranets, content management systems, and enterprise portals have become commonplace, providing employees with opportunities to discover knowledge enshrined in information objects (e.g., electronic documents) (Latham, 2001; WebCKS, 1999). Nonetheless, dealing with information and finding the right content are inefficient actions. Davenport, Harris, and Kohli (2001) stated, "Information management must begin by thinking about how people use information" (p. 63). This is a precondition for using information judiciously.

Although organizations currently have access to various information repositories, the process of knowledge discovery still has major shortcomings, such as the following:

- **Lack of information:** Finding information objects on a topic is frustrating if users know that they exist but cannot trace them.
- **Overload of information:** Knowledge discovery is time-consuming if too many information objects with no or little relevance are found.

One key to successfully minimizing these deficits is by controlling the semantics (i.e., the meaning of terms), making explicated context available, and methodically classifying information objects utilized in business environments (Dale, 2001; Felber & Budin, 1989). Various technological approaches—based on various degrees of context explication, such as discovering implicit meanings and expressing those meanings explicitly—have been proposed to address the lack of relevant information and the overload of remotely related information problems in knowledge discovery. Examples of such approaches are search and classification engines. Nevertheless, there are hardly any criteria available with which to support organizations' choices of an appropriate solution. Consequently, we present a comprehensive overview of several approaches, their underlying principles, advantages, and constraints. To fulfill specific organizations' needs, criteria also are provided in respect of the degree of context explication required.

Objective and Research Approach

The overall objective of this chapter is to propose a continuum of context explication comprised of the relationships between information objects and their contexts in order to foster knowledge discovery. We will demonstrate that the continuum allows organizations to make deductions with regard to the appropriate approach with which to stimulate knowledge discovery through portals.

A review of the knowledge management literature provided a comprehensive overview of the topic and related works. Furthermore, it revealed the gap between research on an information object's context and its explication. Desk and action research (e.g., prototypical implementations of our conceptual approaches, led to logically deduced concepts (Checkland & Holwell, 1998), while the case research allowed the deduction and validation of these concepts. In terms of our research questions, the latter was particularly suitable since the research and theory are still in the early stages of formulation (Benbasat, Goldstein, & Mead, 1987). Consequently, the research and descriptive processes also were influenced by the results of workshops conducted and projects undertaken with our corporate partners during the action research (Gummesson, 2000; Whyte, Greenwood, & Lazes, 1991). We currently are testing and expanding the findings with other partners as well.

Structure of the Chapter

The subsequent section deals with related works in the area of knowledge discovery and portals. It also defines the most important and relevant terms for an understanding of the research field.

In the third section, we describe the challenges facing knowledge discovery. Thereafter, we introduce three major approaches for discovering knowledge through portals by providing a chronological outline of the different development phases. We describe the three approaches—full-text search, attribute-based search, and topic maps—with reference to their characteristics, advantages, and restrictions in relation to context explication.

Knowing the three approaches' capabilities and constraints, and based on given prerequisites, we then propose a continuum of context explication, providing criteria for and advice in respect to choosing an appropriate solution.

In the penultimate section, we provide two examples of how the continuum was applied successfully in a normal work situation. This was done at two institutions where, based on different prerequisites, we chose and implemented different solutions for knowledge discovery. Finally, we draw conclusions and propose directions for further research.

Related Work and Definitions of Terms

In this section, we introduce the theoretical background and define the most relevant terms. We identify related works and explain how they differ from our approach.

Knowledge and Context

Within the literature there are many definitions of knowledge (Biggam, 2001; Davenport & Prusak, 1998; Lai & Chu, 2000; Murray, 1996; Nonaka & Takeuchi, 1995; Polanyi, 1966; Sveiby, 1997) (see Table 1). Some of these references also provide detailed discussions on the differentiation of the terms, data, information, and knowledge, as well as discussing knowledge types and their classification (see Figure 1).

In our view, knowledge comprises both information and person-specific aspects, such as experiences, values, and insights. An important characteristic of knowledge, which simultaneously distinguishes it from information, is its strong affinity to activities (Davenport & Prusak, 1998). Individuals act and react in keeping with their experiences and intrinsic attitudes. Knowledge, on the other hand, is much more than transformed information and, therefore, cannot be represented in the form of information objects or data. Polanyi (1966) developed a concept of implicit knowledge that he described as follows: "We can know more than we can tell" (p. 4). We concur with Polanyi's (1966) basic concept that knowledge's implicit and explicit dimensions are complementary—all knowledge contains both dimensions. Pure explicit or implicit knowledge, or the conversion of one into the other, is thus impossible.

Many of the previous definitions of knowledge have context as an important common aspect. One form of transition from information to knowledge is contextualization. Dey and Abowd (1999) define context as "any information that can be used to characterize the situation of an entity. An entity is a person, place, or object that is considered relevant to the interac-

Figure 1. Data, information, and knowledge (Klemke, 2000)

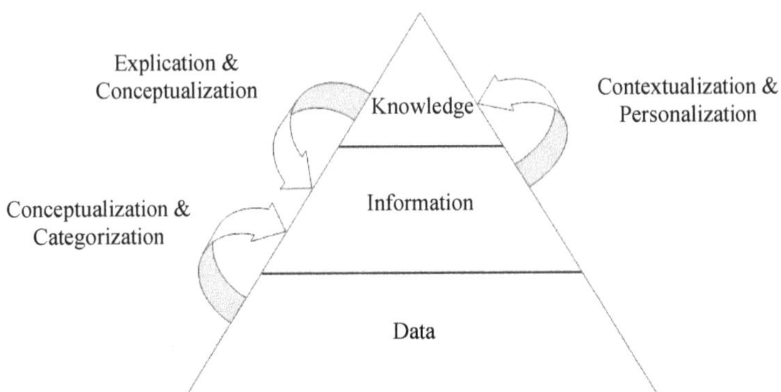

Table 1. Overview of selected definitions of the term knowledge

Author	Definition
Davenport & Prusak (1998)	"Knowledge is a fluid mix of framed experience, values, contextual information, and expert insight that provides a framework for evaluating and incorporating new experiences and information. It originates and is applied in the minds of knowers" (p. 5).
Nonaka & Takeuchi (1995)	"First, knowledge, unlike information, is about beliefs and commitment. … we consider knowledge as a dynamic human process of justifying personal belief toward the 'truth." (p. 58).
Alavi & Leidner (1999) Lai & Chu (2000)	Knowledge is created and organized by the very flow of information, anchored by the commitment and beliefs of its holders. Information becomes knowledge when it is processed in the mind of an individual, and knowledge becomes information when it is articulated or communicated to others in the form of text, computer output, speech or written words, and so forth.
Murray (1996)	"Knowledge solves a problem; it produces competence leading to effective action."(p.5). "Making the tacit explicit often includes the following activities: … Identifying terminology that is clearly understood and using language that is appropriate for the culture and context."(p.4).
Sveiby (1997)	Knowledge is the capacity to act within context.
Polanyi (1966)	Tacit knowledge is personal, context-specific, difficult to express in verbal, symbolic, and written form, and therefore hard to formalize and communicate.
Biggam (2001)	"- It must be true. - The perceiver must believe this to be the case. - The perceiver must be in a position to know this to be the case." (p. 3)

tion between a user and an application, including the user and applications themselves" (p. 3f). Similarly, Sowa (2000) describes context in its nonlinguistic meaning as "situation, environment, domain, setting, background, or milieu that includes some entity, subject, or topic of interest" (p. 275).

Klemke (2000) describes the differentiation of context types by means of a level-based approach. The first level identifies the following context dimensions: organizational, domain-/content-based, personal, and physical. These dimensions are specified in more detail on the second level (e.g., the organizational dimension is subdivided into a process and a structure component). In spite of the common assumption that context consists only of implicit information, the previous definitions allow context to be either explicit or implicit. In this chapter, we reveal that the explication of information objects' implicit context (i.e., the discovery of implicit meanings and expressing those meanings explicitly) supports the creation of new knowledge. Moreover, we describe various approaches with which to achieve this.

Klemke (2000) recommends a holistic understanding of context by means of several dimensions (see Figure 2) and the implementation of an integrated architecture to trace and maintain context models. In addition, the literature regards contexts as having different characteristics and uses different approaches to model these contexts (e.g., workflow process context is modeled by workflow management systems (Wargitsch & Habermann, 1998), while organizational structures are modeled by enterprise ontologies). Conversely, we focus directly on information objects and their contexts and provide approaches with which to discover,

Figure 2. Context typology (Klemke, 2000)

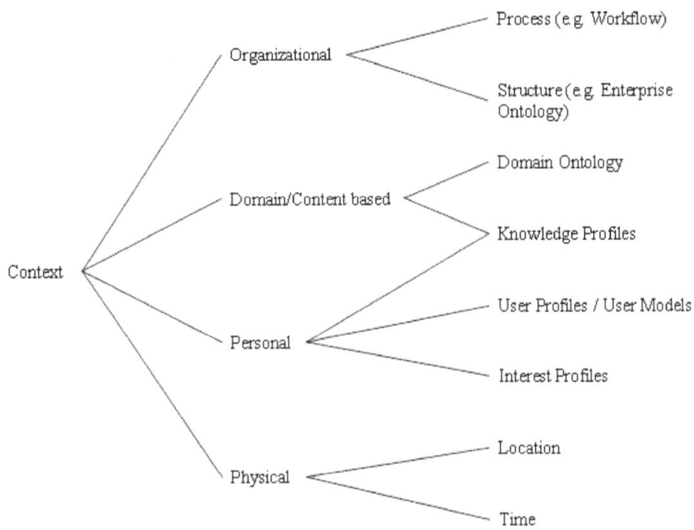

explicate, and use these contexts in various situations in order to increase the information objects' potential value and to stimulate knowledge discovery.

We believe that all documents ultimately are information objects. Users are able to create knowledge by processing and understanding them, although the information objects do not comprise knowledge. However, we recognize that some KM researchers differentiate between information and knowledge object documents (i.e., they acknowledge that documents with context can be knowledge objects). This chapter considers all documents as information objects and requests that the readers accept this viewpoint throughout the rest of this chapter.

Knowledge Discovery as an Important Knowledge Management Activity

Many knowledge management activities, methods, and modules have been discussed in the literature. Lai and Chu (2000) suggest an integrated knowledge management framework that comprises the following activities: initiation, generation, modeling, repository, distribution and transfer, use, and retrospect. Davenport and Prusak (1998) differentiate between specifying a requirement and capturing, distributing, and using knowledge. Probst, Raub, and Romhardt (1999) present a pragmatic approach to the organizationwide management of knowledge. This approach comprises six core processes and two pragmatic modules: the identification, acquisition, development, distribution, use, and preservation of knowledge as well as knowledge's objectives and performance measurement. Nonaka and Takeuchi (1995), Andersen (1996), and Alavi (1997) offer relatively similar classifications of knowledge management activities.

All these approaches have a method in common for the identification or use of knowledge, whether implicit or explicit. Unused knowledge that generally is found within organizations can be uncovered with appropriate methods and, thereafter, utilized. Knowledge discovery methods additionally foster knowledge transparency in organizations as well as supporting users to find relevant information objects. They are a necessary precondition for the core processes of knowledge identification and knowledge use (Probst, Raub, & Romhardt, 1999). They not only improve the organizational use of existing individual and common knowledge but also contribute to the knowledge generation process (i.e., the development or collecting of new knowledge) (Güldenberg, 1996).

Portals

Portals have been discussed as an integration concept for user access to personalized information and applications since 1998 (Bristow, Dickinson, Duke, Henry, & Makey, 2001). Although there are many descriptions of portals (Davydov, 2001; Dias, 2001; Kalakota & Robinson, 2001; Röhricht & Schlögel, 2001; Schwarz, 2000), we focus on them as Web-based, personalized, and integrated access systems to internal and external applications and information repositories.

Portals support knowledge-oriented processes by providing users with a graphical front-end integration of back-end systems that comprises, among others, integration, personalization, and administrative services. Knowledge discovery methods through portals are supported mainly by navigation and search mechanisms (Fleisch & Österle, 2001; Puschmann, 2003). The role of search mechanisms is especially significant in these methods, as the following section shows.

Knowledge Discovery Through Portals

As stated previously, search and retrieval play a vital role in the concept of portals, but knowledge discovery through portals faces special challenges (Baeza-Yates & Schäuble, 2002; Raghavan, 2002).

- **Heterogeneous structures and formats:** Information objects are stored in multiple, roughly structured formats, classified in various ways, and presented in various languages. Portal users, therefore, need a standardized view of all the available information objects.
- **Distributed and redundant information:** Organizations have information objects that reside in a variety of sources (e.g., e-mail, content management, and file server systems) in a partly redundant way. Knowledge discovery processes have to offer mechanisms that connect these repositories to the portal to provide users with a consolidated view.

- **Protected content:** The role of each individual portal user dictates which information objects that individual is able to access. In the process of knowledge discovery, navigation entries and search results have to be filtered in order to display only the information objects accessible to the user (i.e., secure access has to be provided).

Addressing these challenges is fundamental to supporting knowledge discovery methods through portals (Andrews, 2003). The following sections illustrate three major approaches with which to achieve this objective by providing a chronological outline of different development phases.

Full-Text Search

The classic full-text search has been an established retrieval approach since the early 1990s (Rappoport, 2002). A search engine is an information technology component of a portal that acts as a central instance between the user's information need and the available information objects that are stored in one or more repositories. Users convert their information need into a search query and enter it in a search field within a portal. In order to respond to the user queries, a search engine indexes each information object, representing it as a set of weighted words. The search engine compares the entered terms with the previously indexed information objects and provides the users with a result list.

The benefits for users are as follows:

- **Speed:** Searching the content of multiple repositories by means of a single query is faster than searching each application individually by means of separate queries.
- **Ease of use:** Currently, the full-text search is well-known, and most users have some experience in this (Gordon & Pathak, 1999).
- **No prestructuring required:** Since the relevant terms are indexed automatically, no human-driven intervention is necessary.

Since the expressing of an information need in a single query has a strong impact on the quality of the search results, the main restrictions of the full-text search emanate from the following semantic issues:

- **Wrong or too many results:** Receiving search results that refer to information objects with no or little relevance is time-consuming (Cathro, 1997). In this context, the challenge is for users to anticipate the correct terminology (i.e., to match the authors' terminology).
- **Spelling:** A user's query should be orthographically correct.

As shown, the full-text search is dependent on the information object's content, because its context is contained solely within the information object itself. The authors do not provide explication during the information object's creation, nor does a system later do so.

Attribute-Based Search

To overcome the restrictions of the full-text search, the attribute-based search was developed in the mid-1990s (Cathro, 1997). This approach is based on a context explication model that stores an information object's context as metadata (i.e., data about the data) (Berners-Lee, 1997). The metadata are stored with the information object itself and can be viewed and retrieved by users and applications. Common metadata attributes that are associated with information objects include the author's name, date of publication, source of publication, and so forth. The attribute-based search during knowledge discovery through portals would therefore permit structured queries on the context explicated in information objects' metadata (McGovern, 2001). Currently, there are several metadata standards e.g., the Dublin Core Metadata Element Set, which proposes 15 fields or attributes according to which a document can be described (Baeza-Yates & Ribeiro-Neto, 1999; Dublincore, 2003).

The major benefits of the attribute-based search are as follows:

- **Reduced result sets:** Compared to the full-text search, users retrieve relevant information objects more swiftly.
- **Controlled vocabulary:** Users can choose standardized terms from drop-down lists.
- **Personalization capabilities:** Search queries can be enriched automatically with personalized information (user attributes, such as roles, language, and organizational unit).

But there are also certain constraints:

- **Maintenance of controlled vocabulary:** Although this approach is less time-consuming when users want to find relevant information objects, human intervention is required at the time of creation in order to provide them with appropriate context attributes.
- **Metadata are stored with the information object itself:** Since terms could change over time, reclassification may be necessary. Alternatively, reclassification could be avoided by separately mapping old terms with new ones (e.g., with a customized thesaurus).

It is clear that because the information objects contain content and explicated context, both maintained by their author at the time of creation or during maintenance changes, the attribute-based search actually is based on context explication.

Topic Maps

The initial idea behind topic maps, which date back to the early 1990s, arose from the need to model intelligent electronic indexes (of books), tables of content, glossaries, thesauri, or cross references in order to merge them automatically. During many years of discussion and evolutionary development cycles, the topic map model has developed into something very powerful that no longer is restricted merely to the modeling of indexes. The ISO standard ISO/IEC 13250 Topic Maps, adopted in 1999, defines a model and architecture for the semantic structuring of link networks. Topic maps establish an associative network between subjects, which represent information objects, and provide navigation paradigms that allow them to be searched. By applying topic maps to large sets of heterogeneous information repositories, reusable and structured semantic link networks are created on a level above those resources (Rath & Pepper, 1999). The key concepts of topic maps are topics that represent real-world subjects, occurrences of topics, and relationships among topics (topic associations). In addition, the topic map standard provides extended concepts of scope, public subject, and facets. For a comprehensive introduction and reference, refer to Rath and Pepper (1999) and ISO/IEC 13250 (2002).

Topic associations describe the relationships among topics. They are completely independent of the information object itself and, therefore, represent the topic map's essential added value. The addition of topic associations to the concept of topics enables topic maps to model information networks.

Topic maps organize information repositories in a new knowledge space by relating them to topics and structurally associating them. Furthermore, they enable heterogeneous sets of information repositories to be used in an integrated way by interrelating them by means of a unifying conceptual framework. Another characteristic of topic maps is that they are well-suited to represent ontologies. Consequently, they facilitate the description of a shared common understanding (e.g., about the kinds of objects and relationships that are being discussed) (Wrightson, 2001).

The link mechanism between topics and topic occurrences provides a means with which to bridge the gap between knowledge representation and information management fields (Pepper, 1999).

Since the human brain always remembers memorized issues in a specific context (Goldfarb & Prescod, 2000), association is the basic way of thinking. Topic maps support this way of thinking by pointing to related themes when a user searches for a specific theme.

To summarize, topic maps have the following benefits:

- **Creation of knowledge structures:** Applying topic maps to information repositories generates knowledge structures. They form structured, semantically linked networks above large sets of information repositories.
- **Creation of metalayers:** Transparent access to information objects is provided by searching and navigating knowledge structures (i.e., a metalayer above the information objects). Modifications of the metacontext do not affect the information objects or their descriptors. Searching in topic maps can be compared to searching in knowledge structures.
- **Discovery of new context:** Added value is achieved by the creation of new knowledge through the discovery of new contexts.
- **Support of human thinking:** Topic associations support the basic way of thinking by providing interrelating themes.

Some basic constraints are as follows:

- **Effort required for topic map creation and maintenance.** Intense human effort is needed to define, create, and maintain topic maps. Persons who manage topic maps need expertise in both topic map concepts and paradigms as well as in the specific domain to which the topic map applies.
- **New search paradigm.** Users have to learn to use the topic map search concepts and to adopt the associative way of thinking, while they are familiar with the full-text or attribute-based search concepts and their flat result sets.

As has been described, topic maps provide strong concepts and paradigms with which to discover and explicate information objects' contexts, thus relieving authors and users of the need to provide metadata or descriptions. The explicated context does not form part of the information object and even can be used without it. However, specifically skilled persons are required to support the process of context explication. Concepts for the organizational and process integration of such knowledge workers are introduced in detail in Smolnik and Nastansky (2002). In general, they need expertise in managing topic maps as well as in the specific domain to which the topic map applies.

The Continuum of Context Explication

As pointed out in our motivation, context has been recognized as an important aspect to consider when looking at the meaning of information with respect to knowledge discovery and knowledge creation. In the previous section, we presented three approaches with which to find information objects and with which to recognize, represent, and use contextual information through portals. Even though these approaches have supporting users to find relevant information objects in common, they focus on contextual information in different ways and with varying intensity.

The introduction of the continuum of context explication was one of the major results of our research. This continuum focuses on data, information objects, and knowledge as portals' basic subjects as well as on their varying embodied degree of context explication. Furthermore, it describes approaches with which to find and use information objects and contextual information (see Table 2). We define five approaches, each with a differing degree of context and explication simplicity: three approaches relate to information objects and search methods' chronological development, with the other two forming a logical extension of data's transition into information and information into knowledge. Based on given prerequisites, we furthermore provide criteria for and advice in respect to choosing an appropriate solution.

Data Approach

Data are meaningless symbols without content and context that have no context to explicate. Depending on the data quantity and the relevant domain, there are several methods with which to transform data into information objects or even into domain-specific knowledge. For example, in the knowledge discovery in databases and data mining research domains, the identification of patterns in large structured data sets results in the nontrivial extraction of implicit, previously unknown, and potentially useful knowledge (Fayyad, Piatetsky-Shapiro, & Smyth, 1996). Processed and conceptualized data, such as documents created by authors, are defined as information.

The data approach is appropriate for the following situations:

- No or little interaction with users, authors, or knowledge workers
- Large structured data sets
- Possible automatic data generation or collection

Information Approach

Even a simple information object contains some kind of content (e.g., text, audio annotation, or spreadsheet). Although the information object may provide no explicit context like descriptors or other contextual information, it inherently contains context. The context, however, is interwoven with the content and is difficult to conceptualize, which means that the methods implemented to find requested information objects have to rely on the content and cannot access contextual information. A typical method is the full-text search, as described previously. Normal full-text search engines use information objects' indexed contents to respond to a query and do not access contextual information at all. No effort, therefore, is made to explicate context as neither the authors nor the users provide or use explicit contextual information.

The information approach is appropriate in the following situations:

- Many users who have little or no experience with enhanced searching approaches or who are unwilling to use them
- Authors who have no experience describing their information objects
- Numerous unstructured information objects

Descriptor Approach

Information objects often are enriched with metadata (i.e., they contain content and explicit contextual information). Examples are Microsoft Word documents, Adobe PDF documents, or semi-structured documents in a groupware-based office environment.

Table 2. Continuum of context explication

Data approach	Information approach	Descriptor approach	Meta context approach	Knowledge approach

Large structured data sets	Unstructured information objects	Semi-structured information objects Information objects with descriptors	Large sets of heterogeneous information repositories	Information objects in person-specific contexts
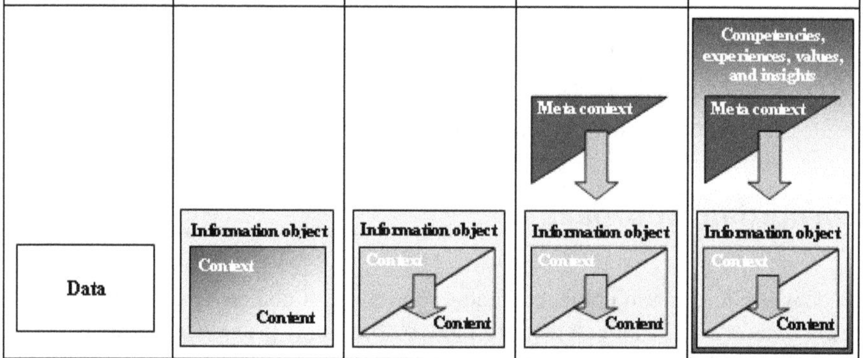				
Pattern identification	Full-text searching	Attribute-based searching Resource Description Framework	Topic maps	Action: Communication Cognition Construction
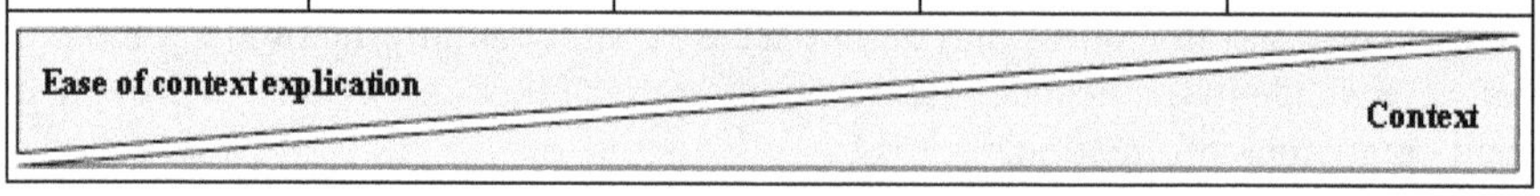				

In contrast to the information approach, information objects contain not only implicit contextual information but also explicit contextual information. As previously explained, a standard for formulating contextual information is the Dublin Core Metadata Element Set that proposes specific attribute classes for the description of an information object. Another concept for structuring and providing metadata is the resource description framework (RDF), which is resource-oriented. Its main objective is the description of resources and their relationships to other resources, with the description mostly residing in the resource.

In contrast to the information approach, some effort is necessary to enrich an information object with explicit contextual information. Authors have to provide this information at the time of creation. In addition, software systems try to maintain some of the contextual information.

The advantage of the attribute-based search as a retrieval method for information objects is dependent on the quality of the provided explicit contextual information (see the introductory section on the attribute-based search). If the metadata are wrong, misleading, or incomplete, the attribute-based search will provide insufficient result sets; if not, the attribute-based search provides more accurate results, which, to a certain extent, will fit the user's context.

The descriptor approach is suitable in the following situations:

- The authors are both trained in and skilled at describing their information objects.
- The information objects contain descriptors.
- There is a large quantity of semi-structured information objects.

Meta Context Approach

When extending the descriptor approach, information objects are described not only by metadata that reside in the information object but also by subjects, concepts, or themes that form contextual information in a metalayer above the information objects. This contextual information is not necessarily stored explicitly within the information object.

Topic maps provide strong paradigms with which to discover, maintain, navigate, and visualize this metacontext and thus explicate the context of an information object (see the introductory section on topic maps).

The semantic relationships among information objects are expressed by associating topics. This semantic network links various information objects' explicated contextual information and discovers new contexts. The discovery of these new contexts supports users in creating new knowledge when they associate known information objects in a new way with other information objects. To achieve these benefits from explicated and new contexts, substantial effort has to be invested to define, create, and maintain a topic map. This effort is disproportionally higher than the definition of metadata in the descriptor approach. In the latter case, authors or software systems explicitly provide contextual information. Authors know what they publish and easily can describe their information objects. In the metacontext approach, knowledge workers are needed to provide and maintain a topic map.

The benefit for users depends on the quality of the knowledge workers' work. If the metacontexts layer covers the entire domain of interest and contains rich and numerous topic associations, users will be able to explore the search domain easily. Thus, they will be enabled to discover new contexts and to leverage and enhance their knowledge.

The meta context approach is suitable in the following situations:

- Knowledge workers who are familiar with both topic map concepts and the domain of interest
- Manageable domains of interest
- Existing taxonomies for the domains of interest
- Users experienced in searching and navigating topic maps
- Large sets of heterogeneous information repositories

Knowledge Approach

So far, we have focused only on the human factor in very specific perspectives, such as authors defining the metadata of information objects or knowledge workers developing topic maps. The human factor plays a decisive role in the conversion of information into knowledge. We subsequently differentiate two facets of the human factor.

First, there are the competencies, experiences, values, and insights that form a rich, person-specific context. This context is a feature of knowledge's implicit dimension and hardly can be explicated (Polanyi, 1966). Within this context, a highly individual and subjective meaning is assigned to an information object. Second, users' active involvement is a necessary precondition to convert information into knowledge. This active involvement comprises actions like communication, construction, or, more intrinsically, cognition. If users experience an "I see!" or epiphanic moment as a result of some action, knowledge is created.

Characteristics of the knowledge approach are as follows:

- Competencies, experiences, values, and insights
- Information objects in person-specific contexts
- Creation of knowledge through human actions (e.g., cognition of information objects)

Small Cases and Lessons Learned

In this section, we present two small cases derived from prototypical implementations at two institutes. They illustrate the benefits and constraints of the previously discussed approaches presented with respect to the discovery of information objects and the subsequent stimulation of knowledge creation.

The first case meets the criteria of both the information and descriptor approaches in the context explication continuum introduced in the previous section. The second case is an example of a solution addressed by the metacontext approach. Motivated by the participatory action research theory (Whyte et al., 1991), our selection of these two cases was based on their significance and the available information in order to achieve an appropriate reliability and validity (Yin, 1994).

A Combination of the Full-Text and Attribute-Based Searches at the IWI

Within the Institute of Information Management (IWI) at the University of St. Gallen, we have several departments, each with two or more competence centers. Project managers lead these competence centers and are responsible for achieving their objectives. Each competence center produces many information objects (e.g., lecture materials, presentations, and publications). These materials are stored in different systems (e.g., file servers, groupware-based office environments, or Web content management systems).

From a terminological point of view, all information objects have one thing in common when contextualizing the content: they all deal with specific topics (e.g., knowledge management, enterprise application integration, business networking, etc.). Since most of the information objects are semi-structured and the maintenance of the metadata is manageable, the continuum of context explication led us to a hybrid approach. In order to reduce the maintenance effort required to achieve the controlled terminology of an attribute-based approach as well as the risk of a misspelled full-text search, we chose a combination of the two.

In an internal project, we proclaimed *topic* as the most important descriptor in contextualizing an information object's content for storage and eventual retrieval. Relevant topics previously had been collected from all the local competence centers and stored in a single parameter database. As far as a specific topic (e.g., portal) is concerned, the following contextual information is embodied in our definition framework: *Item* (preferred term for topic), *Assigned to* (responsible competence center), *Status* (draft, active, or frozen), *Synonyms* (similar terms or different languages), and *Description* (description of the term). A document history supports the traceability of modifications (Kremer, Kolbe, & Brenner, 2003). These topics subsequently are used to classify information objects (e.g., within our team databases or literature and publication applications).

The following challenges motivated us to conduct the previously described project and to implement a combination of the full-text search and the attribute-based search:

- **Availability and access:** Users inside and outside the IWI had to be able to search and to access IWI's information objects in an effective and transparent way, although they had not been provided with any navigation and search mechanisms.
- **Consistent and controlled terminology:** The authors had to be supported by a consistent and controlled terminology, because they had defined information objects' metadata without following any organizational rules or standards, or they had not used metadata at all. This led to an uncontrolled and not utilizable terminology and, thus, to no rational classification of the information objects.

Currently, there are about 350 topics overall, owned by 30 competence centers. Approximately 11,000 documents have been classified according to the introduced topics for eventual retrieval through intranet and extranet portals. Having used this approach for almost a year, we have been able to derive the following success factors from our observations of the effort to solve the challenges:

- **Simplicity:** A lean context explication framework with only a few dimensions reduces the workload associated with information object classification, which increases user acceptance. Authors classify their information objects according to the introduced terminology during the creation phase. Thus, no subsequent reclassification and editorial work is needed.

- **Mixture of centralization and decentralization:** A few simple, centralized rules for topic definitions are helpful (e.g., naming conventions). Decentralized, responsible team members make the detailed decisions regarding terms, thus reducing coordination overheads.

On the one hand, these success factors ensure that the terminology is maintained with very little effort. On the other hand, users benefit from the manageable and consistent terminology during their search.

Topic Maps at the GCC

As pointed out in Smolnik and Nastansky (2002), groupware-based office systems provide an excellent environment for organizational knowledge management. Within the Groupware Competence Center (GCC) of the University of Paderborn, the GCC K-Pool (GCC Knowledge Pool) is used in almost every facet of operative work. It is a groupware-based repository for several kinds of information objects, which chiefly maintains information on books, conferences, links, media objects, contributions, articles, and software. The different information objects are enriched with numerous descriptors:

> *categories are used to set information objects in various contexts, keywords describe the information objects in detail, and publishing information provides further explanation.*

Even though there were many semantic relationships among the information in these databases, it was scarcely possible to navigate among them or to identify knowledge structures. The capabilities to access information objects were restricted to a basic full-text search and navigating through context-sensitive views and categories. Full-text indexes, however, are insufficient when searching for information, while structures such as document types or taxonomies are sometimes too confining to qualify or categorize information objects (Biezunski & Newcomb, 2001). Furthermore, the usage and the scope of these techniques are limited to a single database.

Using the generic approach of applying topic maps to groupware-based organizational memories as described in Smolnik and Nastansky (2002), we applied the search and navigation concepts discussed in the introductory section on topic maps to the GCC K-Pool. We exhaustively defined a topic map template that comprises topic types as well as association types and describes the skeletal structure of the topic map. Typical topic types are *author*, *title*, *publisher*, or *place*; whereas typical association types are *author writes title*, *publisher publishes title*, or *publisher is located in place*. Software agents automatically create and maintain the topic map that is applied to the GCC K-Pool.

The GCC K-Pool topic map facilitates the creation of knowledge structures and metalayers, the discovery of new contexts, and supports users' cognitive capabilities. Furthermore, users are able to search and navigate the GCC K-Pool topic map in several ways. A text-oriented Web browser interface also provides intuitive access. Additionally, users can explore the

GCC K-Pool topic map by using two visualization tools: The K-Viewer, a two-dimensional approach with auto-layout capabilities for restructuring the topic map visualization, and the Sky Surfer, a three-dimensional approach with extensive navigation and search functions. These different topic map visualization approaches are described in detail in Smolnik, Nastansky, and Knieps (2003).

The GCC team consists of highly skilled researchers familiar with topic map concepts and with expertise in the Center's everyday subjects. Users and authors are supported by a slightly distinctive taxonomy. These preconditions meet the criteria that are required for the context explication continuum's metacontext approach. An excellent environment for the deployment of topic maps has therefore been created. We have used this approach for several months now and have observed the following main results:

- **Understanding work contexts:** Users understand better how their work subjects are related when interrelating themes or information objects are utilized. They are able to explore the domain of interest in an intuitive way, and thus, they are able to retrieve relevant and related information objects. The result of both observations is that users' creation of knowledge is stimulated.
- **Low maintenance:** Once configured and created, the topic map is updated automatically. Software agents insert new information objects, topics, and topic associations and delete outdated ones. For the acceptance of such an IT system in an organization or in its subunits, little maintenance effort is important.

Conclusion

As shown, information objects' context and context explication play an important role in the area of knowledge discovery and portals. As there are several possible approaches, the real task for knowledge discovery begins with the selection of the appropriate solution for context explication. Consequently, we have illustrated three approaches—full-text search, attribute-based search, and topic maps—each of which has been described according to their characteristics, benefits, and constraints as far as context explication is concerned, and aligned in our context explication continuum. Successful application of the specified criteria has been illustrated by the implementations at two institutes.

Future Areas of Research

To enrich our proposed model for context explication, we see at least four areas of future research. First, we will evaluate further the distinctness of situations in terms of applying the continuum's elements by adapting GCC's topic map framework to IWI's content. Even if the preconditions are different, this might lead to insights into the various approaches' degree of exchangeability. Second, we have to determine whether patterns can be found that will facilitate transition (e.g., from topic maps to the full-text search or the full-text search to the attribute-based search). Third, we would like to extend our continuum with indications regarding knowledge's implicit dimension by including the explication of skills and skill management. Fourth, we will validate and expand our findings with other external partners. We will focus specifically on industries other than academia to generalize the validity of the continuum. In addition, while we have focused on customer process-oriented portals, we will evaluate the continuum's application in portals that are designed for other purposes. We therefore envision that knowledge discovery through context explication will provide a comprehensive framework with which to support knowledge management processes productively.

References

Alavi, M. (1997). KPMG peat marwick US: One giant brain [Report Nr. 9-397-108]. Boston: Harvard Business School.

Alavi, M., & Leidner, D. (1999). Knowledge management systems: Emerging views and practices form the field. In *Proceedings of the 32th Hawaii International Conference on System Sciences.*

Andersen, A. (1996). *The knowledge management assessment tool: External benchmarking version.* Chicago, IL.

Andrews, W. (2003). *Visionaries invade the 2003 search engine magic quadrant.* Stamford, CT: Gartner.

Baeza-Yates, R. & Ribeiro-Neto, B. (1999). Modern Information Retrieval. Addison Wesley, New York.

Baeza-Yates, R., & Schäuble, P. (2002). Retrieving information: A discipline with a tradition. *UPGRADE (The European Online Magazine for the IT-Professional), III*(3), 3-4.

Benbasat, I., Goldstein, D. K., & Mead, M. (1987). The case research strategy in studies of information systems. *MIS Quarterly, 11*(3), 369-386.

Berners-Lee, T. (1997). *Web architecture: Metadata.* Retrieved May 7, 2004, from http://www.w3.org/DesignIssues/Metadata

Biezunski, M., & Newcomb, S. R. (2001). Topic maps and XTM—A manager's overview. In *Proceedings of the Conference on Knowledge Technologies.*

Biggam, J. (2001). Defining knowledge: An epistemological foundation for knowledge management. In *Proceedings of the 34th Hawaii International Conference on System Sciences.*

Bristow, P., Dickinson, C., Duke, S., Henry, S., & Makey, P. (2001). *Enterprise portals: Business application and technologies.* East Yorkshire: Butler Group.

Cathro, W. (1997). *Metadata: An overview.* Retrieved May 7, 2004, from http://www.nla.gov.au/nla/staffpaper/cathro3.html

Checkland, P., & Holwell, S. (1998). Action research: Its nature and validity. *Systemic Practice and Action Research, 11*(1), 9-21.

Dale, A. (2001). Designing taxonomies at unilever. *Knowledge Management Review, 3*(6), 30-34.

Davenport, T. H., Harris, J. G., & Kohli, A. K. (2001). How do they know their customers so well? *MIT Sloan Management Review, 42*(2), 63-73.

Davenport, T. H. & Prusak, L. (1998). *Working knowledge: How organizations manage what they know*. Boston: Harvard Business School Press.

Davydov, M. M. (2001). *Corporate portals and e-business integration.* New York: Mc-Graw-Hill.

Dey, A. K., & Abowd, G. D. (1999). *Towards a better understanding of context and context-awareness* [GVU Technical Report GITGVU -99–22]. Atlanta, GA: Georgia Institute of Technology.

Dias, C. (2001). Corporate portals: A literature review of a new concept in information management. *International Journal of Information Management, 21*, 269-287.

Dublincore. (2003). *Dublin Core Metadata Initiative.* Retrieved May 7, 2004, from http://dublincore.org

Fayyad, U. M., Piatetsky-Shapiro, G., & Smyth, P. (1996). From data mining to knowledge discovery: An overview. In U. M. Fayyad, G. Piatetsky-Shapiro, P. Smyth, & R. Uthurusamy, *Advances in knowledge discovery and data mining* (pp. 1-36). Menlo Park, CA: AAAI Press (American Association for Artificial Intelligence).

Felber, H., & Budin, G. (1989). *Terminologie in theorie und praxis.* Tübingen: Gunter Narr Verlag.

Fleisch, E., & Österle, H. (2001). Das tor zur IT-welt: Thesen zum erfolgreichen portaleinsatz. *Computerwoche extra, 2*(S), 28-31.

Goldfarb, C. F., & Prescod, P. (2000). *XML handbook—Anwendungen, produkte, technologien.* München: Addison-Wesley Verlag.

Gordon, M., & Pathak, P. (1999). Finding information on the World Wide Web—The retrieval effectiveness of search engines. *Information Processing and Management, 35*, 141-180.

Güldenberg, S. (1996). *Wissensmanagement und wissenscontrolling in lernenden organisationen: Ein systemtheoretischer ansatz*. Wiesbaden: Deutscher Universitäts-Verlag GmbH.

Gummesson, E. (2000). *Qualitative methods in management research.* London: Sage Publications.

ISO, ISO/IEC 13250. (2002). *International Organization for Standardization.* Retrieved May 7, 2004, from http://www.iso.org/iso/en/CatalogueDetailPage.CatalogueDetail?CSNUMBER=38068&ICS1=35&ICS2=240&ICS3=30

Kalakota, R., & Robinson, M. (2001). *M-business.* New York: McGraw-Hill.

Klemke, R. (2000). Context framework—An open approach to enhance organisational memory systems with context modelling techniques. In *Proceedings of the Third International Conference on Practical Aspects of Knowledge Management (PAKM 2000).*

Kremer, S., Kolbe, L. M., & Brenner, W. (2003). Do you know your terms?—A procedure model for terminology management. In *Proceedings of the European Conference of Information Systems (ECIS).*

Lai, H., & Chu, T.-H. (2000). Knowledge management: A review of theoretical frameworks and industrial cases. In *Proceedings of the 33th Hawaii International Conference on System Sciences.*

Latham, L. (2001). *Web content management and portals: Who's doing what?* [Research Note SPA-13-9670]. Stamford, CT: Gartner.

McGovern, G. (2001). *Why metadata is important.* Retrieved May 7, 2004, from http://www.gerrymcgovern.com/nt/2001/nt_2001_10_01_metadata.htm

Murray, P. C. (1996). Information, knowledge, and document management technology. *Knowledge Management Briefs, 1*(2), Retrieved May 9, 2002 from http://www.ktic.com/resource/km2/Information,%20knowledge,%20and%20document%20management%20technology.htm

Nonaka, I., & Takeuchi, H. (1995). *The knowledge-creating company—How Japanese companies create the dynamics of innovation.* New York: Oxford University Press.

Pepper, S. (1999). Navigating haystacks and discovering needles: Introducing the new topic map standard. *Markup Languages: Theory and Practice, 1*(4), 41-68.

Polanyi, M. (1966). *The tacit dimension.* Gloucester: Routledge & Kegan Paul.

Probst, G. J. B., Raub, S., & Romhardt, K. (1999). *Wissen managen—Wie unternehmen ihre wertvollste ressource optimal nutzen.* Wiesbaden: Gabler.

Puschmann, T. (2003). *Collaboration portale—Architektur, integration, umsetzung und beispiele* [dissertation]. Difo-Druck, Bamberg: Universität St. Gallen.

Raghavan, P. (2002). Information retrieval for enterprise content. *UPGRADE (The European Online Magazine for the IT-Professional), III*(3), 5-8.

Rappoport, A. (2002). *Web search engines: Users surprising the experts.* Retrieved May 7, 2004, from http://www.searchtools.com/slides/baychi2002/index.html

Rath, H. H., & Pepper, S. (1999). Topic maps: Introduction and allegro. In *Proceedings of the Conference on Markup Technologies 99.*

Röhricht, J., & Schlögel, C. (2001). *cBusiness—Erfolgreiche internetstrategien durch collaborative business.* München: Addison-Wesley.

Schwarz, J. (2000). Mass customization von prozessen durch unternehmensportale. *Information Management & Consulting, 15*(2), 40-45.

Smolnik, S., & Nastansky, L. (2002). K-discovery: Using topic maps to identify distributed knowledge structures in groupware-based organizational memories. In *Proceedings of the 35th Hawaii International Conference on System Sciences* (p. 10).

Smolnik, S., Nastansky, L., & Knieps, T. (2003). Mental representations and visualization processes in organizational memories. In *Proceedings of the 7th International Conference on Information Visualization (IV03)—International Symposium on Knowledge Domain Visualization (IV03-KDViz).*

Sowa, J. F. (2000). *Knowledge representation: Logical, philosophical, and computational foundations.* Pacific Grove: Brooks/Cole.

Sveiby, K. E. (1997). *The new organizational wealth: Managing & measuring knowledge-based assets*. San Francisco: Berrett-Koehler.

Wargitsch, C., & Habermann, F. (1998). *IMPACT: Workflow-management-system als instrument zur koordinierten prozessverbesserung: Anforderungen*. Nürnberg: Universität Erlangen-Nürnberg Bereich Wirtschaftsinformatik.

Web/CKS. (1999). *Web portals: Present and future positioning in the European marketplace*. [Report].

Whyte, W. F., Greenwood, D. J., & Lazes, P. (1991). Participatory action research: Through practice to science in social research. In W. F. Whyte (Ed.), *Participatory action research* (pp. 19-55). Newbury Park, CA: Sage Publications.

Wrightson, A. (2001). *Topic maps and knowledge representation*. Ontopia AS.

Yin, R. (1994). *Case study research. Designs and methods*. London: Sage Publications.

Chapter XV

Toward the Multidimensional Conceptualization of Knowledge

Mark Nissen, Naval Postgraduate School, USA

Murray Jennex, San Diego University, USA

Abstract

Many taxonomies and definitions of knowledge have been published in the KM literature. This chapter defines knowledge as something that is multidimensional and existent on a continuum. Four dimensions describing knowledge are proposed—explicitness, reach, life cycle, and flow time—and a modeling method is discussed. The chapter concludes with a call for research in the dimensionality of knowledge.

Introduction

Knowledge management (KM) practice continues to mature, and KM research continues to improve in both depth and applicability. Managers and professionals in practice are moving away from near-sole reliance upon technological artifacts, such as databases, document repositories and Web portals, and recognizing the importance of people, organization, communication, trust, and motivation. Edwards, Handzic, Carlsson, and Nissen (2003) contrast this in terms of hard vs. soft KM issues. Researchers in industry and academe finally are distinguishing between knowledge and information, examining the context of KM, and considering dynamic aspects of knowledge as it flows. This maturation of practice and improvement in research reflects the healthy progress of KM as it struggles to separate from information systems research—which has tried vigilantly to usurp the management of knowledge as some variation on managing information—and for concomitant emergence as a stable and fruitful field of its own (Jennex & Croasdell, 2005).

But the KM field as a whole continues to struggle in terms of how it treats the concept knowledge. In particular, researchers and practitioners alike refer broadly to knowledge as a single, monolithic concept. For instance, many otherwise sophisticated and technologically savvy knowledge managers still mistake knowledge as the information technologies used for support (e.g., "it's in the database"); many well-compensated KM consultants still fail to distinguish between tacit and explicit knowledge (e.g., "capture the experience"); many otherwise informed KM researchers still conflate individual and organizational-level knowledge (e.g., "the team learned a lesson"); and even the most thoughtful KM scholars still struggle to characterize knowledge beyond simple, binary contrasts (e.g., explicit/tacit, declarative/procedural, know-what/know-how). As a contrasting instance, it makes little sense to assert that tacit knowledge created by a specific individual, for instance, will behave anything like explicit knowledge shared among organizations. Yet such assertions are widespread and implicit in the bulk of both KM research and practice today. Moreover, in very practical terms, failure to differentiate knowledge is analogous to treating all forms of transportation (e.g., pedestrian, automobile, boat, airplane, teleportation) singly. Clearly, walking, sailing, or flying to China from the U.S. represent qualitatively different behaviors with very different performance characteristics (i.e., in terms of cost and time), but each is part of the concept transportation.

We have begun to call for increased sensitivity and attention to the multidimensional nature of knowledge (Nissen & Jennex 2005), and we draw upon such call here to elaborate our primary points. This call is not new, however. For instance, the ancient Greeks maintained differentiated knowledge concepts (Kane, 2003); 20th century economists measured more than a dozen different kinds of knowledge (Machlup, 1980); and some contemporary researchers are beginning to utilize two-dimensional conceptualizations to characterize knowledge flows and contingency effects (Inkpen & Dinur, 1998; Nonaka, 1994). Nonetheless, this call is timely as KM emerges from fad to necessity in practice and as it transitions from conceptualization to application in research. This call also underpins our ever-clearer realization that KM research and practice face a very real and dangerous risk of stagnation if the many different kinds and behaviors of knowledge cannot be separated out and accounted for.

To overcome such risk, KM practice needs to learn how to manage appropriately—and differently—the many different kinds of knowledge, and which of numerous alternate technologies, organizations, and processes fit best the enterprise mission and context at hand. And KM research needs to sharpen its theoretical frameworks and empirical instruments to interrelate the myriad different states of knowledge and modes of behavior with their differentiated effects on work, people, and performance, particularly where technology is intended to play an important role. This requires a renewed research thrust into understanding the multidimensional nature of knowledge. In this chapter, we argue for development of new empirical instruments to detect and measure the various dimensions of knowledge. And we build upon recent work on multidimensional knowledge flows (Nissen, 2006b) to provide a possible starting point for such research.

Analytical Framework

The first step is to develop an analytical framework to articulate and interrelate the key dimensions of knowledge. Needed are dimensions that are rooted firmly in the KM literature and that offer good potential to support the development of empirical measurement scales. Clearly, many theoretical concepts from a diversity of perspectives in disparate literatures represent candidates for use in developing such an analytical framework. Indeed, the plethora of diverse, nonintegrated concepts and perspectives makes it difficult to compose an integrated, parsimonious framework. In order to address such difficulty, we build upon prior theoretical and empirical research that focuses specifically on modeling and visualizing dynamic knowledge.

This prior research draws from diverse, dynamic knowledge perspectives and proposes a framework to integrate them through four dimensions: explicitness, reach, life cycle, and flow time. Although this prior research remains largely theoretical, each of its four constituent dimensions has some empirical basis reported in the literature. This facilitates our task of operationalizing the corresponding concepts to develop empirical instruments for use in the field. Moreover, this analytical framework has been applied to describe numerous different organizations and processes (Nissen, 2005, 2006a; Nissen & Levitt, 2004, Snider & Nissen, 2003).

The first dimension—explicitness—addresses the type of knowledge. The existence and importance of different knowledge types is noted repeatedly in several literatures (Nonaka, 1994; Postrel, 2002; Saviotti, 1998; Spender, 1996), and the distinction between explicit and tacit knowledge (Polanyi, 1967) is particularly compelling. It cleaves knowledge into one explicit class that is supported well by information systems and a tacit class that is not (Nissen, Kamel, & Sengupta, 2000). We may be able to operationalize this construct as the ability to articulate knowledge; from theory we would expect that explicit knowledge could be articulated readily, whereas tacit knowledge could not. From theory we also would expect explicitness to represent a continuum of knowledge, as what may be tacit to one person may be explicit to another and with various degrees of explicitness in between. This continuum would use the absolutes of tacit and explicit knowledge as endpoints.

The second dimension—reach—addresses the level of social aggregation associated with knowledge. The importance of theory that can cross different levels of analysis is emphasized repeatedly in the organization studies literature (Scott, 2003). Moreover, such levels of social aggregation are interrelated with various types of knowledge by several researchers (Crossan, Lane, & White, 1999), and shown to reveal expressive patterns for visualization (Nonaka & Takeuchi, 1995) and contingency development (Inkpen & Dinur, 1998). The importance of this dimension is in the amount of context of understanding that must be captured with the knowledge in order to make it actionable for users. Jennex (2006) describes how lower levels of social aggregation such as teams and work groups have shared contexts of understanding and readily can share and reuse knowledge with each other, while higher levels of social aggregation may have no understanding of the culture and context in which the knowledge was created and meant to be used, requiring users to have the story behind the knowledge. We may be able to operationalize this construct as the level of social aggregation in the enterprise; from theory, we would expect that different types of knowledge would occur at varying levels of social aggregation. From theory, we also might expect reach to represent a continuum of knowledge, perhaps with intermediate aggregation levels such as individual, group, organization, and inter-organization.

The third dimension—life cycle—addresses the activities associated with knowledge. We note a variety of activities in the literature, ranging from knowledge creation and conversion to sharing and application to reuse and forgetting. We note also how several life-cycle models incorporate these various perspectives into a process view (Nissen et al., 2000). Through work to conceptualize the dynamics of knowledge flows (Nissen, 2002), the dimension life cycle has further been integrated with both explicitness and reach and shown to enable novel visualization of diverse dynamic knowledge patterns such as spirals (Nonaka, 1994) and others. We may be able to operationalize this construct as the activity associated with knowledge; from theory, we would expect that different types of knowledge across multiple levels of organizational reach would involve varying activities of a life-cycle process. From theory, we also might expect life cycle to represent a continuum of knowledge, perhaps with categorical variables such as knowledge creation, sharing, and application.

The fourth dimension—flow time—addresses explicitly the dynamic nature of knowledge, and it enables dynamic knowledge to be conceptualized directly. Incorporating explicitly this temporal dimension departs substantively from the theoretical models already summarized. It makes explicit the dynamic nature of knowledge, and it may support empirical measurement of such dynamics. Through the prior research noted previously, the dimension flow time has been integrated further with all three of the others (i.e., explicitness, reach, life cycle) and has been shown to enable multidimensional visualization of diverse, dynamic knowledge patterns (Nissen, 2006a). We may be able to operationalize this construct as the length of time required for knowledge to flow; from theory, we would expect that different types of knowledge across multiple levels of social aggregation and associated with varying knowledge activities would flow at different rates. From theory, we also would expect flow time to represent a continuum of knowledge, perhaps with relative orders of magnitude (e.g., days, months, years) in order to distinguish among various knowledge flows.

Drawing heavily from Nissen (2006a), Figure 1 helps us to visualize dynamic knowledge in terms of the four dimensions previously outlined. Here, we include axes for the three dimensions—explicitness, reach, and life cycle—and we extend this representation to include the flow time dimension; relatively long vs. short flow times are differentiated by the thickness of arrows depicting knowledge flows. Each arrow in the figure represents a distinct knowledge flow, and we refer to each arrow as a trajectory to connote its dynamic nature. The on-the-job training (OJT) (a euphemism for trial and error) knowledge flows

Figure 1. Dynamic knowledge visualization (Adapted from Nissen, 2006a)

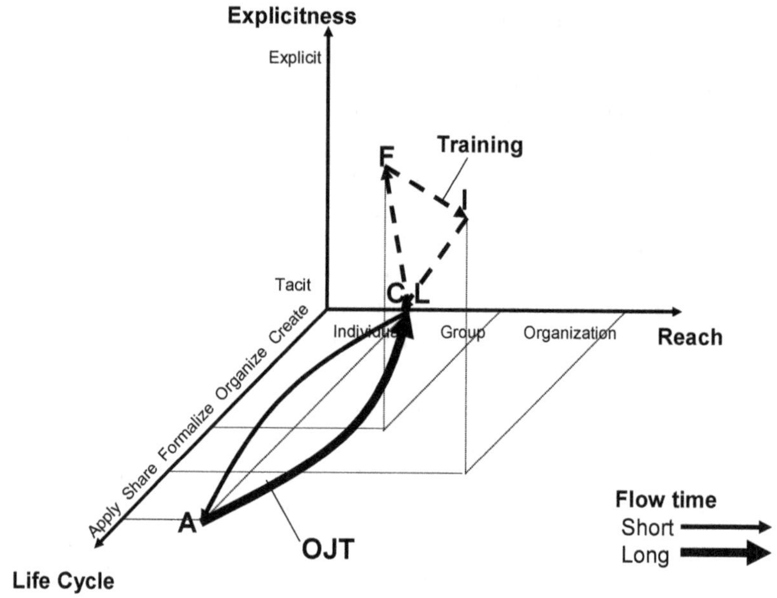

represented in the figure include two corresponding trajectories, one each for the modes of job learning and doing. Likewise, the training (represents classroom instruction) knowledge flows represented in the figure include three trajectories corresponding to the modes of course instruction, development, and learning.

Specifically, the OJT process is delineated as a cycle of two dynamic knowledge trajectories in the tacit plane of the figure. The cycle connects points C and A, which correspond to knowledge creation and application, respectively. Notice that the flow represented by this cycle reflects tacit, individual knowledge flowing at two different speeds along the life-cycle axis. We depict the flow corresponding to knowledge creation at point C using a relatively thick line (i.e., slow flow) and the flow corresponding to knowledge application at point A using a relatively thin line (i.e., fast flow). The training process is delineated by its own flow trajectories that rise up out of the tacit plane in the figure. Beginning at point C, working knowledge is formalized by a group of instructors through course development into an explicit state at point F. We depict classroom instruction as a subsequent flow to point I, through which explicit training material is shared organizationwide. Learning by individual students is denoted by point L, which we depict as an individual process of knowledge creation.

Developing an Instrument

The next step is to operationalize the four dimensions outlined previously and to create empirical instruments that can be taken into the field to measure knowledge in practice. The development of empirical instruments represents an important aspect of research in the physical and social sciences alike. A great many scholarly articles and practical textbooks have been written on the subject, and a great many useful instruments have been developed over a considerable period of time. The KM field needs researchers and practitioners alike to cooperate on the development of one or more instruments in order to measure knowledge multidimensionally. Some research along these lines is underway. For instance, we find work to understand the dynamics of knowledge—and, in particular, their managerial implications—in terms of Inventory Theory (MacKinnon, Levitt, & Nissen, 2005), and advanced, computational modeling of knowledge dynamics is beginning to illuminate—and measure—intriguing interdependencies between different types and behaviors of knowledge (Nissen & Levitt, 2004). Researchers need to ensure that the resulting scales possess characteristics of good validity and can be applied empirically in the field. Practitioners need to ensure that the resulting scales measure aspects of knowledge that are important to managers and for organizational performance. It is not the place of a chapter such as this to specify how such scales should be developed nor to develop them. Rather, we repeat here our call for action to treat knowledge as a multidimensional concept and to develop empirical scales for multidimensional measurement. Given the multidimensional foundation that we outline here and the considerable theoretical and empirical research on which it is based (Nissen, 2006b), researchers with talent should be able to move forward along these lines and to enable multidimensional conceptualization and measurement of dynamic knowledge.

References

Crossan, M. M., Lane, H. W., & White, R. E. (1999). An organizational learning framework: From intuition to institution. *Academy of Management Review, 24*(3), 522-537.

Edwards, J., Handzic, M., Carlsson, S., & Nissen, M. (2003). Knowledge management research & practice: Visions and directions. *Knowledge Management Research & Practice, 1*(1), 49-60.

Inkpen, A. C., & Dinur. A. (1998). Knowledge management processes and international joint ventures. *Organization Science, 9*(4), 454-468.

Jennex, M. E. (2006). Classifying knowledge management systems based on context content. In *Proceedings of the Hawaii International Conference on Systems Sciences, IEEE.*

Jennex, M. E., & Croasdell, D. (2005). Is knowledge management a discipline? *International Journal of Knowledge Management, 1*(1), i-v.

Kane, H. C. M. (2003). Reframing the knowledge debate, with a little help from the Greeks. *Electronic Journal of Knowledge Management, 1*(1), 1-9.

Machlup, F. (1980). *Knowledge: Its creation, distribution, and economic significance* (vol. 1). Princeton, NJ: Princeton University Press.

MacKinnon, D. J., Levitt, R. E., & Nissen, M. E. (2005). Knowledge as inventory: Near-optimizing knowledge and power flows in edge organizations (phase one). In *Proceedings of the International Command & Control Research & Technology Symposium, McLean, Virginia.*

Nissen, M. E. (2002). An extended model of knowledge-flow dynamics. *Communications of the Association for Information Systems, 8*, 251-266.

Nissen, M. E. (2005). Delineating knowledge flows for enterprise knowledge agility. In M. Khosrow-Pour (Ed.), *Encyclopedia of Information Science and Technology, Volume 2 Hershey PA: Idea Group Press (2005), 779-785.*

Nissen, M. E. (2006a). Dynamic knowledge patterns to inform design: A field study of knowledge stocks and flows in an extreme organization. *Journal of Management Information Systems 22*(3), 225-263.

Nissen, M. E. (2006b). *Harnessing knowledge dynamics: Principled organizational knowing & learning*. Hershey, PA: IRM Press.

Nissen, M. E., & Jennex, M. E. (2005). Editorial preface—Knowledge as a multidimensional concept: A call for action. *International Journal of Knowledge Management, 1*(3), i-v.

Nissen, M. E., Kamel, M. N., & Sengupta, K. C. (2000). Integrated analysis and design of knowledge systems and processes. *Information Resources Management Journal, 13*(1), 24-43.

Nissen, M. E., & Levitt, R. W. (2004). Agent-based modeling of knowledge dynamics. *Knowledge Management Research & Practice, 2*(3), 169-183.

Nonaka, I. (1994). A dynamic theory of organizational knowledge creation. *Organization Science, 5*(1), 14–37.

Nonaka, I., & Takeuchi, H. (1995). *The knowledge-creating company: How Japanese companies create the dynamics of innovation.* New York: Oxford University Press.

Polanyi, M. (1967). *The tacit dimension.* London: Routledge.

Postrel, S. (2002). Islands of shared knowledge: Specialization and mutual understanding in problem-solving teams. *Organization Science, 13*(3), 303-320.

Saviotti, P. P. (1998). On the dynamics of appropriability, of tacit and of codified knowledge. *Research Policy 26*, 843-856.

Scott, W. R. (2003). *Organizations: Rational, natural, and open systems* (5th ed.). Upper Saddle River, NJ: Prentice-Hall.

Snider, K. F., & Nissen, M. E. (2003). *Beyond the body of knowledge: A knowledge-flow approach to project management theory and practice.* Project Management Journal, 34(2), 4-12.

Spender, J. C. (1996). Making knowledge the basis of a dynamic theory of the firm. *Strategic Management Journal, 17*, 45-62.

Chapter XVI

Eliciting Tacit Knowledge Using the Critical Decision Interview Method

Hazel Taylor, University of Washington, USA

Abstract

Interest in the capture of tacit knowledge within organizations has risen in recent years. However, while the capture of explicit *knowledge is relatively straightforward, methods for eliciting* tacit *knowledge are less well-developed. This chapter briefly overviews a number of strategies for eliciting tacit knowledge and then provides a detailed examination of one of these strategies: the critical decision interview approach. The critical decision interview method can assist expert respondents to articulate tacit knowledge by probing beyond their espoused theories about their actions to reveal their practice. Tacit knowledge then can be identified by contrasting respondents' practices with theoretical prescriptions for best practice in the field. The application of the method in an investigation of risk management in IT projects is described, and the effectiveness of this method for surfacing tacit knowledge is discussed.*

Introduction

In recent years, there has been rapidly growing interest in the management of organizational knowledge, and significant attention has focused on individual employees' tacit knowledge and the question of how this tacit knowledge can be surfaced and shared or retained within the firm (Alavi & Leidner, 2001; Malhotra, 2000; Nonaka, 1994). While there has been a recognition that employees' tacit knowledge first must be made explicit before it can be managed (Nonaka, 1994; Walsham, 2001), less attention has been paid to methods of eliciting tacit knowledge and helping to make it explicit. Most researchers agree that, at best, tacit knowledge is difficult for its holder to articulate and that respondents' theories of action (in Argyris and Schön's terminology, 1978) may well be different from their actual practice. A key requirement for the capture of tacit knowledge, therefore, is a knowledge elicitation technique that has the potential to prompt and assist a respondent to recall and articulate tacit knowledge and to get beyond the theories or rationalizations that a person may use to explain his or her actions.

In this chapter, I show how the critical decision interview method can aid in eliciting tacit aspects of knowledge from expert practitioners, and I illustrate its use in a research project investigating tacit knowledge in the field of risk management in IT projects. I begin by discussing issues related to tacit knowledge elicitation and the key requirements of a tacit knowledge elicitation method and briefly review possible knowledge elicitation strategies. Then I describe the critical decision interview method, show how it meets the key requirements, and discuss implementation and analysis procedures. Next, I describe the application of the method in a recent research project investigating tacit knowledge in risk management of IT projects. I discuss the results obtained and reflect on the effectiveness of the method as it was applied in that project. Finally, I conclude with a brief discussion of the effectiveness and limitations of the method as a tacit knowledge elicitation tool.

Tacit Knowledge Elicitation

The concept of tacit knowledge has been used by researchers in a wide range of disciplines with a corresponding variety of meanings and characterizations. Consequently, there is some confusion in the literature over the exact definition of tacit knowledge and its relationship to similar concepts, such as implicit learning, procedural knowledge, and practical intelligence (Ambrosini & Bowman, 2001; Berry & Dienes, 1993; Castillo, 2002). While some researchers regard tacit knowledge as completely inarticulable and, therefore, unlikely to be transferable explicitly to other individuals (Tsoukas, 2003), most theorists view the tacit-explicit dimension as a continuum (Ambrosini & Bowman, 2001; Berry & Dienes, 1993; Castillo, 2002; Keane & Mason, 2006; Leonard & Sensiper, 1998; Polanyi, 1966; Reber, 1993) or as two dimensions present in all knowledge (Stenmark, 2002) and, hence, argue that, depending on the degree of tacitness, tacit aspects of knowledge can be surfaced. Nonaka (1994) and Takeuchi (2001) believe that tacit knowledge includes both technical skills and cognitively based knowledge, and argue that cognitive tacit knowledge can be made at least partially explicit by the use of metaphor, analogy, and prototype. Sternberg and

Wagner (1986) and Klein , Calderwood, and MacGregor, (1989) have developed interview techniques based on the use of storytelling approaches to facilitate the elicitation of this type of tacit knowledge.

In applied management studies, there has been a lack of consistency in the operationalization of the tacit knowledge concept and on what distinguishes tacit knowledge from explicit knowledge (Ambrosini & Bowman, 2001; Castillo, 2002). However, one group of researchers working with Sternberg and Wagner (Sternberg et al., 2000; Sternberg & Horvath, 1999; Sternberg & Wagner, 1986) are notable both for developing a clear definition of their concept of tacit knowledge and for gathering a substantial body of empirical evidence to support their concept. For the purposes of this chapter, I have adopted the definition of tacit knowledge developed by Sternberg et al. (2000); namely, that tacit knowledge is quite simply knowledge (rather than technical skills) that is acquired implicitly from everyday experience and that is difficult for the possessor to articulate or explain.

Tacit knowledge, as defined by Sternberg et al. (2000), has three key features. First, it typically is acquired implicitly; that is, by experience, observation, or trial and error, without systematic support from other people or media such as books. Second, tacit knowledge tends to be procedural knowledge that guides behavior (i.e., knowing how rather than knowing what). And third, it is knowledge that has a direct practical outcome for the person acquiring it (i.e., in the context of business-related research; it is knowledge that respondents have acquired that is directly applicable in the course of their work).

Requirements of a Tacit Knowledge Elicitation Method

The key features of tacit knowledge delineated by Sternberg et al. (2000) impose specific requirements on any method that aims to elicit such knowledge. In particular, elicitation of tacit knowledge requires a method that can assist respondents in surfacing and articulating their knowledge about the situations of interest and that takes into account the contextual and experience-based nature of the knowledge. It is necessary to ensure that respondents focus on what they actually did in certain situations rather than on their theories about what they ought to have done, while still capturing information about meaning and purpose ascribed by the respondents to their actions. The context of the situations examined is of paramount importance in gaining an understanding of the dynamics of the process and of the triggers that the respondent looked for to provide clues about what to do in a particular situation. There are three major requirements to be addressed in selecting a suitable tacit knowledge elicitation technique; they are discussed next.

First, one facet of individuals' tacit knowledge is that it is learned by experience and is not commonly known (Sternberg et al., 2000; Sternberg & Horvath, 1999). Thus, any attempt to capture tacit knowledge rather than just explicit knowledge must draw out the respondents' definitions of which situations meet the area of interest rather than imposing a definition on them. By allowing respondents to be the judge of which situations to discuss within the broad field of interest, the results are more likely to capture the participants' interpretations of what is important rather than simply reflecting the researcher's biases.

Second, since respondents' theories of action may be different from their actual practice (Argyris & Schön, 1978), it is important to ensure that the tacit knowledge elicitation method is focused specifically on exploring what actually happens in practice rather than on simply reporting what respondents thought they ought to do in practice. These actual practices are likely to involve both explicit and tacit knowledge about the tasks in question. However, tacit knowledge tends to be context-specific, and what might appear to be the same task may be solved quite differently on different occasions because of variations in the context or the environment. Thus, the research technique must aid in surfacing the environmental clues that experts observe when formulating the issues or problems related to a particular situation and when devising solutions or responses to those situations.

Focusing respondents on what actually happened in practice also helps to address the third requirement; namely, to prompt and assist recall of the underlying tacit knowledge, and since it is typically difficult for respondents to fully explain or articulate the tacit aspects of their knowledge and interpretation of their responses and actions (Berry & Dienes, 1993; Sternberg et al., 2000), the method also should provide a means of inferring or deducing the tacit knowledge from possibly incomplete or unclear articulations provided by respondents.

Possible Tacit Knowledge Elicitation Strategies

The acquisition of expert knowledge, which includes both tacit and explicit components, has been the focus of extensive research (see, for example, reviews by Hoffman, 1992; Hoffman, Shadbolt, & Burton, 1995; Olson & Biolsi, 1991; Shadbolt, O'Hara, & Crow, 1999). Many of the techniques described in these reviews are more appropriate for revealing knowledge that once was known explicitly to the holder and has become tacit simply through years of practice. For example, protocol analysis (Ericsson & Simon, 1993), in which an expert is asked to think aloud while performing a task, has been used extensively, particularly in problem-solving and decision-making research. It is most effective for analyzing tasks that are relatively easy to verbalize (Olson & Biolsi, 1991), since the requirement that subjects give a running commentary to accompany their execution of a task is difficult for many people to meet, particularly when the knowledge associated with the performance of the task has become tacit through years of experience (Wagner, Najdawai, & Chung, 2001). In particular, when respondents are asked to verbalize those aspects of their processes that are more tacitly held, instead they may attempt to rationalize the process and verbalize their theories of action (Gordon, 1992). Respondents also may have difficulty reporting salient cues that enable them to recognize and respond to certain stimuli (Ericsson & Simon, 1993), and these cues are often related to tacit knowledge that respondents have acquired through experience.

There are, in fact, few well-established methods for empirically investigating those aspects of tacit knowledge that were acquired implicitly and, hence, are likely, at best, to be imperfectly articulated by the holder (Ambrosini, 2003). It is also important to note that this key characteristic of tacit knowledge—difficulty of articulation by the holder—implies that tacit knowledge elicitation methods are likely to require a substantial amount of interpretation and deduction of tacit knowledge items from the raw data. Hence, the methods described next typically are applied within an interpretive research framework.

In the workplace, the more technical dimensions of implicitly acquired tacit knowledge often are transferred from expert to novice via observation or apprenticeship (Blackler, 1995; Nonaka, 1994). However, such transfer processes in the workplace typically do not render the tacit knowledge explicit, and there is no clear assessment of the worth of the tacit knowledge being gained by the apprentice. It is quite possible that the knowledge transferred may actually perpetuate inferior work practices (Ambrosini, 2003). In the research arena, these observational and apprenticeship approaches are encompassed in the contextual inquiry method (Beyer & Holtzblatt, 1995), which combines in-depth interviewing and observation techniques. The researcher spends extended periods of time in the workplace with each respondent, and in some cases, actually acts as an apprentice to learn to perform the task in question (Nonaka & Takeuchi, 1995). By using judicious questioning about observed behaviors, this method can effectively meet the three key requirements discussed previously of allowing the respondent to determine the key areas of interest, focusing on what actually happens, and prompting recall of tacit knowledge about the activities of interest. However, this approach is very time-consuming, and often it is difficult in the research context to gain entry to suitable situations for extended periods of observation.

The more cognitive dimensions of tacit knowledge can be transferred effectively in the workplace by mentoring and storytelling techniques (Swap, Leonard, Shields, & Abrams, 2001). This approach has been developed further by researchers to provide a direct method of eliciting tacit knowledge through specialized interview approaches combined with detailed content analysis in order to identify or infer the tacit knowledge items (Klein et al., 1989; Sternberg et al., 2000). These direct approaches are discussed in more detail in the following section.

Cognitive aspects of tacit knowledge also have been examined with indirect methods designed to investigate human cognition, such as causal mapping and rep grid techniques (Ambrosini & Bowman, 2001; Olson & Biolsi, 1991). These indirect methods typically rely on effective interview techniques to elicit raw data and use various inference or analysis techniques to infer underlying tacit knowledge. For example, the causal mapping technique uses a series of probe questions to reveal layers of the underlying causes of a specified outcome in order to prepare a graphical representation or map of the processes. These maps then can be analyzed in a number of different ways, either by the researcher or with the active involvement of the participants (Ambrosini, 2003; Eden, Ackermann, & Cropper, 1992). In particular, tacit knowledge can be inferred by comparison of maps elicited from experts and novices or by sharing individual causal maps with a group of respondents all working at the same tasks in order to identify any aspects of one individual's map not commonly known to the whole group. Similarly, probe questions are used in the rep grid technique to uncover respondents' perspectives on the similarities and differences between sets of key constructs or situations in the area of expertise under investigation (Olson & Biolsi, 1991). The answers to the questions then are subjected to further structured analysis in order to infer the tacit knowledge, if any, being applied to distinguish between different categories in the domain of interest.

Other indirect knowledge elicitation techniques have been used to demonstrate that experts hold (possibly) tacit knowledge that novices in the field do not have, without explicitly articulating the particular knowledge items. Various card sort, rating, and clustering techniques, for example, have been used in studies comparing novices and experts to reveal differences in how experts structure their knowledge about the subject area (Olson & Biolsi, 1991). However, while these techniques can demonstrate that experts make different distinctions from novices about their area of expertise, it can be difficult to give meaningful names to the distinctions or clusters identified, and further investigation and analysis is needed to reveal tacit knowledge aspects of these distinctions (Hoffman et al., 1995).

In terms of the three key requirements for a tacit knowledge elicitation method discussed earlier, the success of these indirect techniques depends on the effectiveness of the interview techniques used as well as on the skills of the researcher in the subsequent analysis. The critical decision interview method is a very effective technique for meeting the elicitation requirements, as discussed next in detail, and can be used to provide data for direct content analysis to identify tacit knowledge or to underpin a causal map or rep grid analysis.

The Critical Decision Interview Method

The critical decision interview method is a detailed approach to the elicitation of tacit knowledge with the aim of revealing aspects of expertise such as the critical cues that form the basis for judgment decisions. It is based on the use of the critical incident technique, which is a method for obtaining specific, behaviorally focused descriptions of job performance (Flanagan, 1954). In the critical incident technique, respondents are asked to provide specific examples with detailed contextual and behavioral information of situations that they consider to be important and relevant to good or poor performance in the area under question. Thus, the critical incident approach focuses on what actions the respondents took, rather than on why they decided on a certain action in a specific situation, and so helps to reveal respondents' actual practices rather than their theories of action.

Building on Flanagan's (1954) critical incident technique by adding a storytelling component, Sternberg et al. (2000) (Sternberg & Horvath, 1999; Wagner, 1987; Wagner, Sujan, Sujan, Rashotte, & Sternberg, 1999) examined domain experts' tacit knowledge in several different settings, including academia, military leadership, sales, and business management. By encouraging the respondents to tell stories that are illustrative of good or poor performance in a particular area, Sternberg et al. (2000) argue that respondents are better able to recall specific and relevant details about the particular context of the story and to identify actual behaviors rather than reporting their own theories about their behaviors. Thus, it is more likely that the story will reveal or uncover underlying tacit knowledge, and the emphasis on actual behaviors also will keep the respondents focused on their actual practice rather than on what they think they ought to have done.

Researchers from a quite different research field, naturalistic decision making (Klein, 1992; Klein et al., 1989) have developed a very similar variation of the critical incident technique—the critical decision method—which has been used extensively for eliciting expert knowledge in situations in which the experts have difficulty accessing their knowledge. It differs from the more general critical incident techniques of Flanagan (1954) and Sternberg et al. (2000) in that it allows more cognitive probing to encourage respondents to reflect on their own strategies and bases for decisions. One of the key strengths of the critical decision method is its concentration on identifying the detailed context of a given critical situation, the specific environmental and situational cue usage, and the reflection by respondents on similarities between a specific situation and other situations in their experience. As Klein et al. (1989) note, focusing the respondent on similarities and differences between situations in their experiences can illuminate the real reasons for taking a particular action in a specific situation and so get beneath respondents' rationalizations about their actions in order to reveal underlying tacit knowledge.

The critical decision interview method described in the present chapter combines Sternberg et al.'s (2000) storytelling and Klein et al.'s (1989) cognitive probing refinements in a semi-structured protocol that focuses the respondent on actual events and draws on the respondent's judgment about which specific incidents and aspects of incidents are important. The method meets the key requirements discussed in the preceding section as follows. First, by asking respondents to determine which situations are important and challenging, the approach ensures that the identification of critical situations emerges from the data itself rather than being imposed by the researcher. Second, the use of careful follow-up questioning, such as asking for another similar story in which the expert did something different, helps to identify context-specific details and environmental clues that may be important factors in aiding the expert to determine the specific action to take. Third, the effectiveness of the method for facilitating verbalization of difficult-to-articulate tacit knowledge has been shown by both Sternberg et al. (2000) and Klein et al. (1989), who have used their variations of the method extensively in research for eliciting expert knowledge in situations in which the experts have difficulty accessing their knowledge.

Implementing the Critical Decision Interview Method

Critical decision interviews typically follow a semi-structured protocol, providing enough guidance to ensure overall consistency across interviews but retaining enough flexibility to allow developing concepts to be well-grounded in the data with complete and thorough contextual descriptions (Walsham, 1995). Respondents are chosen for their expertise in the field of interest, with expertise being established on criteria such as number of years in the key role under investigation and recognition by peers and/or superiors of their expertise in the role.

Typically, respondents are asked to focus on a recent specific situation or incident that they have experienced directly in the field of interest. They are encouraged to tell the story about this situation without interruption in order to help to activate their memory of the incident as a context for the following questions and to establish a rapport. This storytelling process also helps to create a context for the interviewer's understanding of the situation from the perspective of the respondent and to guard against the interviewer's own biases and preconceptions. A structured series of follow-up probe questions then are used to elicit information about the situational cues surrounding the incident, the strategies and options considered, the factors or triggers that determined one response rather than another, detail about the action taken and the consequences of the action, and why this situation could have been difficult for novices. Once respondents have exhausted their recall of key incidents for the specific situation chosen, they are asked to consider whether those incidents and the actions and consequences were typical of other similar situations in their experience. Follow-up questions are used again to identify any key differences between the situations identified.

Analytic Procedures

A content analysis approach is used to identify details of each incident, the contextual cues, and actions taken. In order to determine what, if any, of the knowledge captured in the critical decision interview process is actually tacit knowledge rather than explicit knowledge, the analysis of the interview transcripts must include a comparison with some kind of explicit knowledge benchmark. Previous studies have sought to achieve this benchmarking by comparing expert and novice respondents (Klein et al., 1989; Sternberg & Horvath, 1999; Wagner, 1987; Wagner et al., 1999; Walsham, 1995), based on the assumption that novices to the field of interest will have gained a level of explicit knowledge but will not have had time to develop any tacit knowledge from their actual experiences. While this approach has some merit, it typically involves using graduate students as the novice respondents. These students may have a good grasp of the academic requirements in the field of interest, but they may not yet have gained a full appreciation of the body of explicit practitioner knowledge referred to as best practice in their field. Thus, the alternative proposed in this chapter is to benchmark expert responses against a recognized and well-established best-practice manual for the field. By seeking the congruencies and variances between the actual practice described by respondents and the prescribed best practice, tacit knowledge items can be deduced.

Illustrative Research Project—Tacit Knowledge in Risk Management of IT Projects

This interpretive study concerned the practice of risk management in IT projects and, in particular, aimed to identify knowledge, both tacit and explicit, that experienced project managers used in order to plan for and address critical risk situations that arose during the course of their projects. Two key aims of the research were to examine the congruence between theory and practice for risk management in IT projects and to uncover tacit knowledge related to practice. A key requirement, therefore, was to use a research technique that could surface respondents' underlying tacit knowledge. The critical decision method was used to guide the interview approach in this research, and details of its application are described next.

Interviews

A semi-structured interview protocol (see Appendix A) was developed using a set of core procedures based on the critical decision method (Klein et al., 1989). The interviews consisted of the following six parts:(a) introduction and explanation of the process; (b) general description of a specific project the participant had worked on recently and his or her role in the project; (c) description of the risk management processes applied in the project; (d) elicitation of specific incidents during the course of the project that were risky and challenging and for which the respondent considered a less experienced person might have made different decisions; (e) exploration of the typicality of this project in the respondent's experience; and (f) collection of demographic data.

Sample Selection

The study was carried out in Hong Kong. Organizations that had a significant involvement in IT software implementation were identified on a reputational basis. Once I had made an initial contact with a senior project manager within a firm, I used a snowball strategy to obtain further contacts, both within the firm and in other firms. Within each firm, I asked the senior initial contact to identify IT project managers who would be recognized as proficient in their domain, based on their years in the profession and on the number of IT implementation projects with which they had been involved. In general, I used a guideline of seeking respondents with a minimum of five years experience with at least five projects. Since I was interviewing respondents who had had a long career in IT project management, most of the interviewees were able to talk about their experiences with two or three different projects that they considered worthy of discussion (i.e., interesting) from a risk management perspective. I interviewed 25 respondents who discussed a total of 60 projects. By the completion of the 25 interviews, the preliminary data analysis was showing a convergence of themes with no new themes emerging, even though the sample included a very wide range of respondents in terms of type and extent of experience.

Data Analysis

Since tacit knowledge is difficult (but not impossible) to articulate, to some extent its existence must be deduced from a detailed examination of the situations described by respondents and a comparison of their descriptions of their actions with what they might have been expected to do if they had applied only explicit knowledge gained, for example, from formal training courses. Thus, as discussed next in more detail, the analysis focused on identifying

Figure 1. NVivo coding structure

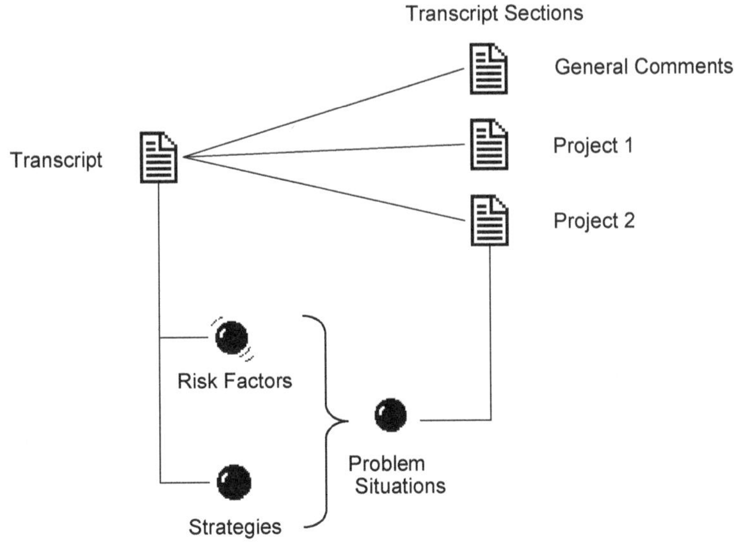

specific risk-related situations in each transcript, classifying the risk involved in the situation according to a risk category framework derived from the literature, and comparing the actions described by the respondent in response to the risk with prescribed practices and strategies from the literature for dealing with that risk. Respondents' tacit knowledge about risk management practices then was deduced from the variances in their actions from prescribed practices and strategies.

A qualitative content analysis procedure was used to analyze the interview transcripts, supported by the NVivo version 2.0 software package. I set up an overall structure for each transcript, illustrated in Figure 1, by splitting the transcript into sections corresponding to each of the projects discussed by the respondent, plus a section for the respondent's general comments. I split each of these sections further into passages encompassing specific risk or problem situations that the respondent had discussed. Once this overall structure had been established, I used two stages of coding for the detailed analysis.

The first stage was to identify the specific risks being discussed by each respondent, and to do this, I used a framework of risk factors identified from the literature (Schmidt, Lyytinen, Keil, & Cule, 2001). The development of this risk factor framework is described in more detail elsewhere (Taylor, 2004a). When I had completed this stage of the coding I returned to the first transcripts coded and recoded them without reference to my original coding in order to check whether there had been any shifts in my interpretation or application of codes over time. This recoding helped to ensure consistency in the application of codes for the risk factors. The second stage involved analyzing each of the risk-related or problem-related passages identified earlier to establish the specific actions or strategies described to address the risk, together with any environmental or situational pointers that managers attended to as they dealt with the particular problem facing them. These signals, together with contextual detail relating to the project circumstances, were of particular interest, as they were likely to represent key tacit knowledge that the managers have learned about what to watch for during the progress of their projects.

In conjunction with this second stage of coding, I used an iterative process to compare the actual practices in response to context-specific problems with the commonly accepted and promoted best-practice recommendations provided in the literature in order to identify any practices that seemed to characterize tacit knowledge about aspects of addressing critical risk situations. I used the Project Management Institute's practitioner guide (PMI Standards Committee, 1996; Project Management Institute, 2000) as the benchmark for this comparison and established the extent of congruence with and variance from these prescriptions in the respondents' descriptions of their projects.

Results and Discussion

The variances among respondents' practices and literature prescriptions enabled me to deduce tacit knowledge about IT risk management practice held by this group of Hong Kong project managers. One key aspect of the managers' tacit knowledge was the use of a few broad general strategies to manage wide ranges of risk in their projects, rather than the recommended prescription of developing and applying risk-specific strategies. This approach reflected managers' tacit knowledge about how best to manage their own time and resources and about how to be best prepared for whatever problems might arise, whether or not they were anticipated at the start of the project. Managers also applied rules of thumb to guide their use of typical project management tools and generally approached client interactions from a negotiation standpoint rather than from the contract enforcement point of view that often predominates in the literature, particularly with reference to issues of change control.

Some of the items of tacit knowledge seemed very commonplace and, on the surface, appeared to add little to the body of explicit knowledge. However, these items reflected very practical rules of thumb developed through experience, which address the question of how to enact the basic steps found in any project management guide. For example, development and control of the work breakdown structure (WBS) are fundamental steps in all prescriptive literature about project management, and yet, respondents felt it important to highlight what they had *not* learned from the basic instruction courses but had found out through their own experience; namely, rules of thumb on developing the WBS (break down tasks to a maximum of one week duration for practical control) and on controlling progress on the WBS (probe for specific evidence to support claims about the percentage completion of tasks). Similarly, while standard prescriptions recommend strict adherence to specified requirements, the application of change control, and the use of escalation to quickly resolve any problems, the respondents in this study described a much more flexible approach with a rule of thumb of trying to view the situation from the customer's perspective in order to build and maintain trust with their clients. These managers' experiences had led them to place high importance on the need to build and maintain a trusting relationship from the start, and in order to achieve this, they showed a reluctance to rely on the contract specifications to enforce progress, preferring to negotiate agreement and being prepared to waive contract conditions at times in order to maintain the overall relationship with their clients. Full details of the tacit knowledge items deduced are reported elsewhere (Taylor, 2004b).

Reflections on the Critical Decision Interview Method

On reflection, as I worked through the analysis of transcripts, I did not find that items of tacit knowledge immediately became obvious to me. Rather, I was aware of the extensive and detailed scrutiny of the transcripts that was required in order to tease out the often subtle variations between theory and practice. No doubt, this was due in part to the fact that the respondents, for the most part, followed rigorous processes in their risk management. However, as I continued to iterate between prescriptions and descriptions of practice, the gaps and variations did become clear.

The effectiveness of the method also varied from respondent to respondent. Some respondents seemed to grasp very quickly the idea of recounting a specific project to tell the story of that project. They were eager to talk about what actually had happened, and it was easy to keep them focused on what they actually did at each stage. Other respondents, however, were more uncomfortable with the idea of simply talking about what had happened in a project and seemed to think that I would want more from them than just a description of what had happened in a particular project. They wanted to share their opinions about what ought to be done or to speculate about things that might go wrong in IT projects, based on the conclusions they had drawn from reflections on their experiences over a number of years. I had to work hard with these respondents to focus them in the direction I wanted, and with some of them, I found it useful to allow the first part of the interview to follow their agenda before drawing them back to the specific areas in which I was interested.

Indeed, the experience with the more reluctant respondents convinced me that just asking respondents to identify key risk factors and the strategies they used to manage risk would have been unlikely to reveal anything new or anything more than the respondents' espoused theories. As noted in the section on data analysis, I coded and analyzed passages relating to respondents' general comments separately. I found little evidence in these general comments of any variation from the prescriptions, with the exception of knowledge about flexible application of change control processes, which emerged both from respondents' general comments and specific project descriptions. However, the process of focusing respondents on what they had actually done in specific projects did enable the surfacing of tacit knowledge and the identification of contextual and environmental details that provided important pointers to how project managers approach real situations in practice rather than what they say they would do in hypothetical situations.

Conclusion

The question of how to elicit tacit knowledge from expert employees has become increasingly important to address with the advent of increasing interest in knowledge management. In this chapter, I have described a useful tool—the critical decision interview method—which can assist researchers in the task of drawing out tacit knowledge from expert respondents, and I have illustrated the effective use of the tool in a recent tacit knowledge research project. One of the strengths of the method, as used in that research, was that it encouraged participants to report on a specific project rather than simply to discuss their general opinions, and I found that by maintaining respondents' focus on what actually happened, I was able to draw out the gaps between the typical risk management prescriptions, of which all respondents were aware, and what they actually did with respect to risk management.

Two key limitations should be mentioned in closing. First, the critical decision interview method has the limitation that it relies on managers' self reports and recollections of their actions and carries with it the assumption that these self reports will provide an accurate picture of respondents' actual practices, including the tacit knowledge they applied. While this limitation is acknowledged, extensive use of critical incident and critical decision methods (Klein et al., 1989; Sternberg et al., 2000) for eliciting tacit knowledge has demonstrated that these techniques can effectively tease out the tacit knowledge applied by respondents in performance of key tasks.

The second limitation of the method is that, even though a tacit knowledge item may be deduced, there is no assessment of the worth of this tacit knowledge item—it may be something from which all practitioners in the area would benefit, or it may be evidence of poor practice that the respondent has got away with in the past. The use of expert respondents mitigates this limitation to some extent, but ultimately, the usefulness of the tacit knowledge captured during the research is judged best by another independently selected set of expert practitioners.

References

Alavi, M., & Leidner, D. E. (2001). Knowledge management and knowledge management systems: Conceptual foundations and research issues. *MIS Quarterly, 25*(1), 107-136.

Ambrosini, V. (2003). *Tacit and ambiguous resources as sources of competitive advantage.* Basingstoke, UK: Palgrave Macmillan.

Ambrosini, V., & Bowman, C. (2001). Tacit knowledge: Some suggestions for operationalization. *Journal of Management Studies, 38*(6), 811-829.

Argyris, C., & Schön, D. A. (1978). *Organizational Learning II.* Reading, MA: Addison-Wesley.

Berry, D. C., & Dienes, Z. (1993). *Implicit learning: Theoretical and empirical issues.* Hove, UK: Lawrence Erlbaum.

Beyer, H. R., & Holtzblatt, K. (1995). Apprenticing with the customer. *Communications of the ACM, 38*(5), 45-53.

Blackler, F. (1995). Knowledge, knowledge work and organizations: An overview and interpretation. *Organization Studies, 16*(6), 1021-1046.

Castillo, J. (2002). A note on the concept of tacit knowledge. *Journal of Management Inquiry, 11*(1), 46-59.

Eden, C., Ackermann, F., & Cropper, S. (1992). The analysis of cause maps. *Journal of Management Studies, 29*(3), 309-324.

Ericsson, K. A., & Simon, H. A. (1993). *Protocol analysis: Verbal reports as data* (rev. ed.). Cambridge, MA: MIT Press.

Flanagan, J. C. (1954). The critical incident technique. *Psychological Bulletin, 51*, 327-358.

Gordon, S. E. (1992). *Implications of cognitive theory for knowledge acquisition.* In R.R. Hoffman (Ed.), The psychology of expertise: Cognitive research and AI (pp. 99-120). New York: Springer-Verlag.

Hoffman, R. R. (Ed.). (1992). *The psychology of expertise: Cognitive research and empirical AI.* New York: Springer-Verlag.

Hoffman, R. R., Shadbolt, N. R., & Burton, A. M. (1995). Eliciting knowledge from experts: A methodological analysis. *Organizational Behavior and Human Decision Processes, 62*(2), 129-158.

Keane, B. T., & Mason, R. M. (2006). On the nature of knowledge: Rethinking popular assumptions. In *Proceedings of the 39th Annual Hawaii International Conference on System Sciences.*

Klein, G. A. (1992). *Using knowledge engineering to preserve corporate memory.* In R. R. Hoffman (Ed.), The psychology of expertise: Cognitive research and empirical AI (pp. 170-187). New York: Springer-Verlag.

Klein, G. A., Calderwood, R., & MacGregor, D. (1989). Critical decision method for eliciting knowledge. *IEEE Transactions on Systems, Man, and Cybernetics, 19*(3), 462-472.

Leonard, D., & Sensiper, S. (1998). The role of tacit knowledge in group innovation. *California Management Review, 40*(3), 112-132.

Malhotra, Y. (Ed.). (2000). *Knowledge management and virtual organizations.* Hershey, PA: Idea Group Publishing.

Nonaka, I. (1994). A dynamic theory of organizational knowledge creation. *Organization Science, 5*(1), 14-37.

Nonaka, I., & Takeuchi, H. (1995). *The knowledge-creating company: How Japanese companies create the dynamics of innovation.* New York: Oxford University Press.

Olson, J. R., & Biolsi, K. J. (1991). *Techniques for representing expert knowledge.* In K. A. Ericsson, & J. Smith (Eds.), Toward a general theory of expertise: Prospects and limits (pp. 240-285). New York: Cambridge University Press.

PMI Standards Committee. (1996). *A guide to the project management body of knowledge.* Upper Darby, PA: Project Management Institute.

Polanyi, M. (1966). *The tacit dimension.* Garden City, NY: Doubleday.

Project Management Institute. (2000). *A Guide to the project management body of knowledge* (PMBOK guide). Newton Square, PA: Project Management Institute.

Reber, A. S. (1993). *Implicit learning and tacit knowledge: An essay on the cognitive unconscious.* New York: Oxford University Press.

Schmidt, R., Lyytinen, K., Keil, M., & Cule, P. (2001). Identifying software project risks: An international Delphi study. *Journal of Management Information Systems, 17*(4), 5-36.

Shadbolt, N.R., O'Hara, K., & Crow, L. (1999). The experimental evaluation of knowledge acquisition techniques and methods: History, problems, and new directions. *International Journal of Human-Computer Studies, 51*(4), 729-755.

Stenmark, D. (2002). Information vs. knowledge: The role of intranets in knowledge management. In *Proceedings of the 35th Hawaii International Conference on System Sciences, Hawaii.*

Sternberg, R. J. Forsythe, G. B., Hedlund, J., Horvath, J. A., Wagner, R. K., Williams, W. M., et al. (2000). *Practical intelligence in everyday life.* Cambridge, UK: Cambridge University Press.

Sternberg, R. J., & Horvath, J. A. (Eds.). (1999). *Tacit knowledge in professional practice: Researcher and practitioner perspectives.* Mahwah, NJ: Lawrence Erlbaum.

Sternberg, R. J., & Wagner, R. K. (Eds.). (1986). *Practical intelligence: Nature and origins of competence in the everyday world.* NY: Cambridge University Press.

Swap, W., Leonard, D., Shields, M., & Abrams, L. (2001). Using mentoring and storytelling to transfer knowledge in the workplace. *Journal of Management Information Systems, 18*(1), 95-114.

Takeuchi, H. (2001). *Towards a universal management of the concept of knowledge.* In I. Nonaka, & D. J. Treece (Eds.), Managing industrial knowledge: Creation, transfer and utilization. London: Sage.

Taylor, H. (2004a). Risk factors in vendor-driven IT projects. In *Proceedings of the Tenth Americas Conference on Information Systems, New York.*

Taylor, H. (2004b). *Risk management and tacit knowledge in IT projects: Making the implicit explicit* [unpublished doctoral dissertation]. Brisbane: Queensland University of Technology.

Tsoukas, H. (2003). Do we really understand tacit knowledge? In M. Easterby-Smith & M. A. Lyles (Eds.), *Handbook of organizational learning and knowledge* (pp. 410-427). London: Blackwell.

Wagner, R. K. (1987). Tacit knowledge in everyday intelligent behavior. *Journal of Personality and Social Psychology, 52*(6), 1236-1247.

Wagner, R. K., Sujan, H., Sujan, M., Rashotte, C. A., & Sternberg, R. J. (1999). Tacit knowledge in sales. In R. J. Sternberg & J. A. Horvath (Eds.), *Tacit knowledge in professional practice: Researcher and practitioner perspectives* (pp. 155-182). Mahwah, NJ: Lawrence Erlbaum Associates.

Wagner, W. P., Najdawai, M. K., & Chung, Q. B. (2001). Selection of knowledge acquisition techniques based upon the problem domain characteristics of production and operations management expert systems. *Expert Systems, 18*(2), 76-87.

Walsham, G. (1995). Interpretive case studies in IS research: Nature and method. *European Journal of Information Systems, 4*, 74-81.

Walsham, G. (2001). Knowledge management: The benefits and limitations of computer systems. *European Management Journal, 19*(6), 599-608.

Appendix: Sample interview protocol

(An outline of the research project and the interview protocol was provided to respondents by e-mail when the interview appointment was confirmed.)

Interview Introduction

Briefly review the purpose of the research with the respondent—to understand more about actual risk management practices for software package implementation projects and, in particular, to learn about the subtle practical problems faced by managers of these projects and the judgments they exercise during the course of the project. Explain the focus on the practical skills and judgment that managers acquire and apply with experience rather than the official view we tend to teach in formal classes. Explain the interview process, assure confidentiality, and seek permission to record the interview.

Choice of Specific Project

Ask the respondent to focus on a particular recent project in which he or she was involved so that he or she is reporting on events that actually have occurred rather than talking about general conceptions of rules and procedures. Encourage the interviewee to choose a project in which he or she faced some risky or challenging situations. Ask for a general description of the project, including information about the type of project, client organization, personnel involved, budget, and schedule.

Risk Management Processes

Ask the respondent to describe the risk management processes, if any, that were applied to the chosen project. Let this account proceed without interruption, except for clarification.

Elicitation of Specific Incidents

Ask respondent to describe specific incidents from the project that were challenging from a risk perspective and that might have been dealt with differently by a novice manager. If necessary, ask prompting questions to elicit situations that illustrate aspects of risk management processes. Use probe questions (see Table 1) to elicit information about the situational cues surrounding the incident, the strategies and options considered, the factors or triggers that determined one response rather than another, detail about the action taken and the consequences of the action, and why this situation could have been difficult for novices.

Exploration of Typicality

Once the respondent has finished recalling key incidents for the specific project chosen, ask whether those incidents and the actions and consequences were typical of other projects in his or her experience. Use follow-up questions to identify any key differences between projects identified.

Collection of Demographic Data

Collect demographic information about the respondent and details (i.e., type, duration, effort in terms of person-months, team size, budget) about the projects, if not already captured.

Table 1. Probe questions (Adapted from Klein, Calderwood, & MacGregor, 1989)

Area	Questions
Problem description	Describe the situation.
	What happened leading up to the situation? (context, environment)
	What did you do?
	What was the outcome?
Planning	Had you anticipated the possibility of this problem at the planning stage?
	Did your plans include contingency measures for a problem like this?
	Did the contingency measures work?
	Did you deviate from the plan? How, what factors caused the deviation?
Cues	What key points alerted you to …? How did you know that …?
Options	What alternatives did you consider …? What limitations did you face regarding possible actions?
Interactions	Did you have direct control? Who were the key players?
Analogues	Were you reminded of any previous experience …?
Goals	What were your specific goals at this point?
Basis	How did you decide on your choice of action/reject other options?
Knowledge	What information did you use for deciding …? What training or experience was useful in making this decision? How did you learn about …?
Hypotheticals	With hindsight, what would you have done differently? What training or experience would have helped? What do you think a novice might have done in this situation?
Exceptions	Can you think of another situation in which you would have done things differently?
Results of actions	Did your action work as expected? If not, why do you think that was? If so, what might have caused it not to work? What would have happened if your action hadn't worked? What would you have done?

Chapter XVII

Knowledge Acquisition and Transfer in Developing Countries: The Experience of the Egyptian Software Industry

Ahmed Seleim, University of Alexandria, Egypt

Ahmed Ashour, University of Alexandria, Egypt

Omar Khalil, University of Kuwait, Kuwait

Abstract

This investigation explored knowledge acquisition and transfer practice in Egyptian software firms. It used a combination of a cross-sectional field survey of 38 firms and an in-depth qualitative analysis of 14 firms. Although most of the firms in the sample recognize the importance of knowledge, their idiosyncrasies appear to affect the way knowledge is acquired and transferred. The firms were found to have a limited use of their software developers' initiatives, R&D, and the academic and research institutions as sources for knowledge acquisition. They also were found to have limited capabilities in transferring and sharing knowledge. The Egyptian culture is rich in social and emotional capital, which can play an important role in building relationships, facilitating the exchange of knowledge, and sharing experience. The Egyptian software firms should develop and implement KM strategies that attract expert software developers, capitalize on trust and social relationships, and build IT-based KM systems in order to enable knowledge acquisition and transfer.

Introduction

Organizational knowledge accumulates over time and enables firms to attain deeper levels of understanding of their business. Knowledge is a critical factor that can be used to explain the growth of a firm, which is viewed as a repository of knowledge and experience (Penrose, 1959). Knowledge production is an economic activity (Machlup, 1962, 1983), and knowledge importance is on the rise in the post-capitalist society (Drucker, 1993). Consequently, knowledge management (KM) has become one of the major challenges facing today's organizations.

In spite of the varying views of KM in the literature, knowledge acquisition, knowledge documentation, knowledge transfer, and knowledge application are believed to be four interdependent basic dimensions of the KM process. KM is a cross-functional, multifaceted phenomenon (Bontis & Fitz-enz, 2002; Lee & Choi, 2003), and a considerable variation in KM literature and KM processes and practices exists. Effective KM requires approaching organizational knowledge as a process rather than a resource (Alavi & Leidner, 2001; Alavi & Tiwano, 2002; Davenport & Prusak, 1998; Lee & Choi, 2003; Spender, 1996; Wiig, 2000).

The coordination of KM dimensions in organizations is critical, since the shortage of any dimension may result in less than optimum outcomes of KM processes and systems (Bhatt, 2001; Darroch, 2003). KM success models such as those of Bots and de Bruijn (2002), Massy et al. (2002), Lindsey (2002), Maier (2002), and Jennex and Olfman (2005) suggest that effective KM processes (e.g., acquisition, documentation, transfer, and application) is essential to the successful development and implementation of knowledge management systems (KMS) and the adoption of KM strategies. Improving KM processes, in turn, necessitates understanding how organizations practice and manage such processes.

On the other hand, KM is a relatively new research area (Zhang & Zhao, 2006), and most of the earlier research on KM, especially in software firms, is case-based (Carter, 2000; Dingsoyr & Conradi, 2003; Hellstrom, Malmquistm, & Mikaelsson, 2001; Kautz, Thaysen, & Vendelo, 2002) and nearly limited only to developed countries. Given the inadequate external validity of the findings of such research, empirical cross-sectional investigations of KM practices in developed and developing countries are wanted. In addition, Mathiassen and Pourkomeylian (2003) assert that it is far from clear how knowledge-intensive organizations such as software firms can practically take advantage of KM insights.

The objective of this study is to understand how knowledge acquisition and transfer are practiced in a number of relatively young and small software firms from Egypt, which is viewed as a developing country. These two KM dimensions are selected for investigation, because software firms may not sustain competitive advantages without constantly learning from experience and developing and transferring new knowledge (Brown & Woodland, 1999; Garvin, 1993).

This chapter is organized accordingly. A background on KM is presented first, followed by the research method, research results, implications, and conclusion.

Background

Knowledge Types

Since knowledge is too complex and vague of a term to be defined precisely, Davenport and Prusak (1998) propose only a working definition. They view knowledge as a "fluid mix of framed experience, values, contextual information, and expert insight that provides a framework for evaluating and incorporating new experiences and information."(p. 5). They add that individual knowledge is generated and applied in the minds of people and that organizational knowledge generally is embedded in the organizational documents, repositories, routines, processes, practices, and norms.

Knowledge takes different forms and types. Perhaps the most familiar distinction in the KM domain is between explicit and tacit knowledge (Nonaka, 1991). While tacit knowledge is personal and remains in the human mind, behavior, and perception (Karhu, 2002; Nonaka, 1991; Sveiby, 1997), explicit knowledge easily can be formalized and documented through different tools such as information technology (IT), rules, and procedures.

Other examples of knowledge include objective and experience-based knowledge (Penrose, 1959), organizational routines knowledge (Ashour, 2000), procedural knowledge (Kogut & Zander 1996; Nickolas, 2001; Winter, 1987), general (e.g., theoretical) knowledge and domain-specific (e.g., customer, technical, competitor, supplier, product, and market) knowledge (Demsetz, 1991; Grant, 1996), individual and organizational knowledge (Bhatt, 2001), external knowledge (i.e., produced by people outside the organization as resides in books, journals, magazines, etc.), and internal knowledge (i.e., created primarily within the organization, largely through experience and experimentation) (Jalote, 2003). Knowledge also can be viewed in a hierarchy of data, information, knowledge, and wisdom (Alavi & Leidner, 2001; Davenport & Prusak, 1998).

In addition, Quinn, Anderson, and Finkelstein (1998) classify knowledge in software engineering into four types: cognitive knowledge (know-what), advanced skills (know-how), system understanding (know-why), and self-motivated creativity (care-why). Software domain (e.g., software development and management) knowledge is of particular interest in this investigation.

Knowledge Management (KM)

KM is rooted in three schools of thought: (1) the resource-based theory of the firm (Aker, 1989; Amit & Schoemaker, 1993; Barney, 1991; Dierickx & Cool, 1989; Hall, 1992, 1993; Itami, 1987; Penrose, 1959; Prahalad & Hamel, 1990; Teece, Pisano, & Shuen, 1997; Wernerfelt, 1984); (2) the knowledge-based theory of the firm (Bierly & Chakrabarti, 1996; Choi & Lee, 2003; Demsetz, 1991; Grant, 1996; Kogut & Zander, 1992,1996; Nonaka, 1994; Nonaka &

Takeuchi, 1995; Spender, 1994); and (3) the organization learning theory (Crossan, Lane, & White, 1999; Davenport & Prusak 1998; DiBella & Nevis, 1998; Garvin, 1994; Huber, 1991; Senge,1990; Slater & Narver, 1995; Tippins & Sohi, 2003).

The essence of these schools of thought is that the organization's ability to develop, use, and benefit from its knowledge through learning is the only source of sustainable competitive advantage. Huber (1991) identifies four KM-related constructs in organizational learning, including knowledge acquisition, information distribution, information interpretation, and organizational memory. Jennex and Olfman (2002) add that organizational learning uses organizational memory as its knowledge base.

KM researchers have adopted different views of what KM entails. The differences, however, appear to be in the number and labeling of KM processes rather than in the underlying concepts (Alavi & Leidner, 2001). Stewart (1997), for example, views KM as a process of creating, maintaining, and exploiting all the possibilities of the knowledge that each organization can use. Liebowitz (2000) defines KM as the process of creating value from an organization's intangible assets. KM also is viewed as the process of capturing an organization's knowledge and using it to foster innovation through organizational learning (Nonaka 1991, 1994; Nonaka & Takeuchi, 1995; Wiig, 2000). Zack (1999a) asserts that KM includes the acquisition, refinement, storage, retrieval, distribution, and presentation of knowledge. Prusak (1997), however, uses KM to refer to what organizations know, how they use what they know, and how fast they can know something new.

Drawing on KM literature, KM is defined as the processes or dynamics that an organization adopts to acquire, document, transfer, and apply knowledge in order to achieve its goals. KM practice focuses on the tactical and operational implementation of knowledge-related activities (Wiig, 1997). Therefore, KM possibly is practiced differently when organizations adopt variant processes or dynamics to acquire, document, transfer, and apply their knowledge. Only knowledge acquisition and knowledge transfer are addressed in this investigation.

Knowledge Acquisition

Tiwana (2000) defines knowledge acquisition as the process of developing and creating insights, skills, and relationships in the organization. Edwards (2003) uses knowledge creation and acquisition interchangeably. Davenport and Prusak (1998) define knowledge acquisition as only one of five modes—acquisition, dedicated resources, fusion, adaptation, and knowledge networking—of knowledge generation. To Davenport and Prusak (1998), knowledge generation includes acquisition of external knowledge as well as development of internal knowledge.

For the purpose of this investigation, knowledge acquisition includes all activities required to add to an organization's knowledge resources from external and internal knowledge sources. Accordingly, knowledge acquisition encompasses selection and acquirement of knowledge from external sources (e.g., journals, magazines, books, Web sites, professional meetings, etc.) as well as the creation of knowledge internally (e.g., through experience, experimentation, special groups such as R&D, etc.). Knowledge acquisition is a continuous process in which individuals and groups within an organization build the organization's knowledge by developing new knowledge content or by replacing existing knowledge within the organization's tacit and explicit knowledge (Pentland, 1995).

Knowledge acquisition strategies vary. Zack (1999b) distinguishes between knowledge exploration and exploitation as two strategies for knowledge acquisition. Knowledge exploration includes activities that focus on importing new knowledge into the organization, while knowledge exploitation involves applying existing knowledge to new uses. Bhatt (2001), however, provides a more pragmatic approach for knowledge acquisition by reconfiguring and recombining existing pieces of knowledge, developing competence by focusing on capabilities and limiting shortcomings, strengthening research and development (R&D) capabilities, scanning and monitoring external environments, and borrowing and employing external technologies.

Knowledge Transfer

Knowledge transfer, on the other hand, refers to the ability of an organization to diffuse and share knowledge. It includes the exchange of different types of knowledge, including tacit and explicit knowledge, among individuals, groups, and units, and at the different organizational levels. It is the process through which one unit (e.g. group, department, or division) is affected by the experience of another unit (Argote & Ingram, 2000). Furthermore, knowledge transfer can occur explicitly—when an individual or a unit communicates with another individual or another unit—or implicitly through norms and routines (Argote & Ingrom, 2000). Davenport and Prusak (1998), however, argue that since the goal of knowledge transfer is to improve an organization's ability to perform, knowledge transfer should include not only the transmission (sending) of knowledge but also the absorption of the knowledge by the recipient.

Bartol and Srivastave (2002) identify four mechanisms for individuals to share their knowledge in organizations: contribution of knowledge to organizational databases, sharing knowledge in formal interactions, sharing knowledge in informal interactions, and sharing knowledge within communities of practices. Abou-Zeid (2002) points out that knowledge can be transferred within an organization through personal communication (e.g., electronic mail, groupware, telephone, videoconferencing, face-to-face meetings, training seminars and courses, and communities of practice), codified communication (e.g., written reports, databases, and faxes), and embodied knowledge transfer (e.g., rules, procedures, and directives). Also, IT-based KMS provide appropriate means for intraorganizational knowledge transfer (Abou-Zeid, 2002; Jennex & Olfman, 2005).

Knowledge transfer also includes the exchange of knowledge externally with other individuals, groups, and organizations. The international business studies provide some insights for understanding how knowledge is transferred from one organization to another (Inkpen, 1998; Mowery, Oxley, & Silverman, 1996). Mechanisms of knowledge transfer among organizations include arrangements such as technology transfer, R&D collaboration (Amabile et al., 2001), strategic alliances (Borys & Jemison, 1989; Das, Sen, & Sengupta, 2003), learning alliances (Prahalad & Hamel, 1990), and marketing alliances. Channels through which knowledge is transferred from universities and research centers to organizations and industries include publications, patents, consulting, informal meeting, recruiting, licensing, joint venture, research contracts, and personal exchange. The choice of transfer mechanism, however, depends on the nature of knowledge (tacit vs. explicit), intended business uses, and the target subsidiary.

Nevertheless, culture and trust are critical to an effective acquisition and transfer of knowledge (Abou-Zeid, 2002; Davenport & Prusak, 1998; Goh, 2002; Lee & Choi, 2003; Mason & Pauleen, 2003; Stoddart, 2001; Yahya & Goh, 2002). De Long & Fahey (2000) point out that culture influences knowledge acquisition and transfer through shaping the assumptions around the importance of knowledge, determining how knowledge is distributed, utilized, controlled, shared, and hoarded, creating the context for social interaction that determines how knowledge is used in particular situations and determining the processes by which new knowledge is applied. Culture obstacle occurs when employees feel possessive about their knowledge and may not be forthcoming in sharing it (Lindvall & Rus, 2003; Okunoye & Bertaux, 2006; Usoro & Kuofie, 2006).

Trust is viewed as an element of the organizational context (Bartlett & Ghoshal, 1998) and as an antecedent of cooperation (Gulati, 1995; Ring & Van de Ven, 1992). Trust is critical for collaboration and *interaction within the firm as well as between the firm and its coalitions such as customers, experts, associations, and competitors* (Fukuyama, 1995; Gillis, 2003). Trust facilitates a culture of openness that allows knowledge and information exchange (Eppler & Sukowski, 2000). Withholding knowledge because of lack of trust is detrimental to the processes of knowledge acqisition and transfer (Hedlund & Nonaka, 1993).

Importance of Knowledge Acquisition and Transfer Practice in the Software Industry

Effective knowledge acquisition and transfer is particularly important in the software industry. The essence of software development is pure knowledge (Grant, 2000), as 95% of software business is intangible capital (Hoch, Roeding, Purkert, Linder, & Muller, 2000). Software engineering literature suggests viewing software firms as learning organizations (Argyris, 1998; Lennselius & Wohlin, 1987). Mathiassen and Pourkomeylian (2003) view KM practice in software firms as a significant issue for both researchers and practitioners. Software firms are expected to adopt effective KM in order to inform practices. Findings from empirical KM research should enrich the KM literature and help knowledge managers in software firms to develop effective KM strategies and systems.

In addition, organizational knowledge systems consist of a combination of three possible forms of paper documents, computer documents, and self-memory (Jennex & Olfman, 2002). Successful IT and non-IT-based KMS that aim to manage organizational knowledge should perform the functions of knowledge acquisition, storage, transfer, and application (Alavi & Leidner, 2001; Edward, Feng, & Liou, 2005; Jennex & Olfman, 2005). KMS improvement decisions necessitate a thorough understanding of how KM is practiced. Investigating knowledge acquisition and transfer practices should provide insights on how to improve the efficiency and coherence of knowledge acquisition and transfer processes, which ultimately strengthens KMS. Moreover, empirically based descriptions and analyses of knowledge acquisition and transfer processes should contribute to the ongoing attempts to assess KMS success factors and models (Jennex & Olfman, 2005) and the building and testing of KMS within a global context.

Methodology

Research Strategy

The selection of a research strategy depends on the research methods employed in the relevant prior research and the nature and objectives of the study. For the purpose of this investigation, a combination of a cross-sectional field survey of a number of Egyptian software firms along with an in-depth qualitative analysis of multiple cases have been adopted. Such a research design should provide a rich data set that can be used to accurately understand and describe KM practices in the investigated firms.

The use of the survey method generally does not provide the detailed information needed to deeply understand a complex phenomenon such as KM practice. Quantitatively based research seldom captures the subjects' perspectives, because they have to rely on more remote, inferential empirical methods and materials (Denzin & Lincoln, 2000). Alternatively, the expanded use of qualitative methods, such as ethnography, ethnology, participation observation, unstructured interviewing, and case study methods, has been one of the most important developments in recent social research, including information systems (IS) research (Olson, 1981; Robey, 1981, 1983; Shadish, Cook, & Campbell, 2002). A qualitative research is valuable when conducting investigations that seek processes-related information and explore new phenomena (Marshall, 1985). In such situations, the qualitative method allows researchers to capture reality in more detail compared to the survey method.

Qualitative research relies on the integration of data from variant sources of information that allow a better assessment of the validity and generality of the explanations made about the investigated phenomenon (Balogun, Huff, & Johnson, 2003; Maxwell, 1998). These methods have unrivaled strengths for the elucidation of meanings, the in-depth description of cases, and the discovery of new hypotheses (Denzin & Lincoln, 2000). Therefore, the qualitative research should provide researchers with a rich, deep knowledge that facilitates the interpretation of the survey results, which provide only generic attributes of the practice of knowledge acquisition and transfer in the software firms.

Sampling

The study focuses on software firms in the Egyptian private sector as the primary population. The Egyptian software industry is a promising one. In 2000, software exports were estimated at $50 million (El-Rashidi, 2002), and the total Egyptian software market was estimated at $140 million. Also, Egypt has thousands of qualified software professionals and has become a major supplier of software products, services, and software developers to its neighboring Arab countries.

The research population consisted of 107 software companies that were members of the Egyptian Chamber of Software Industry. These software firms had no formal KM initiatives and have made no significant efforts in this direction. Therefore, they were considered a suitable setting for conducting this research. Upon an initial contact with these firms, which are located in the Cairo and Alexandria areas, 38 firms agreed to participate in this

Table 1. Sample profile: Size and age (N = 38)

Characteristics	Minimum	Maximum	Average	St. Dev.
Number of developers	2	80	25.55	22.30
Other employees	3	230	35.97	51.53
Total number of employees	5	300	62.10	67.91
Age of firm	2	17	7.65	4.32

investigation, and 14 of these firms agreed to take part in the in-depth qualitative analysis of KM practices.

The research sample (n = 38) represents 35.5% of the total population. Twenty-one firms are incorporated, 14 are individually owned, and four are branches of international firms. Twenty-six firms employ Egyptian capital, four employ foreign capital, and eight employ joint capital. Table 1 provides a description of the sample in terms of size (number of employees) and age. Most of the firms in the entire sample are relatively small, with averages of 62 employees and approximately 26 software developers. The largest firm has 300 employees, 80 of whom are software developers, and the smallest firm has five employees, two of whom are software developers. The firms in the sample are relatively young (average age in the sample is less than eight years); the oldest is 17 years old, and the youngest is two years old. The relatively high standard deviations suggest wide dispersions in the firms' number of employees and ages.

Table 2 further describes the sample in terms of business type, target markets within Egypt, export markets, exported software, existence of branches inside and outside Egypt, and platform type. Twenty-eight firms produce packaged Arabic and English software, 27 develop custom software, 13 Arabize packaged software, 12 are local agencies for international product, and 14 are engaged in other software-related businesses. In addition, 26 of the firms produce business software (e.g., accounting, finance, etc.), 11 produce educational software, eight produce children's software, seven produce religiously oriented software, five produce cultural software, and 14 produce other types of packaged software.

Regarding the targeted local Egyptian market, 33 firms target the private sector with their products, 21 target the government sector, 16 sell software products to the public sector, 12 produce and market software for individual users, and nine target civil institutions such as unions and clubs. However, 31 firms reported software exports to Arab countries, 14 exported software to North America, 11 exported software to Europe, seven exported software to Asia (non-Arab states), and five exported software to other foreign markets.

Twenty-five firms in the sample exported packaged (standardized) software, 18 exported custom software, nine exported Arabized software, and nine exported other custom and packaged software to foreign markets. Twenty firms reported to operate on multiple hardware platforms, and 18 were operating on a single hardware platform. Finally, 17 firms of the sample had multiple branches inside and outside Egypt.

Table 2. Characteristics of the software firms in the entire sample (N = 38)

Sample Characteristics	No. of Firms	Percentage
Types of Business		
Custom software	27	71%
Arabic and English packaged software	28	74%
Arabization of packaged software	13	34%
Local agencies for international products	12	32%
Others	14	37%
Target Market in Egypt		
Government sector	21	55%
Private sector	33	87%
Public sector	16	42%
Individual user	12	32%
Civil institutions (e.g., unions and clubs)	9	24%
Export Market		
Arab world	31	82%
Europe	11	29%
North America	14	37%
Asia (non-Arab states)	7	18%
Other areas	5	13%
Type of Exported Software		
Packaged software	25	66%
Arabized software	9	24%
Custom software	18	47%
Others	9	24%
Type of Produced Package Software		
Religion-oriented software	7	18%
Cultural software	5	13%
Educational software	11	29%
Children's software	8	21%
Business software (e.g., accounting, sales)	26	68%
Others	14	37%
Firms That Have Branches Inside and Outside Egypt	17	45%
Type of Platform		
Single platform	18	47%
Multiple platforms	20	53%

Measurement and Data Collection

The instrument, which consists of quantitative and qualitative measures, was tested and revised a number of times in order to fit the context of the study. The final instrument (in Arabic) includes three sets of questions (an English translated summary of the instrument is presented in the Appendix). The first set consists of 11 probing, open-ended questions designed to guide the interviews and the gathering of data from the 14 firms that agreed to participate in the qualitative analysis of KM practices. The second set comprises 23 questions designed to gather cross-sectional data from the entire sample (38 firms) on the mechanisms used in KM practice in general and knowledge acquisition and transfer in particular, along with the barriers and challenges of KM in the software firms. The third set comprises 16 questions designed to gather data on each firm's background (e.g., age, number of software developers and other employees, type of capital, type of ownership, products and services, local and foreign markets, types of exported software, number of foreign branches, software and hardware platforms, quality certificates, and rewarding systems.

The interviews and questionnaires were used to supplement each other in data collection. Data on KM methods were collected from the entire sample (38 firms) by having the firms' CEOs or their representatives respond to the second set of questions (23 questions) in a structured interview setting. In addition to the CEOs, a number of software developers, project managers, and quality managers in the 14 firms that agreed to participate in the qualitative study were interviewed, and the first set of 11 questions was used to further explore KM practices. The CEOs of the 14 firms arranged for additional interviews with other employees who were considered to play active roles in software development in their firms. Semi-structured interviews, personal observations, and investigation of relevant documents were the main methods used for data collection from the 14 firms.

The interviews were conducted in Arabic, and each lasted between one and four hours. All interviews were recorded in order to maintain accuracy and proper use of evidence. Forty interviews were completed in the 14 firms, with an average of approximately three interviews per firm. In order to discover important insights from the data, the contents of the interviews were analyzed in accordance with the two dimensions of knowledge acquisition and knowledge transfer. The researchers utilized their knowledge of the literature on KM and other related fields as well as their familiarity with the Egyptian culture and business environment to analyze the qualitative data. In addition, descriptive statistics were used to present the quantitative data pertinent to knowledge acquisition and knowledge transfer.

Validity and Reliability

Yin (1989) suggests the use of multiple sources of evidence and reviewing by key informants as methods to enhance validity. Qualitative data were gathered by conducting 40 interviews with key informants in the 14 firms (three interviews per firm in 12 firms and two interviews per firm in two firms). In addition, the design of the questionnaire (in Arabic) in 2002 was guided by the research methods and findings of prior KM research. In addition, earlier drafts of the instrument were developed and revised based on the feedback received from a number of experts from the software industry in Egypt.

Reliability can be improved by reducing the errors and biases that influence the ability to obtain the same results, if the study is replicated. To minimize any possible bias in our data, follow-up phone calls were made to obtain additional information and/or to seek clarification of information that was collected already. Moreover, the raw qualitative data were translated from Arabic to English and then from English to Arabic in order to ensure having the same contents (converted translation). All data gathered from the main sources were consolidated and synthesized in order to obtain a full picture of knowledge acquisition and transfer in the investigated firms.

Results

The key findings on the two KM dimensions of knowledge acquisition and knowledge transfer are presented in this section.

Knowledge Acquisition

Terms such as *acquire*, *seek*, *generate*, *create*, and *capture* often were used by the CEOs in the interviews. All these terms have the common theme of focusing on building and accumulating knowledge. Most of the interviewed CEOs recognize the fact that their firms operate in a fast-changing environment, which requires a continuous acquisition of information and knowledge. They indicated that knowledge acquisition was essential to their capability to innovate and face the threats of the aggressive local and international competition. Also, the CEOs asserted that business core knowledge (e.g., customer knowledge) and pure technical knowledge are two critical types of knowledge in the software industry. Each of the two types of knowledge is acquired and created in the firms through different methods.

Customer Knowledge

Customer knowledge is considered one of the most valuable assets in organizations and one of the core competences (Glazer, 1991; Day, 1994; Wright & Ashill, 1998), and software firms are no exception. It involves the activities that generate and acquire knowledge regarding the customer's current and potential needs (i.e., business core knowledge). In addition, software developers may not have access to application domain knowledge, and therefore, a relationship with users/customers is important in order to provide an understanding of why specific requirements are included or excluded from the system specification. Such knowledge is essential to software product specifications, product innovation ideas, and product performance requirements.

Customers' business core knowledge is acquired through different sources and activities in order to identify customers' needs. Business core knowledge is acquired through different sources and activities in order to identify customers' needs. Table 3 depicts the methods commonly used for the creation and acquisition of customer knowledge and information in the entire sample.

The frequency distributions in Table 3 suggest that the most commonly used strategies of acquiring customer knowledge and information were visitation with customers, market research, means for receiving customers' suggestions and ideas, and global demand for and trends in software. The formal and informal meetings, such as conferences, parties, and friendships, were the least commonly used methods by the firms in acquiring knowledge on customers and their business core. However, the 22 firms that reported market research as a strategy for customer knowledge acquisition are among the firms that produce packaged (standardized) software. In this sector, marketing research is an important source of knowledge about the general needs of target customers.

Table 3. Frequency distribution of customer knowledge and information acquisition methods (N = 38)

Forms of Company Attention to Customers	Number of Firms	Percentage
On-site visitation with customers	28	74%
Market research	22	58%
Formal and informal gatherings (e.g., conferences, parties)	10	26%
Friendship	9	24%
Means for receiving suggestions and ideas from customers	22	58%
Global demand for and trends in software	17	45%
Others	8	21%

The basic objective in identifying the customer's needs in custom and packaged software firms is to understand the business core knowledge. A number of the CEOs of the 14 firms that participated in the qualitative research pointed out that the real challenge in this field lies in the business core knowledge rather than in the technical knowledge. For this reason, software firms rely on two types of system analysts: user analysts and systems analysts. The user analyst meets with customers to identify their requirements and specifications for the new system and conveys them to the systems analyst, who translates the requirements into a codified knowledge from the computer system perspective.

The firms that produce packaged software usually conduct market research in the target market in order to identify the basic and common requirements for the largest number of potential customers. However, customer knowledge is particularly important for the firms that produce customized software. For these firms, business core knowledge, such as customer knowledge, is acquired in the course of identifying customers' needs through on-sight visitation with customers, direct observation of how work is done, and participation with customers while performing their jobs. Such visits enable software developers to understand the needs of their customers and to satisfy their operations-related requirements (e.g., speed, accuracy, complexity, interdependence, and growth). There was a general agreement among the CEOs of the 14 firms participating in the qualitative study that the biggest problem facing custom software firms is the customers' lack of awareness of their actual needs.

Technical Knowledge

Software firms were found to acquire the needed technical knowledge through a number of external and internal sources.

As to the external knowledge sources, knowledge exploration includes different activities to import new knowledge into the firm (Zack, 1999b). Academic sources (e.g., universities, research centers), the Internet, and products of the local and international software firms were found to be among the sources used for external knowledge acquisition in the firms. Table 4 includes comparisons between the importance of universities/research centers and the Internet as two external technical knowledge sources for the firms in the sample.

Table 4. The importance of universities/research centers and the Internet as two sources of external knowledge (N = 38)

Knowledge Source / Relative Importance	Universities and Research Center		Internet Network	
	Number of Firms	Percentage	Number of Firms	Percentage
Very important	2	5%	27	71%
Important	3	8%	8	21%
Moderately important	13	34%	3	8%
Slightly important	4	11%	0	.0%
Not important	16	42%	0	.0%
Total	38	100%	38	100%

Only five firms agreed that establishing links and relationships with academic sources such as universities and research centers is either important or very important. Thirteen firms perceived such links and relationships to be moderately important. However, 20 firms perceived the academic institutions as a source of external knowledge to be slightly important or absolutely not important.

From the interviews with the CEOs in the 14 firms, only four firms were found to have actual connections with the faculty in the engineering departments in a number of local universities, such as Alexandria University, Cairo University, the American University at Cairo, and Al-Azhar University. The working relationships take different forms, such as the offering of seminars and workshops, occasional meetings between faculty members and software developers for the purpose of exchanging ideas and solving problems, having faculty members working as consultants for the firms, firms' sponsorship of students' graduation projects, and internship programs.

Nevertheless, only one firm (H Logic Company) was found to actually sponsor research projects at Alexandria University. The rest of the firms were found to have no significant connection with local universities and research centers. The CEOs in these firms commented that the universities are not in a position to support the Egyptian software industry. They believe that the computer curricula and courses offered at these universities are out-dated. As an example, one CEO stated that the Sadat Academy in Cairo continues to offer ADA, which is a programming language of the 1980s. Another CEO bitterly complained about his firm's failure to establish a relationship with Mubark City for Scientific Research in Amria due to the city's lack of responsiveness.

Perhaps the lack of connection between these firms and the universities is attributed to the firms' inabilities to exploit the knowledge that is available in such universities. Cohen and Levinthal (1989, 1990) assert that a firm's ability to apply or use university research for a commercial purpose depends on the amount of investments in the firm's R&D, which affects its absorptive capacity. According to World Bank's World Development Report (1998/1999) and the United Nation report (Arab Human Development Report, 2002), the percentage of the business sector's expenditure on R&D is generally very low in Egypt, resulting in a negative impact on the absorptive capacity of its firms, including software firms.

Table 5. Frequency distribution of products of local and international firms as sources for external knowledge (N = 38)

External Sources for Knowledge	Number of Firms	Percentage
Products of local (competitors) software firm	6	16.2%
Products of international software firms	9	24.3%

As to the Internet, the majority of the sample (27 firms) pointed out its high importance as a source of very low-priced and accessible information and knowledge (Table 4). Eight firms perceived the Internet to be important as a source of external knowledge, and only three firms perceived it to be moderately important. In addition, when interviewed, a number of software developers in the 14 firms that participated in the qualitative research highlighted the fact that some of the Web sites do provide free services and consultancy to solve real problems when developing software.

Furthermore, production of software that imitates what the international and local software firms introduced to the market is an example of external knowledge acquisition. A significant portion of the software production introduced to the Egyptian and Arabian markets was found to be a result of reverse engineering and imitation, not innovation. The number of firms that imitate the products of international and local firms were found to be nine and six, respectively (Table 5).

In their interviews, four of the 14 CEOs asserted that most of the software firms in Egypt engage in activities aimed at leveraging the transferring and sharing of ideas and knowledge among each other in the form of what the economists call utilization of knowledge spillovers (Cockburn & Henderson, 1994). Their evidence for this claim is that most of the software products that were introduced to the Egyptian and Arabian market were similar. This finding is not surprising in an industry like software in which many aggressive firms exist and knowledge flows among firms relatively quickly (Staples, Greenaway, & McKeen, 2000).

Also, the interviews with the CEOs of 14 firms revealed that software firms acquire knowledge and information through memberships in professional associations such as the Chamber of Software Industry, the Egyptian Software Association, and the Association of Exporters. The Chamber of Software Industry, for example, offers its members information and knowledge-related services through sharing industry-related information and knowledge, opening communication channels with similar firms, holding conferences and seminars, establishing centers for training and product quality, and participating in international exhibits. A number of the CEOs pointed out that they had the opportunity through the Chamber of Software Industry to export their software to foreign markets. Another CEO admitted that his firm had to cancel a major software project as a result of information they received while participating in one of the Chamber's sponsored exhibits.

As to internal knowledge sources, R&D and individual initiatives of developers and employees were the most commonly used sources for the acquisition of internal technical knowledge (Table 6). In order to create new technical knowledge, 25 firms were found to rely on their R&D activities, and 20 firms were found to rely on the individual initiatives of their software developers and other employees.

Table 6. Frequency distribution of the internal knowledge sources (N = 38)

Knowledge Strategies	Number of Firms	Percentage
Research and development (R&D)	25	66%
Individual initiatives of developers and employees	20	53%

However, the results of the interviews with the CEOs in the 14 firms indicate that independent R&D units exist only in large software firms such as Sakhar, Harf, RDI, and Kollay Engineering. Kollay Engineering, for example, views R&D as one of its most important success factors. The firm depends on its R&D activities to ensure that software is developed according to the market's latest business practices and the accounting legal requirements.

Also, Sakhr Software Company has a pioneering R&D program in its main center in Egypt to help develop Arabic software solutions and products. Sakhr has embarked on an ambitious R&D project that involves more than two million employee hours and is funded entirely by private investment in order to develop Arabic-based software for its individual, corporate, and government customers. In addition, H-LOGIC Company has an R&D unit that enabled the company to produce a new range of compact electronic solutions using state-of-the art micro controllers and integrated systems technologies.

Data on R&D expenditure were available from only eight of the 14 firms. The average R&D expenditure as a percent of total income was 3.8%, which is much lower than the reported similar average of 21% among the international software firms (Hoch et al., 2000). Given that the eight firms are among the most successful firms in the Egyptian software industry, the actual ratios of the R&D expenditures in the Egyptian software industry are likely to be even lower than the reported ratio of 3.8%. This finding is not a surprise, since the reported percentage of the R&D expenditure in Egypt's business sector is generally very low (Arab Human Development Report, 2002; World Development Report, 1998/1999).

More than half of the entire sample (20 firms) was found to rely on the individual efforts of their software developers and other employees to enrich their technical knowledge. The dependence on the individual initiatives as a source for knowledge creation in these 20 firms may reflect their adoption of personalization strategies, which focus on the employees and their efforts and abilities to develop knowledge (Hansen, Nohria, & Tierney, 1999). However, the lack of individual initiatives as a source of internal technical knowledge in the other 18 firms is surprising in an industry in which success depends heavily on innovation and creativity.

One reason for the lack of individual initiatives as a source of technical knowledge is that software firms may find external sources more feasible and affordable. For example, the interviews with the CEOs in the 14 firms revealed that most of the Egyptian software firms have licenses from international software firms (e.g., Microsoft, Oracle, and Sun Corporation) that are considered a source of technical knowledge. Another study of the intellectual capital in the same software firms revealed that the average number of software licenses from international software firms is 5.97 (Seleim, Ashour, & Bontis, 2004), a rather significant number, given the relatively small software production operations of these firms.

Scientific and professional references in the companies and projects libraries are considered an important source of explicit knowledge and information (Gunnlaugsdottir, 2003; Motwani, Gopalakrishna, & Subramanian, 2003). Many of the 14 firms were found to have their own libraries as a source for knowledge, especially the technical knowledge required to develop software products. The number of specialized books in these libraries ranged from five in the relatively small firms with only two developers to 900 books in the relatively large firms. Also, almost all the firms reported that they use object-oriented programming languages (e.g., Visual Basic and C++) in systems development and Web-based programming languages (e.g., HTML, XML, Java script) in their Internet application developments.

In addition to R&D and the individual initiatives of software developers, the interviews with the 14 CEOs revealed that a number of firms have adopted other strategies for knowledge acquisition, including hiring bright and experienced software developers and other employees, providing training and development programs, and rotating developers on different projects and tasks. Subsidization of both employees' professional certifications, such as Microsoft Certified Systems Engineer (MCSE), Microsoft Certified Security Administrator (MCSA), Certified Internet Webmaster (CIW), and enrollment in academic programs (e.g., the MBA program at the Arab Academy for Science and Technology and Maritime, MBA program at American University in Cairo) were among the initiatives taken by the firms to help their employees develop and acquire knowledge.

One of the firm's list of its training programs during the last six months of 2002 included programs on internal quality auditing, teamwork, project management, effective meeting and presentation skills, technical report writing for engineers, and time and pressure management. Ten of the 14 firms participating in the qualitative investigation indicated that human capital development was and continues to be a major part of their missions.

Sakhr Company, the largest software developer in the Middle East, stressed that software developers are their most important assets, essential to their continuous success in the software market. The company's management team believes that creative thinking and innovation only can be achieved by sustaining staff growth and development. Their ongoing training programs take place both locally and internationally to ensure employees' advancement and growth and to provide ample opportunities for innovation. To emphasize the importance of its intellectual capital, the CIT Company names its human resource department the Human Capital Department. This finding is consistent with those of Simon (1991), who asserts that organizations learn through informing/educating their existing members and adding new members who have new knowledge that the organization did not have before.

The firms that were found to rely heavily on the individual initiatives and efforts of their employees for knowledge acquisition were found to adopt favorable human resource management practices. The CEOs in these firms indicated that the only sustainable competitive advantage for a firm is its ability to learn faster than its competitors, an assertion that confirms the views of a number of scholars (Senge, 1990). In addition, the CEOs emphasized the importance of their human capital and the need to hire bright and experienced software developers. This hiring practice, common in Western organizations, is unusual in a country like Egypt in which nepotism is a common practice.

At the national level, the government took a number of initiatives in support of the transfer and adoption of information technology in Egypt. The Cabinet Information and Decision Support Center was established in 1992, and the Information and Communication Ministry was established in 1999. In addition, the government has allocated 55 million Egyptian pounds to a five-year (2001-2005) training program aimed at professionally qualifying 20,000-30,000 college graduates to enter the labor market in the software industry. Each selected graduate registers in one of the professional six-month training programs in order to be professionally certified by one of the international companies such as Microsoft, Oracle, and Sun. Such a program should help software firms to improve their intellectual capital, which is essential to the building of their knowledge absorptive capacity. This knowledge absorption capacity enables the firms to recognize the value of the new information and to apply it in order to create new knowledge and capabilities (Cohen & Lenvinthal, 1990).

Knowledge Transfer

Knowledge may be transferred and shared internally among members of an organization and with entities external to the organization. A significant part of the knowledge transferred at the individual and departmental levels in a software firm is technical, which is needed to perform specific tasks. This knowledge helps to design, build, sell, and support software products and services, which, in turn, result in experience accumulation over time. This section presents how knowledge is transferred and shared among software developers at the individual and departmental levels as well as knowledge transfer through training, education, and cooperation with competitors.

The interviews with the 14 firms revealed that they use different strategies—e-mail, voice-mail, forums, traditional meetings, groupware, on- and off-the-job training, education—for internal knowledge transfer. A number of the interviewed CEOs stressed the important role of information technology (IT) tools (e.g., e-mail, groupware, intranets and portals, video-conferencing, bulletin boards, etc.) in building relationships, facilitating the exchange of information, and sharing experience and knowledge. In particular, IT was found to provide convenient ways for knowledge transfer among software developers, who are always busy and have no time to share their knowledge in person. These findings are in agreement with those of Daveport and Prusak (1998), who reported that lack of time inhibits knowledge transfer among individuals.

Training and education were found to be important ways for knowledge transfer and diffusion among software developers and other employees in the firms in the sample (Table 7). Hiring and training employees who received education from recognized foreign institutions is essential to the renewal of the human capital stocks of the firms in the developing countries. Five percent of the total employees in the entire sample received education from foreign institutions that are considered superior to their domestic counterparts.

Also, 22% of the total employees (developers and other employees) received training on soft skills (e.g., managerial, group interaction, and decision-making skills), 21% of the total employees received training on project management, and 28% of software developers were certified. In addition, 14% of software developers in the firms were recognized as in-field experts in the Egyptian software industry. These findings are indicative of the firms' commitments to investing in human resource development and its efforts in recruiting such experts as means for knowledge transfer and diffusion. Based on estimations given by the CEOs of nine of the 14 firm samples, the expenditure on off-the-job training and education was approximately 6.75% of their total revenues.

Table 7. Frequency distribution of training and education as means for knowledge transfer (N = 38)

Training and Education	Number	Percentage
Number of employees (developers and other employees) who received foreign education	72	5%
Number of employees (developers and other employees) who received training on soft skills	248	22%
Number of developers who received training on project management	198	21%
Certified developers	224	28%
Developers who were recognized as experts in software business in Egypt	127	14%

Furthermore, the majority of the 14 firms that participated in the qualitative study practice rotation of employees on different projects for learning purposes and to ensure projects' continuity in light of a high turnover rate among software developers. A number of firms was found to offer orientation programs to their new developers and employees. The quality manager at one of the firms commented that his firm has a policy to guide and mentor its new developers and employees. According to such a policy, at least two of the oldest technicians are assigned the task of mentoring, guiding, advising, and auditing the performance of the new hires.

Also, most of the firms asserted that they are making sincere efforts to develop a knowledge-sharing culture while at the same time maintaining the strategic interest of their firms. However, there was a general agreement among the CEOs that knowledge sharing continues to be an issue. Software developers are reluctant to share their knowledge. In separate interviews with a number of software developers, they commented, "Competition in the labor market often inhibits sharing knowledge and experience." They added that "tacit knowledge is the most precious asset we have." They felt that knowledge is power. As one CEO put it, "When asked to collaborate in solving problems in a project, the developers' reactions were commonly negative, and their responses were rather slow."

A highly competitive Egyptian software market, combined with a relatively high unemployment rate, makes software developers hesitant to share knowledge in an attempt to ensure job security. Also, knowledge was not found to be effectively transferred and shared among software developers because much of the needed knowledge was not codified, often distributed throughout the firm, and embedded within its routines. This finding confirms Tsoukas' (1996) contention that tacit and unstructured knowledge and information are difficult to transfer and share. However, this finding is inconsistent with those of Navaretti and Tarr (2000), who assert that knowledge is mobile and can be used repeatedly across applications within and without a firm.

When interviewed, a number of software developers stressed the fact that the Egyptian culture plays a critical factor in knowledge transfer. It affects the willingness of the developers to transfer and share their experience with others. It was pointed out that the informal and

Table 8. Frequency distribution of compensation systems in the sample (N = 38).

Compensation System	Number	Percent
Fixed salary	35	92%
Workgroup incentives	22	58%
Individual incentives	23	61%
Profit sharing	6	16%
Stock sharing	8	21%
Nonmonetary recognition (e.g., gifts, trips, dinner)	14	37%

personal relationships among developers play a major role in knowledge exchange. This finding is inconsistent with those of Davenport and Prusak (1998), who investigated KM in a number of organizations and found that cultures negatively may affect knowledge transfer. However, this finding is consistent with the characterization of the Egyptian national culture as a culture that is rich in its social capital, which encourages people to assist each other and exchange experience and knowledge merely because they are colleagues in a workplace.

One interesting finding is that a significant portion of knowledge transfer and sharing among developers is incidental and takes place in informal, social functions. Knowledge sharing occurs when one needs to know something in order to do the job and takes a positive action to learn it. One important question raised by a number of software developers was how one is promoted when he or she shares knowledge and experience with others. Many of the firms that claim to be active in KM still reward people for hoarding rather than sharing knowledge.

Individually based compensation systems, which may impede knowledge sharing within organizations, were found prevalent in the firms in the sample (Table 8). Thirty-five firms pay their developers fixed salaries. Twenty-three firms employ individually based incentives for their developers; 22 firms provide their employees with workgroup incentives plans; 14 firms adopt individually based nonmonetary compensations (e.g., recognition, gifts, free trips, and dinner invitations); and eight firms provide their employees with stocks and profit sharing. However, when asked to elaborate on these group-based incentives systems, the CEOs were not sure whether the systems actually encourage knowledge sharing among group members, since the performance evaluation systems are still very much individually based.

Finally, although knowledge also can be transferred through collaboration with other firms in the industry, only limited cooperative efforts were found to exist among the software firms in the sample (Table 9).

Fifteen firms were found to have subcontractual relationships with other software firms; and only 12 firms indicated that they cooperate with other firms in the industry for the purpose of knowledge and information exchange. Limited cooperation means limited knowledge exchange among local software firms. Also, no serious cooperation initiatives with key software firms at the international level were attempted. Nonetheless, 17 firms in the sample have branches inside and outside Egypt (Table 2). The branches in the other countries are likely to be channels through which firms transfer foreign knowledge. Firms with foreign links have higher productivity growth rates than firms without foreign links (Navaretti & Tarr, 2000).

Table 9. Frequency distribution of knowledge transfer through cooperation with competitors (N = 38)

Forms of Cooperation with Competitors	Number	Percentage
Knowledge and information exchange	12	32%
R&D	2	5%
Subcontractors	15	39%
Joint Investment	2	5%
Others	7	18%

Also, the Egyptian Chamber of Software was reported to help establish consortiums among software firms based on their markets (e.g., education, industry, banks, and healthcare). These consortiums are expected to focus on activities such as R&D, marketing research, and knowledge and information exchange in order to create and share a knowledge infrastructure that can be used in the development of software products and solutions. Clearly, the success of such efforts depends, among other factors, on the extent of trust among the firms, their desire for corporation, a supportive culture, and an appropriate institutional infrastructure.

Conclusion and Implications

Software development entails various forms of explicit and implicit knowledge, which is dynamic and evolves with technology, organizational culture, and the changing needs of the organization's software development practices (Aurum, Jeffrey, Wohlin, & Handzic 2003; Lindvall & Rus 2003; Mathiassen, Pries-Heje, & Ngwenyama, 2002). In such a knowledge-intensive work environment, effective KM processes lie in developing organizational capability to acquire, document, transfer, and apply knowledge. Using quantitative and qualitative research approaches, this investigation aimed at exploring knowledge acquisition and knowledge transfer in software firms in Egypt.

Most of the firms in the sample seem to be cognizant of the importance of knowledge as a valuable resource affecting their capabilities to remain competitive. In addition, the software firms' idiosyncrasies appear to affect the way knowledge acquisition and knowledge transfer are practiced. Sources found for knowledge acquisition include visitation with customers; secondary information on demand; and trends of software, market research, the Internet, individual initiatives of software developers, and R&D activities. However, the firms' reliance on the relationships with academic institutions as a knowledge source is rather limited.

In addition, the investigated software firms seem to have a limited capability for effectively transferring and sharing the knowledge developed in one project to other projects. The rather limited knowledge sharing often occurs incidentally and informally through social networking and is based on trust. Formal IT-based KMS either did not exist in the surveyed firms or simply were not recognized and reported by the participants. Although IT does and potentially can play an important role in building relationships, facilitating information exchange and the sharing of experience and knowledge among software developers, knowledge transfer will continue to be a challenge that Egyptian software firms need to deal with.

Practices of knowledge acquisition and transfer in the software firms are scattered, and cohesive KM strategies are lacking. Therefore, KM strategies (e.g., codification, personalization, or both) must be adopted in order to guide knowledge acquisition and transfer practices in the software firms. Codification strategies emphasize the systematization and storage of knowledge and making it available for software developers and other employees in the firm (Hansen et al., 1999). Codification strategies, which utilize IT applications such as intranets and knowledge repositories, are more appropriate for capturing, storing, and transferring software development technical knowledge.

Personalization strategies, on the other hand, support the flow of information in a firm by storing information about knowledge resources like "yellow pages" of who knows what in the firm (Edwards, 2003; Hansen et al., 1999). Personalization strategies, which often favor people-based solutions such as communities of practice and storytelling, are particularly effective for knowledge acquisition and transfer in a culture that is rich in the emotional and social capitals like the Egyptian culture.

Software developers are always busy and are reluctant to share their knowledge in order not to jeopardize their job security. This finding was expected, since lack of knowledge culture frequently was cited as a critical obstacle to an effective KM practice (Agresti, 2000; Edwards, 2003; Lindvall & Rus, 2003). Culture obstacle occurs when individuals are not willing to transfer and share their knowledge and experience. They feel their knowledge is why they are valuable to their firms (Lindvall & Rus, 2003). Therefore, software firms should stimulate and support a collaborative knowledge culture that allows software developers to share and reuse their development experience and knowledge. Such a culture must encourage a bottom-up buy in to KM activities that matches the KM strategies employed from the top down (Edwards, 2003)

In order to create a collaborative knowledge culture that allows knowledge transfer among software developers, managers in the software firms should capitalize on the Egyptian national culture, which is a family-collective and humane-oriented culture (House, Javidan, Hanges, & Dorfman, 2002; Javidan & House, 2001). In such a culture, the process and content of communication are expected to help group cohesion and harmony. Human relations, sympathy, and support for others are highly valued characteristics; and people are generally friendly, sensitive, tolerant, and they value harmony (Javidan & House, 2001).

Accordingly, the Egyptian culture appears to be rich in its social capitals—the interpersonal relationships of a person and the resources embedded in those relationships (Nahapiet & Ghoshal, 1998). The social capital plays a central role in knowledge transfer as it encourages people to assist each other and to exchange experience and knowledge merely because they are colleagues in a workplace or classmates in a school. It also encourages openness, trust, and mutual respect, which are vital to enable knowledge transfer and learning to take place (Kautz et al., 2002). Software firms need to leverage such a supporting culture in order to reduce developers' resistances to sharing knowledge and information.

Once KM strategies are clearly defined, software firms should find effective mechanisms and practices for knowledge transfer among software developers in order to allow a better utilization of knowledge and experience. These mechanisms may include formulating teamwork, learning networks, knowledge sessions, tutoring and mentoring, brainstorming, effective communications systems, on-the-job training, job enlargement, job rotation, broadly defined jobs, job redesign, reward systems, motivational techniques, and change programs. These mechanisms should enable the firms to transfer an important part of the tacit knowledge of the experienced software developers to their colleagues and to become organizationwide competencies.

As advocated in the KM literature (Davenport & Prusak, 1998; Goh, 2002; Lee & Choi, 2003; Mason & Pauleen, 2003; Prusak, 1997), trust plays an essential role in knowledge acquisition and transfer within and without software firms. The shared values and beliefs of the employees of an organization influence the way the organization's members feel, think, and behave (Schein, 1996). Openness, trust, and mutual respect are vital in enabling knowledge transfer within the firm. In addition, collaborative and trusty relationships with local academic institutions and research centers are crucial to external knowledge acquisition and transfer, especially when the financial and human resources available for R&D in the relatively young, small software firms are rather limited.

Furthermore, software firms should develop human resource strategies and policies in order to attract and retain expert software developers as a source for knowledge acquisition. Human resource practices that emphasize team-based performance evaluation and compensation, continuous training and learning, social and informal interactions, membership in professional organizations, and the promotion of a climate of trust are expected to support knowledge acquisition and transfer in the Egyptian software firms. Institutional and human learning mechanisms along with a supportive culture of shared values, norms, and beliefs ensure a productive learning environment that increases the stock and flow of knowledge in these firms.

Finally, the findings of this investigation on knowledge acquisition and transfer practices and mechanisms in a number of Egyptian software firms should contribute to the growing empirically based literature on KM in general and KM in developing countries in particular. However, future research designs are needed to investigate factors (e.g., organization's maturity, size, age, type, leadership style, and culture) that are believed to influence KM practices in general and knowledge acquisition and transfer in particular. These factors should be investigated within different contexts and at variant levels of analysis in order to develop a better understanding of KM practice.

References

Abou-Zeid, E. (2002). An ontology-based approach to inter-organizational knowledge transfer. *Journal of Global Information Technology Management, 5*(3), 32-47.

Agresti, W. (2000). Knowledge management. *Advances in Computers, 53*, 171-283.

Aker, D. (1989, Winter). Managing assets and skills: The key to a sustainable competitive advantage. *California Management Review, 32*(1), 91-106.

Alavi, M., & Leidner, D. E. (2001). Knowledge management and knowledge management systems: Conceptual foundations and research issues. *MIS Quarterly, 25*(1), 107-136.

Alavi, M., & Tiwano, A. (2002). Knowledge integration in virtual teams: The potential role of KMS. *Journal of the American Society for Information Science and Technology, 53*(12), 1029-1037.

Amabile, T. M., Nasco, C. P., Mueller, J., Wojcik, T., Odomirok, P. W., Marsh, M., et al. (2001). Academic-practitioner collaboration in management research: A case of cross-profession collaboration. *Academy of Management Journal, 44*, 418-431.

Amit, R., & Schoemaker, P. J. (1993). *Strategic assets and organizational rent. Strategic Management Journal, 14*(1), 33-46.

Arab Human Development Report. (2002). *United Nation Development Program.* Regional Bureau for Arab States. Jordan (Arabic Version).

Argote, L., & Ingram, P. (2000). Knowledge transfer: A basis for competitive advantage in firms. *Organizational Behavior and Human Decision Processes, 82*(1), 150-169.

Argyris, C. (1998). Teaching smart people how to learn. In P.F. Drucker, D. Grvin, D. Leonard, S. Strauss & T. Brown (Eds.), *Harvard business review on knowledge management* (pp. 81-108). Harvard Business School Press.

Ashour, A. S. (2000). Knowledge capital management. Reinventing management paradigm in the 21st century. In *Proceeding of the Twelfth International Conference on Training and Management Development Towards Arab Learning Organization,* Cairo.

Aurum, A., Jeffery, R., Wohlin, C., & Handzic, M. (Eds.) (2003*). Managing software engineering knowledge.* Berlin: Springer.

Balogun, J., Huff, A. S., & Johnson, P. (2003). Three responses to the methodological challenges studying strategizing. *Journal of Management Studies, 40*(1), 197-224.

Barney, J. B. (1991). F*irm resources and sustained competitive advantage. Journal of Management, 17*(1), 99-120.

Bartlett, C. A., & Ghoshal, S. (1998). Beyond strategic planning to organization learning: Lifeblood of the individualized corporation. *Strategy & Leadership, 26*(1), 34-39.

Bartol, K. M., & Srivastave, A. (2002). Encouraging knowledge sharing: The role of organizational reward systems. *Journal of Leadership and Organization Studies, 9*, 64-76.

Bhatt, G. (2001). Knowledge management in organizations: Examining the interaction between technologies, techniques, and people. *Journal of Knowledge Management, 5*(1), 68-75.

Bierly, P., & Chakrabarti, A. (1996). Generic knowledge strategies in the U.S. pharmaceutical industry. *Strategic Management Journal, 17*, 123-135.

Bontis, N., & Fitz-enz, J. (2002). Intellectual capital ROI: A current map of human capital antecedents and consequent. *Journal of Intellectual Capital, 3*(3), 223-247.

Borys, B., & Jemison, D. B. (1989). Hybrid arrangements as strategic alliances: Theoretical issues in organizational combinations. *Academy of Management Review, 14*, 234-250.

Bots, P. W. G., & de Bruijn, H. (2002). Effective knowledge management in professional organizations: Going by the rules. In *Proceedings of the 35th Hawaii International Conference on Information Sciences.*

Brown, R. B., & Woodland, M. J. (1999). Managing knowledge wisely: A case study in organizational behavior. *Journal of Applied Management Studies, 18*(2), 175-199.

Carter, B. (2000). The expert's opinion: Knowledge management. *Journal of Database Management, 11*, 42-43.

Choi, B., & Lee, H. (2003). An empirical investigation of KM styles and their effect on corporate performance. *Information & Management, 40*(5), 403-417.

Cockburn, I., & Henderson, R. (1994). Absorptive capacity, coauthoring behavior, and organization of research in drug discovery. *The Journal of Industrial Economics, XLVI*(2), 157-182.

Cohen, W. M., & Levinthal, D. A. (1989). Innovation and learning: The two faces of R&D. *The Economic Journal, 99*(397), 569-597.

Cohen, W. M., & Levinthal, D. A. (1990). Absorptive capacity: A new perspective learning and innovation. *Administrative Science Quarterly, 35*(1), 128-152.

Crossan, M., Lane, H., & White, R. (1999). An organizational learning framework: From intuition to institution. *Academy of Management Review, 24*(3), 522-537.

Darroch, J. (200). Developing a measure of knowledge management behaviors and practices. *Journal of knowledge management*. 7(5), 41-54.

Das, S, Sen, P. K, & Sengupta, S. (2003). Strategic alliances: A valuable way to manage intellectual capital. *Journal of Intellectual Capital, 4*(1), 10-19.

Davenport, T. H., & Prusak, L. (1998). *Working knowledge: How organizations manage what they know*. Boston: Harvard Business School Press.

Day, G. S. (1994). The capabilities of market driven organization. *Journal of Marketing, 58*(4), 37-52.

De Long, D. W., & Fahey, L. (2000). Diagnosing cultural barriers to knowledge management. *The Academy of Management Executive, 14*(4), 113-127.

Demsetz, H. (1991). The theory of the firm revisited. In O. E. Williamson, & S. Winter (Eds.), *The nature of the firm* (pp. 159-178). New York: Oxford University Press.

Denzin, N. K., & Lincoln, Y. S. (2000). Introduction: The discipline and practice of qualitative research. In N. K. Denzin, & Y. S. Lincoln (Eds.), *Handbook of qualitative research* (2nd ed.) (pp. 1-29). Thousand Oaks, CA: Sage Publications.

DiBella, A. J., & Nevis, B. C. (1998). *How organizations learn: An integrated strategy for building learning capability*. San Francisco: Jossey-Bass.

Dierickx. I., & Cool, K. (1989). Asset stock accumulation and sustainability of competitive advantage. *Management Science, 35*(12), 1504-1513.

Dingsoyr, T., & Conradi, R. (2003). Usage of intranet tools for knowledge management in a medium-sized software consulting company. In A. Aurum, R. Jeffery, C. Wohlin, & M. Handzic (Eds.), *Managing software engineering knowledge* (pp. 49-68). Berlin: Springer.

Drucker, P.F. (1993). *Post-Capitalist Society*. New York: Harper Collins

Edward, C., Feng, K., & Liou, W. (2005). Implementation of *knowledge management systems* and firm performance: An empirical investigation. *Journal of Computer Information Systems, 45(2), 92-104.*

Edwards, S. E. (2003). Managing software engineers and their knowledge. In A. Aurum, R. Jeffery, C. Wohlin, & Handzic (Eds.) (2003). *Managing Software Engineering Knowledge.* Berlin: Springer, 5-27.

El-Rashidi, Y. (2002). Silicon Valley, silicon pyramids. *Al-Ahram Weekly, Jan 31-Feb 6.* (571), 2.

Eppler, M. J., & Sukowski, O. (2000). Managing team knowledge: Core processes, tools and enabling factors. *European Management Journal, 18*(3), 334-341.

Fukuyama, F. (1995). *Trust: The social virtues and the creation of prosperity*. New York: Free Press.

Garvin, D. A. (1993, July-August). Building learning organizations. *Harvard Business Review,* 78-91.

Garvin, D. A. (1994). Building a learning organization. *Business Credit, 96*(1), 19-28.

Gillis, T. (2003). More than a social virtue: Public trust among organizations most valuable assets. *Communication World, 20*(3), 10-11.

Glazer, R. (1991). Marketing in an information-intensive environment: Strategies implications of knowledge as an asset. *Journal of Marketing, 55*(4), 1-19.

Goh, S. C. (2002). Managing effective knowledge transfer: An integrative framework and some practice implications. *Journal of Knowledge Management, 6*(1), 23-30.

Grant, R. M. (1996, Winter, Special Issue). Toward a knowledge-based theory of the firm. *Strategic Management Journal, 17*, 109-122.

Grant, R. M. (2000). Shifts in the world economy: The drivers of knowledge management. In C. Despres, & D. Chauvel (Eds.), *Knowledge horizons: The present and promise of knowledge management.* Butterworth-Heinemann.

Gulati, R. (1995). Does familiarity breed trust? The implications of repeated ties for contractual choice in alliances. *Academy of Management Journal, 38*(1), 85-112.

Gunnlaugsdottir, J. (2003). Seek and you will find, share and you will benefit: Organizing knowledge using groupware systems. *International Journal of Information Management, 23*(5), 363-380.

Hall, R. (1992). The strategic analysis of intangible resources. *Strategic Management Journal, 13*(2), 135-144.

Hall, R. (1993). A framework linking intangible resources and capabilities to sustainable competitive advantage. *Strategic Management Journal, 14*(8), 607-618.

Hansen, M. T., Nohria, N., & Tierney, T. (1999). What's your strategy for managing knowledge? *Harvard Business Review, 77*, 106-116.

Hedlund, G., & Nonaka, I. (1993). Models of knowledge management in the West and Japan. In P. Lorange, B. Chakravarthy, J. Roos, & A. Van de Ven (Eds.), *Implementing strategic processes: Change, learning and cooperation* (pp. I17-44). Oxford, UK: Basil Blackwell.

Hellstrom, T., Malmquist, U., & Mikaelsson, J. (2001). Decentralizing knowledge: Managing knowledge work in a software engineering firm. *Journal of High Technology Management Research, 12*, 25-38.

Hoch, D. J., Roeding, C. R., Purkert, G., Linder, S. K., & Muller, W. R. (2000). *Secrets of software success: Management insights from 100 software firms around the world.* Boston: Harvard Business School Press.

House, R., Javidan, M., Hanges, P., & Dorfman, P. (2002). Understanding cultures and implicit leadership theories across the globe: An introduction to Project GLOBE. *Journal of World Business, 37*, 3-10.

Huber, G. P. (1991). Organizational learning: The contributing processes and the literatures. *Organization Science, 2*(1), 88-115.

Inkpen, A. C. (1998). Learning and knowledge acquisition through international strategic alliances. *Academy of Management Executive, 12*(4), 69-80.

Itami, H. (1987). *Mobilizing invisible assets.* Cambridge, MA: Harvard University Press.

Jalote, P. (2003). Knowledge infrastructure for project management. In A. Aurum, R. Jeffery, C. Wohlin, & M. Handzic (Eds.), *Managing software engineering knowledge* (pp. 361–375). Berlin: Springer.

Javidan, M., & House, R. (2001). Culture acumen for the GLOBE manager: Lessons from project GLOBE. *Organizational Dynamics, 29*(4), 289-305.

Jennex, M. E., & Olfman, L. (2002). Organizational memory/knowledge effects on productivity: A longitudinal study. In *Proceedings of the 35th Hawaii International Conference on System Sciences.*

Jennex, M. E., & Olfman, L. (2005). Assessing knowledge management success. *International Journal of Knowledge Management, 1*(2), 33-49.

Karhu, K. (2002). Expertise cycle—An advanced method for sharing expertise. *Journal of Intellectual Capital, 3*(4), 430-446.

Kautz, K., Thaysen, K., & Vendelo, M. T. (2002). Knowledge creation and IT systems in a small software firm. *OR Insight, 15*, 11-17.

Kogut, B., & Zander, U. (1992). Knowledge of the firm, combinative capabilities and replication of technology. *Organization Science, 3*, 383-397.

Kogut, B., & Zander, U. (1996). What firms do? Coordination, identity, and learning. *Organization Science, 7*(5), 502-518.

Lee, H., & Choi, B. (2003). Knowledge management enablers, processes, and organizational performance: An integrative view and empirical examination. *Journal of Management Information Systems, 20*(1), 179-229.

Lennselius, B., & Wohlin, C. (1987). Software metrics: Motivation and fault content estimation. *Microprocessors and Microsystems, 11*, 365-375.

Liebowitz, J. (2000). *Building organizational intelligence: A knowledge management primer.* Boca Raton, FL: CRC Press.

Lindsey, K. (2002). Measuring knowledge management effectiveness: A task-contingent organizational capabilities perspective. In *Proceedings of the Eighth American Conference on Information Systems,* 2085-2090.

Lindvall, M., & Rus, I. (2003). Knowledge management for software organizations. In A. Aurum, R. Jeffery, C. Wohlin, & Handzic (Eds.), *Managing software engineering knowledge* (pp. 73-94). Berlin: Springer.

Machlup, F. (1962). *The production and distribution of knowledge in the United States*. Princeton, NJ: Princeton University Press.

Machlup, F. (1983). *Knowledge, its creation, distribution and economic significance*. Princeton, NJ: Princeton University Press.

Maier, R. (2002). *Knowledge management systems: Information and communication technologies for knowledge management*. Berlin: Springer-Verlag.

Marshall, C. (1985). Appropriate criteria of trustworthiness and goodness for qualitative research on education organizations. *Quality on Quantity, 19*, 353-373.

Mason, D., & Pauleen, D. J. (2003). Perceptions of knowledge management: A qualitative analysis. *Journal of Knowledge Management, 7*(4), 38-48.

Massey, A.P., Montoya-Weiss, M.M., & O'Driscoll, T.M., (2002). Knowledge management in persuit of performance: Insights from nortel networks. *MIS Quarterly, 26*(3), 269-289.

Mathiassen, L., & Pourkomeylian, P. (2003). Managing knowledge in a software organization. *Journal of Knowledge Management, 7*(2), 63-80.

Mathiassen, L., Pries-Heje, J., & Ngwenyama, O. (2002). *Improving software organizations: From principles to practice*. Boston: Addison-Wesley.

Maxwell, J. A. (1998). Designing a qualitative study. In L. Bickman, & D. Rog (Eds.), *Handbook of applied social research methods* (pp. 69-100). Thousand Oaks, CA: Sage Publications.

Motwani, J, Gopalakrishna, P., & Subramanian, R. (2003). Sources of knowledge acquisition by U.S. managers: An empirical analysis. In A. Gunasekaran, O. Khalil, & S. M. Rahman (Eds.), *Knowledge and information technology management: Human and social perspective* (pp. 14-28). Hershey, PA: Idea Group Publishing.

Mowery, D. C., Oxley, J. E., & Silverman, B. S. (1996). Strategic alliances and interfirm knowledge transfer. *Strategic Management Journal, 17*, 77-91.

Nahapiet, J., & Ghoshal, S. (1998). Social capital, intellectual capital, and the organizational advantage. Academy of Management Review, 23(2), 242-266.

Navaretti, G. B., & Tarr, D. G. (2000). International knowledge flows and economic performance: A review of the evidence. *The World Bank Economic Review, 14*(1), 1-15.

Nickolas, F. (2001). What is in the world of work and working: Some implication of the shift to knowledge work. In J. Crtada, & J. Woods (Eds.), *The knowledge management yearbook 2000–2001* (pp. 12-21). Boston: Butterworth-Heinemann.

Nonaka, I. (1991, November-December). The knowledge creating company. *Harvard Business Review,* 96-104.

Nonaka, I. (1994). A dynamic theory of organizational knowledge creation. *Organization Science, 5*(10), 14-37.

Nonaka, I., & Takeuchi, H. (1995). *The knowledge creating company—How Japanese companies create the dynamics of innovation*. Oxford: Oxford University Press.

Okunoye, A., & Bertaux, N. (2006). KAFRA: A context-aware framework of knowledge management in global diversity. *International Journal of Knowledge Management, 2*(2), 26-45.

Olson, M. (1981). User involvement and decentralization of the development function: A comparison of two case studies. *Systems, Objectives, Solutions*, *1*(2), 59-69.

Penrose, E. T. (1959). *The theory of the growth of the firm*. Oxford: Basil Blackwell.

Pentland, B. T. (1995). Information systems and organizational learning: The social epistemology of organizational learning systems. *Accounting, Management and Information Technologies*, *5*, 1-21.

Prahalad, C. K., & Hamel, G. (1990, May-June). The core competence of the corporation. *Harvard Business Review*, 79-91.

Prusak, L. (1997). *Knowledge in organizations*. Boston: Butterworth-Heinemann.

Quinn, J. B. P., Anderson, P., & Finkelstein, S. (1998). Managing professional intellect: Making the most of the best. In P.F. Drucker, D. Grvin, D. Leonard, S. Straus, & T. Brown (Eds.), *Harvard business review on knowledge management* (pp. 181-205). Harvard Business School Press.

Ring, P. S., & Van de Ven, A. (1992). Structuring cooperative relationships between organizations. *Strategic Management Journal, 13*(7), 483-498.

Robey, D. (1981). Computer information systems and organization structure. *Communications of the ACM, 24*(10), 679-687.

Robey, D. (1983). Information systems and organizational change. *Systems, Objectives, Solutions, 3*(3), 145-154.

Schein, E. H. (1996, Fall). Three cultures of management: The key to organizational learning. *Sloan Management Review, 83*(2), 3-20.

Seleim, A, Ashour, A, & Bontis, N. (2004). Intellectual capital in Egyptian software firms. *The Learning Organization, 11*(4,5), 332-346.

Senge, P.M. (1990). *The fifth discipline: The art and practice of the learning organization*. New York: Doubleday.

Shadish, W. R., Cook, T. D., & Campbell, D. T. (2002). *Experimental and quasi-experimental designs for generalized causal inference*. Houghton Mifflin Company.

Simon, H.A. (1991). Bounded rationality and organizational learning. *Organization Science, 2*(1), 125-134.

Slater, S. F., & Narver, J. C. (1995). Market orientation and the learning organization. *Journal of Marketing, 59*(3), 63-74.

Spender, J. C. (1994). Organizational knowledge, collective practice and Penrose rents. *International Business Review, 3*(4), 353-367.

Spender, J. C. (1996, Winter Special Issue). Making knowledge the basis of dynamic theory of the firm. *Strategic Management Journal, 17*, 45-62.

Staples, D. S, Greenaway, K, & McKeen, J. D. (2000). *Research opportunities relevant for managing knowledge based enterprises* [framework paper]. Queen's School of Business.

Stewart, T. A. (1997). *Intellectual capital: The new wealth of organizations*. New York: Doubleday Publishing Group.

Stoddart, L. (2001). *Managing intranets to encourage knowledge sharing: Opportunities and constraints. Online Information Review, 25*(1), 19-28.

Sveiby, K. E. (1997). The intangible asset monitor. *Journal of Human Resource Costing and Accounting, 2*, 73-97.

Teece, D. J., Pisano, G., & Shuen, A. (1997). Dynamic capabilities and strategic management. *Strategic Management Journal, 18*(7), 509-533.

Tippins, M. J., & Sohi, R. S. (2003). IT competence and firm performance: Is organizational learning a missing link. *Strategic Management Journal, 24*(8), 745-761.

Tiwana, A. (2000). *The knowledge management toolkit: Practical techniques for building a knowledge management system*. Upper Saddle River, NJ: Prentice-Hall.

Tsoukas, H. (1996, Winter Special Issue). The firm as a distributed knowledge system: A constructions approach. *Strategic Management Journal, 7*, 11-25.

Usoro, A., & Kuofie, M. H. S. (2006). Conceptualization of cultural dimensions as a major influence on knowledge sharing. *International Journal of Knowledge Management, 2*(2), 15-25.

Wernerfelt, B. (1984). A resource based view of the firm. *Strategic Management Journal, 5*(2), 171-180.

Wiig, K.M. (1997). Integrating intellectual capital and knowledge management. *Long Range Planning, 30*(3), 399-405.

Wiig, K. M. (2000). Knowledge management: An emerging discipline rooted in a long history. In C. Despres, & D. Chauvel (Eds.), *Knowledge horizons: The present and the promise of knowledge management*. New York: Butterworth-Heinemann.

Winter, S. G. (1987). Knowledge and competence as strategic assets. In D. J. Teece (Ed.), *The competitive challenge: Strategies for industrial innovation and renewal* (pp. 156-184). Cambridge, MA: Ballinger.

World Development Report: Knowledge for Development, 1998/1999. The World Bank.

Wright, M., & Ashill, N. (1998). A contingency model of marketing information. *European Journal of Marketing, 32*(1-2), 125-145.

Yahya, S, & Goh, W. K. (2002). Managing human resources toward achieving knowledge management. *Journal of Knowledge Management, 6*(5), 457-468.

Yin, R. K. (1989). *Case study research: Design and methods*. Newbury Park, CA: Sage Publications.

Zack, M. H. (1999a). Managing codified knowledge. *Sloan Management Review, 40*(4), 45-58.

Zack, M. H. (1999b). Developing a knowledge strategy. *California Management Review, 41*(3), 125-145.

Zhang, D., & Zhao, J. L. (2006). Knowledge management in organizations. *Journal of Database Management, 17*(1), I-VIII.

Appendix

An English Summary of the Data Gathering Instrument

[The original instrument is in Arabic]

Open-ended probing questions used to gather qualitative data on knowledge acquisition and transfer:

1. What are the most important driving forces for acquiring and transferring knowledge that your firm considers important to software development?
2. What types of knowledge does your company need in order to develop software products? How does your company develop and acquire such knowledge? How importance is tacit knowledge for software development?
3. What sources does your company use to acquire external knowledge that is considered important to software development? What problems, if any, prevent your company from effectively utilizing each source?
4. Does your company have an R&D unit? How effective is it in creating useful knowledge for software development? What programming languages and development tools does your company use in software development?
5. What strategies and mechanisms does your company use to facilitate knowledge transfer and knowledge sharing among software developers and employees in your company?
6. What IT applications and tools does your company use to facilitate the acquisition and transfer of software development knowledge?
7. To what extent is the knowledge accumulated from past projects reused in developing new software projects? What problems prevent such a reuse?
8. To what extent do you cooperate with your competitors in knowledge exchange?
9. How do software developers and other employees in your company practice knowledge transfer and sharing take place in your company?
10. What are the problems of and barriers to knowledge transfer and sharing in your company?
11. What are the barriers and challenges that face your company in managing its knowledge in general?

Questions used to gather quantitative data on knowledge acquisition and transfer practice:

1. Does the company depend on R&D for acquiring new software development knowledge?
2. What is the percentage of R&D expenditure to annual sales?
3. Does the company depend on market research as a source for acquiring knowledge?

4. Does the company depend on the individual initiative of developers and employees to create knowledge?
5. Does the company depend on global demand for and trends in software development as sources for knowledge?
6. Does the company imitate the products of local competitor software firms as a source for software development knowledge?
7. Does the company imitate the products of international software firms as a source for software development knowledge?
8. Does the company use on-site visitation to gather customer-related knowledge?
9. Does the company use formal and informal gatherings such as conferences and parties to collect customer knowledge?
10. Does the company use means of receiving suggestions and ideas from the customers?
11. Does the company depend on friendship as a means for gathering customer knowledge?
12. Does the company exchange knowledge and information with competitors?
13. Does the company collaborate on R&D projects with competitors?
14. Does the company have joint investments with competitors?
15. Does the company have any cooperation with competitors in the form of subcontracts?
16. Does the company have any cooperation with competitors in the form of sharing experience?
17. Does the company identify any other forms of cooperation with competitors for the purpose of knowledge acquisition and transfer?
18. How many employees (developers and other employees) have received foreign education?
19. How many employees (developers and other employees) have received training on soft skills?
20. How many software developers have received training on project management?
21. How many developers have received professional certificates in software development?
22. How many software developers were recognized as experts in the software business in Egypt?
23. What is the relative importance of universities, research centers, and the Internet as sources of external knowledge?

The questions on the software firm's background were designed to gather the following information:

1. Starting date of the firm
2. Type of capital (e.g., Egyptian, foreign, or joint)
3. Type of ownership (e.g., individual, incorporated, branch of an international company)
4. Types of developed software products
5. The target markets (e.g., users) inside Egypt
6. Whether the firm exports software products
7. The markets for exported software products
8. Types of exported software products
9. Types of packaged software products
10. Whether the firm has branches in other countries
11. The number of the firm's branches and the countries in which they are located
12. Software development platform(s) in use
13. Number of employees
14. Number of software developers
15. Number of other employees (not directly involved in software development)
16. Employee compensation and reward systems in use

Section V

Experience with Knowledge Management

Chapter XVIII

Adopting Knowledge-Centred Principles in Innovation Pursuits: The Case of Singapore Airlines

Andrew Goh, Management Development Institute of Singapore, Republic of SIngapore

Abstract

With the emergence of the knowledge economy, organizations are beginning to see a need to apply knowledge management (KM) practices to their business activities. While knowledge management (KM) has gathered considerable momentum to be a vital source of competitive advantage, how its role could harvest knowledge assets for innovation has yet to be firmly established. This chapter aims to address this issue by examining how innovation can be fostered through knowledge-centered principles. It first describes the globalization of economies and the coming of the new knowledge age as the backdrop to Singapore's vision of transiting into a knowledge economy. Then it discusses how knowledge management (KM) practices can be harnessed better for innovation management and explains why organizations should foster innovation by adopting an evolving set of knowledge-centered principles. Next, based on the case of Singapore Airlines (SIA), it provides a theoretical review of these principles. Finally, it outlines the future challenges of exploiting knowledge for innovation.

Background

Globalization of Economies

The world today is far more interconnected and interdependent than before. As globalization speeds up and cross-border barriers between nations are dismantled, economic development will depend less on physical resources and more on developments that require nations to be more global in their approach to trade and investment activities. Coupled with the fast pace of technological advancements, national economies not only have to be global in order to stay competitive, but they also have to be ready to embrace the demands of innovation (Giget, 1997; Goh, 2002; Grossman & Helpman, 1992). This inadvertently has placed immense pressure on emerging economies to accelerate the process of innovation through knowledge acquisition and application. If the experience of some developed nations that went through such a wave of economic globalization could offer some insights, then a sea of change in employment trends, industrial transformation, and economic revolution should be expected in Asia's emerging economies. Three phenomena seem imminent: (1) outsourcing of transportable jobs to countries offering the most competitive labor cost is now a ubiquitous trend; (2) migration of existing industries up the value chain to new knowledge-intensive ones is increasingly prevalent; and (3) economies are forced to respond to rapid technological changes and constant industrial renewal in order to remain competitive and relevant.

In Asia, a new economy termed *knowledge economy*[a] has arrived (OECD, 1996). Organizations are pressured to be knowledge-intensive in their activities. As advanced technologies proliferate and new products become obsolete faster than before, organizations that are able to capitalize on opportunities arising from the availability of knowledge assets[b] and derive the most value from them will be the industry winners, while those who cannot will be the industry losers. Since innovations constitute the embodiment of knowledge assets in new products and services, innovation pursuits are centered on leveraging the value of knowledge. Corporate leaders thus are taking a keen interest in effective means of harvesting knowledge to foster the pursuit of innovation and are differentiating themselves from competitors based on new management initiatives (Malhotra, 2001; Nonaka, 1991; Skyrme, 1991). This is fast becoming pervasive in today's knowledge-intensive enterprises and soon will be mandatory for the economic survival of all organizations.

Emergence of Knowledge Economy: Singapore's Opportunities

In less than one century, the world has gone through several stages of economic transformation—from agricultural economy to industrial economy, then information economy, and now, the knowledge economy. Economists have argued that national economies as well as advocates of centralized planning should attribute their economic problems to the utilization of knowledge and not on the allocation of resources. Strong evidence also exists to support that economies, which are poor in natural resources but skilled in knowledge creation and utilization, generally outperform those economies that have abundant natural resources but are lacking in knowledge competence and skills. Knowledge has emerged as the primary resource for economic development; land, labor and capital—the economist's traditional factors of production—have become secondary (Drucker, 1988; Miller & Morris, 1999). It is also argued that traditional factors of production are limited by a threshold of scale and scope as every marginal increase in land, labor, or financial capital results in diminishing returns on additional investment. In contrast, a different law of economic returns seems to govern the returns arising from knowledge, and investment in every additional unit of information or knowledge created and utilized results in much higher returns[c].

For more than a decade, Singapore has viewed the emergence of the knowledge economy as offering a wellspring of opportunities. But what seems most immediate is to identify and nurture strategic industries that would spur steady and sustainable growth based on the economics of knowledge or, put simply, transforming the economy into one that exploits the commercialization of knowledge (Amidon, 1997; Goh, 2004; Grossman & Helpman, 1992). As a small nation devoid of vast natural resources, Singapore's hope to eventually become a developed nation[d] may be attained via knowledge-based imperatives such as (1) shifting industries to value-added, technology-based, and knowledge-intensive sectors beyond mere production and manufacturing of foreign goods; (2) building effective communications and information infrastructures that encourage leading-edge knowledge transfer as a lever to increase the value of existing goods and services; (3) investing in a good education system and scientific and applied research and development (R&D) and promoting lifelong learning to enhance the quality of innovation pursuits and to foster a vibrant and entrepreneurial business environment. Yet, in order to leverage these imperatives, one needs to understand the challenges faced by organizations in the new knowledge age.

The New Knowledge Age: Singapore's Vision

Singapore, which attained independence from Britain on August 9, 1965, has a short history of industrial development. Due to the absence of a sizable domestic market, Singapore's industrialization strategy relies mainly on offshore manufacturing and services. The backdrop of this strategy was the promotion of foreign direct investments (FDI) into Singapore, based on an inflow of multinational corporations (MNCs) into the country. This transformed the nation from an entrepôt into a global economy with diversified industries, including logistics, petrochemicals, electronics, and finance (Goh, 2004; Liao & Chew, 2000). As Singapore's economy matures, the need to be a knowledge hub has become more urgent. Hence, in the early 1990s, the country embarked on envisioning a wired island—a vision for the new knowledge age wherein a high-bandwidth infrastructure enables free and rapid flows of information and knowledge. Indeed, Singapore's three national strategic plans—IT2000 for the information and communication technologies (ICT) sector, Industry21 for the nation's economic blueprint, and Library2000 for the National Library Board—were part and parcel of the country's efforts to support this vision. Common in these plans was the emphasis on the preparedness for a new knowledge economy, highlighted as follows:

1. A clear vision of a knowledge economy and how the nation is heading and forging toward the vision

2. A strong emphasis on lifelong learning, knowledge-based activities, and knowledge-intensive businesses
3. An emergent role of knowledge management in high-performing and successful organizations

With a commitment to changing the city state to a knowledge hub, the Singapore government focused on building three interdependent pillars of economic development: technology, innovation, and capability. First, technology areas such as life sciences, multimedia, microelectronics, data storage, wireless communications, and manufacturing technologies were emphasized strongly. Second, to blaze new trails of innovations, as the developed world has succeeded in doing for several decades, an Innovation Development Program was initiated to intensify innovation projects in support of the vision of a knowledge economy. Third, to develop capabilities in industry generation and job creation, the Technopreneurship 21 Plan was launched to promote a culture of entrepreneurial spirit by supporting startups with focused initiatives on technology and innovation development.

The Knowledge Revolution: Singapore's Challenges

Emerging economies now realize that the information revolution has been superseded by the knowledge revolution. The *what* that creates innovations has shifted from the tangibles and physical assets to processes wherein various forms of knowledge are assimilated, shared, and utilized with the objective of creating new knowledge innovations. For a country like Singapore with no natural resources on which to depend, a knowledge economy poses a gold mine to be unearthed. With the global economic paradigm becoming highly knowledge-intensive, which dictates the way businesses compete, innovation-driven value creation is the best bet to secure high economic growth. Singapore's future economic challenges will depend heavily on its ability to develop and utilize knowledge in the innovation landscape. Singapore's economic development has to be centered on innovation-driven initiatives, which would enhance its global competitiveness through areas such as product development, high-tech venture creation, and the like.

Since its GDP reached US$93 billion in 2003, the country is now able to provide a vibrant business environment for promoting innovation as a source of economic growth. As a fitting test-bed for knowledge innovations, the country decided to establish first-rate ICT infrastructures and, thus, became the world's first country to be wired up in every home, school, and business via the Singapore ONE. This infrastructure is a high-speed, high-capacity broadband network capable of transmitting 622 megabits of data per second for most generations of advanced technological applications. Besides ICT infrastructures, other initiatives included developing areas such as video-on-demand services, value-added television, and online education services, whose purpose is to transform digital nervous system and Web lifestyle into reality. Another initiative is the NETtv, which is a proprietary, picture-perfect, Web-casting solution that allows all of the color, sound effects, and media presentations of a large-scale event or entertainment program to be delivered from one place to another. The solution comes with an extensive suite of high-quality interactive multimedia services, live or on-demand on a wide-reaching basis, to anyone anywhere in the world. Additionally, since R&D activities constitute an important component of the knowledge economy, the government realizes that legislation and enforcement measures of patents, industrial designs, trademarks, and copyrights also should be put in place and, thus, its intellectual property laws[e] updated to match those of developed countries.

Data to Information to Knowledge Management

From the 1970s to the early 1980s, the acquisition of effective data management tools was the *raison d'être* for business growth. Then, from the 1980s to the 1990s, the focus shifted from data management to information management. Now, the use of knowledge and its management—generally termed knowledge management[f] (KM)—has emerged as such a critical area of functional management that developing it as a core competence is believed to provide a sustainable competitive advantage. Increasingly, a key corporate strategy is the use of knowledge to speed up innovation. The centrality of KM in most corporate strategies bears testimony that adopting the best KM practices would result in better, higher-quality, and cost-effective innovations. It is not surprising that managers are widely employing KM techniques in innovation programs; and it seems that failure to do so may impede innovation performance and, thus, undermine corporate competitiveness (Fahey & Prusak, 1998; Lindgren & Henfridsson, 2002; Storey & Barnett, 2000). However, the actual role of KM is still far from being fully understood. For instance, while recognizing the importance of knowledge, the apparent confusion between knowledge and information has caused organizations to sink huge investments in information technology (IT) that yielded marginal corporate performance (Malhotra, 1997, 2000; Strassmann, 1997). This is because IT expenditures may not be related directly to corporate performance, and this lack of understanding may be attributed to the economic transition from an era based primarily on information to one dependent on knowledge[g].

Nevertheless, the focus on knowledge management resulted in more resources being directed at acquiring and utilizing knowledge assets for innovation programs. In today's corporate world, KM functional roles have emerged at a global scale with designations such as chief knowledge officers (CKOs) or vice-president of knowledge management being established in MNCs in the United States, Japan, Europe, and Canada, and the trend has caught on in some parts of Asia. In Singapore, for instance, the role of knowledge management is fast becoming a vehicle for institutionalizing innovation in high-tech and knowledge-based industries (e.g., information and communications technologies and biomedical industries). To transit into a knowledge economy, Singapore by 2000 had positioned itself as one of the world's largest manufacturers of disk drives, tape drives, and proprietary pharmaceuticals with more than 6,000 multinational corporations (MNCs) located in the country. Other leading industries include computer peripherals and petroleum refineries[h].

Strategic Management Perspectives

KM Practices Harnessed for IM Processes

Currently, the breadth of KM interest stretches across a wide range of business areas. While its early interest was concentrated largely on information technology, competitive strategy, and business development, the field of KM gradually has extended to areas like operations management, human resources, accounting, and finance (Gupta & MacDaniel, 2002; Mullin, 1996). Though there may be extensive literature written on knowledge management (KM) and innovation management (IM) as separate management areas of concern, limited research has dealt singly on the management issues relating to the management of knowledge for innovation. Yet, the strategic concern encountered by organizations appears to be more than just dealing with KM or IM issues separately; instead, it involves the underpinning issue of how KM practices can be harnessed for IM processes as a corporate strategy. In the past, organizations that rely on new innovations for survival often ask, "How does innovation improve organizational performance?" With KM as a source of competitive advantage, knowledge-intensive enterprises now ask, "How does managing knowledge for innovation enable us to sustain our long-term competitive advantage in the knowledge age?"

Integrating KM and IM: A Strategic Management Framework

The rising interest in integrating KM and IM is not entirely new, with support having already been prevalent in the mid-1990s. For instance, employees in large organizations often are encouraged to participate in all forms of knowledge processes in the pursuit of innovation as a means of enhancing corporate performance. For example, in Singapore, the opportunities offered by the services sector (predominantly knowledge-intensive) should be exploited, as the sector's GDP contribution has risen from US $52.4 billion in 1999 to US$58.4 billion in 2003, making up half of the nation's GDP growth within the four-year period. In the last few years, the importance of KM practices in innovation has reached a significant proportion that warrants a closer look at the strategic issues involving knowledge innovation (KI), an area of managerial concern in a time of change from an information economy to a knowledge economy. However, the two management areas—KM and IM—each with its own theoretical foundation, have yet to be drawn closer into one singular focus in order to provide useful insights into how KI could be managed better. To this end, a framework integrating common KM and IM issues that impact KI materially would be of relevance.

In order to offer an integrative view of KM issues, a strategic management framework[i] for KI is proposed, as depicted in Figure 1. Three strategic aspects—people, processes, and products—that would lead to human, structural, and intellectual capitals, respectively, are identified as critical areas of concern (Blumentritt & Johnston, 1999; Davenport & Prusak, 1998; O'Dell, 1996).

Given that organizational success lies more in knowledge assets than in physical assets, the strategic management of knowledge innovation is now an important area of interest. With

Figure 1. A strategic management framework

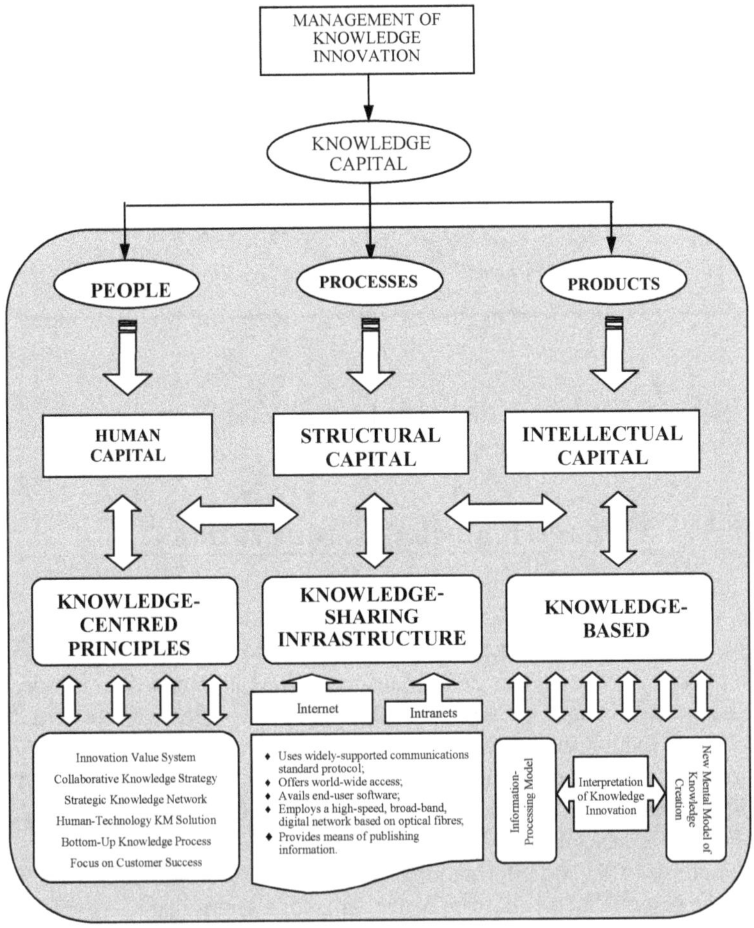

less attention accorded to it so far, three questions are addressed to broaden the strategic management perspectives, and the SIA case is analyzed to shed light on practical considerations. The three questions are discussed as follows:

What Are Knowledge Innovations?

Knowledge innovation is described as the creation, evolution, exchange, and application of new ideas into marketable goods and services, leading to the success of an enterprise, the vitality of a nation's economy, and the advancement of society (Amidon, 1997; Gold, Malhotra, & Segars, 2001). The definition implies that knowledge innovation consists of two key components: knowledge use and the actions associated with managing the flow, use, and handling of knowledge. Although both KM and IM have dealt separately with diverse management issues, the current state of literature on KI has been delineated less clearly as

Figure 2. Management of knowledge innovation

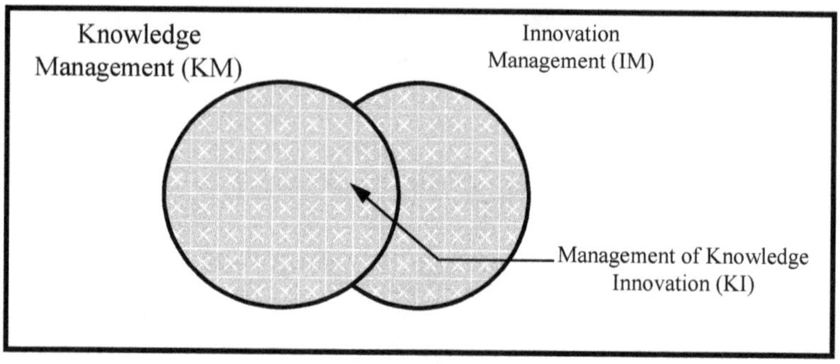

to what extent the literature between KM and IM has overlapped to provide insights into strategic management issues (see Figure 2).

How Do We Manage Knowledge Innovations?

The difficulties encountered in managing knowledge innovation are characterized by increased complexity of knowledge, rapid technological changes, and dynamic forces of competition. Limited literature has covered this area. Mainstream KM literature does not deal specifically with the management of knowledge innovation (KI) as a topic of concern. Most KM writers tend not to exhibit the same degree of understanding on the economic significance of innovation. Conversely, innovation writers are less able to articulate how KM practices can be applied as an effective strategic management tool. Understanding the management of KI requires one to discern how KM should be employed in IM and vice versa. Yet the prevailing emphasis of KM on information technology (IT) capabilities appears less able to explain how the potential of KI could be better realized. It seems that in order to harness the most value from knowledge for innovation, it is more appropriate to understand knowledge-centered principles first and then to deploy suitable KM tools[j] accordingly.

Why Knowledge-Centered Principles?

In order for KM practices to be incorporated into innovation management (IM) processes, organizations should view their roles within the context of an integrative management framework of knowledge-centered principles. Like any strategic issue, effective KM should be built on overarching principles, regardless of industry context or business domain, which constitute broad and yet important thrusts that offer explicit, systematic, and coordinated approaches to directing a theory of actions in order to realize the potential of KI. Organizations, therefore, should adopt these principles collectively as an underpinning belief system in the pursuit of knowledge innovation. In order to make headway along this line of thinking, principles that typify effective approaches of fostering knowledge Iinnovation are identified through a comprehensive review on literature domains, including the fields of KM itself, information systems, library science, IT strategy, information economics, and organization behavior. Based on a distillation of contemporary strategic management issues, six knowledge-centered principles stand out in contrast to generic KM or IM principles or other conventional approaches of management (Davenport, 1993; Davis & Botkin, 1999; Harkema & Browaeys, 2002; Miller & Morris, 1999; Nonaka & Takeuchi, 1995; Skyrme & Amidon, 1997). The six knowledge-centered principles are (1) innovation value system, (2) collaborative knowledge strategy, (3) strategic knowledge network, (4) human-technology KM solution, (5) bottom-up knowledge process, and (6) focus on customer success. While actual KM implementation in firms may be organization-specific and technology-dependent, these knowledge-centered principles empower individuals with the ability to develop their own effective KM strategies for fostering innovation.

The Case of Singapore Airlines (SIA)

Introduction

While Singapore, as indicated by a World Economic Forum (2003) study, is ranked the 10th spot in national innovative capacity, it has much room for improvement in terms of how KM practices are employed for innovation. However, KM assimilation in Singapore generally has increased over the years. According to a study undertaken by the Singapore Human Resources Institute (SHRI) on the status and role of KM practices amongst Singaporean firms, it was found that KM implementation has reached a pronounced level with more than 44% of these firms involved in one or more KM projects. Moreover, these firms supported the country's vision of a knowledge economy and have invested substantial resources in order to improve business processes through KM practices.

The case of Singapore Airlines (SIA) is selected for analysis for three reasons. First, the company's businesses operate in an emerging economy of Asia, which offers an ideal context to discuss the challenges encountered in an evolving knowledge economy. Second, SIA is hailed popularly as a leading knowledge enterprise with a relentless drive for knowledge-driven strategies in order to achieve sustainable performance and long-term growth. Third, the company is recognized as an innovation-based organization that places strong emphasis on value creation through knowledge.

About SIA

SIA's origin can be traced to the government-owned Malaysia-Singapore Airlines (MSA), the national airlines of the Malaysian federation of states to which Singapore belonged until its separation in the mid-1960s. Formed in 1972, SIA has come a long way from its humble beginnings as a small regional airline with a modest fleet of 10 aircraft and a route network spanning 22 cities in 18 countries. For more than three decades, SIA has prided itself as a carrier with a reputation for superior service (i.e., in-flight services)[k]. Internationally recognized as one of the world's youngest and leading passenger and cargo carriers,

SIA now boasts a modern fleet of more than 90 aircraft with a sophisticated route network to more than 60 cities at 89 destinations in 40 countries. Since SIA is a national carrier, the Singapore government owns 57% of SIA through Temasek Holdings, the national investment vehicle. SIA's units also include regional carrier SilkAir, a pilot school, and other repair and maintenance facilities.

SIA generally is considered one of the more successful airlines in Asia in terms of financial performance and service delivery. Prudent investment, judicious planning, and continuous product innovation have propelled the national carrier to excel in the hypercompetitive aviation business environment under tough conditions (e.g., global security threats, political instability, and public health crises). In order to achieve a growing international reputation and a high level of profitability, SIA has identified innovation as one of the most important strategic thrusts for the airline in the coming years. SIA has received many accolades,[1] as reflected by the numerous awards given to the company (e.g., Asian Most Admired Knowledge Enterprises (MAKE) Award by Teleos, Asia's Best Managed Company of the Decade by Asiamoney, Asia's Most Admired Company for five successive years by Asian Business Magazine, Best Airline for the 11th time in 12 years by Conde Naste Traveller, to name only a few). Besides these generalist awards, SIA also has won specialist awards, such as the Passenger Service award by Airline Transportation World and the Marketing Category award by Asia Inc. and Arthur D. Little.

SIA's Innovation Efforts through Knowledge-Centered Principles

SIA's KM capabilities have made the company a model of success in the airline industry, which has managed to maintain its superior customer service, constant differentiation of products and services, and consistent delivery of quality client solutions. Given the turbulence that is plaguing the world's aviation market, SIA's business model is constantly being challenged. As a result, the demands placed on the company's human resources, structures, and systems have increased tremendously. Hence, SIA continuously has to create new value for the markets it serves. If one analyzes the sources of value creation in a knowledge economy, it would not be difficult to observe that its center of gravity has migrated from resource allocation of physical assets toward the pursuit of knowledge innovation. The former offers less value, and the latter is concerned with the use of scientific, technological, organizational, and managerial assets, which are related to knowledge and are key to enhancing corporate competitiveness. SIA has no choice but to continue pursuing innovation through KM practices. The case analysis of SIA, based on a theoretical overview of six knowledge-centered principles, offers insights into how the company has assimilated these principles in its core business activities. In order to provide richer insights and more action-based perspectives, initiatives, projects, and schemes employed by SIA are highlighted.

Innovation Value System

Some organizations link the value of innovation to mechanistic actions, with a lack of emphasis on what can be derived from an innovation value system. These organizations are overly preoccupied with innovation value chain ideas about knowledge processes, which are linear and static. One example is innovative portfolio management, a linear and static process on which some organizations rely to identify innovation projects by adhering closely to an organization's value chain in order for knowledge to be reused. In such projects, the idea behind an innovation value system of knowledge creation resulting from networked resources and external relationships is not fully employed. Knowledge assets, in contrast to tangible assets, are nonlinear and dynamic with respect to the origins from which they are derived or the effects that they produce. Knowledge innovation is the outcome of leveraging on a system view of value creation from knowledge instead of a chain view. The former recognizes innovation as comprising a system of knowledge assets originating from virtually anywhere with knowledge processes intertwined in a complex web of interrelationships that add value to it through knowledge capital.

In order to harness knowledge capital for innovation, organizations must understand how an innovation value system works. Toward this end, SIA has displayed strong dedication in implementing innovation value systems in different situations in order to harvest knowledge from a system of value enablers comprising a whole spectrum of prospective stakeholders rather than depending merely on chain-type information like industry trends or static consumer needs. One case in point was SIA's innovation project with IBM's Business Innovation Services in 2000 to jointly develop an electronic commerce solution to provide more in-flight options as a means of improving its service quality. The project team approached the solution from the perspective of an innovation value system. The team members conducted extensive research and in-depth studies with business partners, air-travelers, employees, and SIA's senior management in order to explore ideas like offering mobile services via WAP and PDA and other in-flight innovations. By conceiving onboard services, cabin amenities, entertainment programs, and other prospective value enablers as an innovation value system, it took into account the diverse knowledge gathered from customer information, industry dynamics, and travelers' preferences, and was able to determine how its legendary SIA Girl could further enhance the company's brand name.

Collaborative Knowledge Strategy

To innovate effectively, organizations should adopt KM practices based on a collaborative knowledge strategy that which encourages win-win situations through symbiotic relationships by knowledge sharing and growing the knowledge pie for all. In today's complex business environment, good innovations require the melding of knowledge from diverse disciplines, which draws upon a variety of functional expertises such as engineering, packaging, and marketing, just to name a few. In contrast, competitive information strategy creates win-lose scenarios vying for the same information pie. The former strategy should be employed in the knowledge age, while the latter strategy is relevant only for the information age. Organizations that do not collaborate in knowledge sharing often fall victim to suboptimal results (e.g., longer timelines or frequent reworks in innovation projects). Without knowledge collaboration, innovations tend to be inferior in quality, impact, and foresight. In comparison, a collaborative knowledge strategy that cuts across all boundaries, whether cultural, organizational, or geographical, adds new dimensions to innovation ideas. Organizations that collaborate through forging knowledge communities with technology partners, suppliers, and specialists often produce better and more successful innovations.

SIA consistently has aligned its product development goals and innovation investments with a collaborative knowledge strategy. Being committed to building a knowledge-driven culture, SIA also is reputed to be a world-class enterprise in collaborative knowledge sharing with external parties. For example, the company works very closely with partners in various communities to develop destinations into world-class event hubs through festivals, performances, and exhibitions. In order to realize the potential of a collaborative knowledge strategy, SIA created a department called Commercial Partnerships—Associate Airlines to explore initiatives that facilitate knowledge collaboration in areas such as code-sharing services, frequent-flyer programs, and network optimization and sales policy. Professional KM facilitators skilled in building teams also are hired to catalyze knowledge sharing with external parties, including conflict management, as innovations frequently are created through collaboration among members with opposing interests. In order to transform enterprise knowledge into world-class innovative products, services, and solutions, resources also are invested heavily to foster a strong knowledge-sharing culture, such as the introduction of Web-based technologies and a companywide focus on developing team-based competencies in order to improve the company's capabilities in knowledge collaboration.

Strategic Knowledge Network

In the 1990s, there was much interest on strategic business units whose objective was to manage businesses functionally in order to improve organizational performance. With the popularity of virtual organizations, there is heightened awareness that networked forms are more relevant to knowledge innovation. In a knowledge economy, it is more critical to develop strategic knowledge networks than strategic business units. The former exploits information and communication technologies (ICT) efficiently, reduces distances between knowledge communities, and enables organizations to achieve full information power and economies of scale for the pursuit of innovation. In comparison, the latter is far more limited, since it conceives isolated islands of information assets for functional business purpose only. A networked resource view of the world's knowledge enables organizations to adapt to the chaos, uncertainty, and complexity of innovation, and if these networks are managed well, they constitute influential agents for superior innovation performance. In this view, knowledge creation evolves mostly from strategic knowledge networks, wherein innovation ideas occur between and within organizations at knowledge exchanges due to cross-fertilization and accelerated transfer among knowledge communities.

Over the years, SIA's technology strategy has involved huge investments in many state-of-the-art technologies in order to strengthen its knowledge networks. Suppose, for instance, that the company invests heavily in building strategic knowledge networks for predicting the demand and supply of airline seats. Because the company's airline seats are sold to a variety of markets with different traffic mix and seasonal travel capacity, a networked knowledge resource is vital for a best match between supply and demand to optimize system load factor and, thus, to keep seat wastage to a minimum. With a commitment to continuous improvement, SIA exploited knowledge networks like the PROS Revenue Management system, which, in 2000, supplied the company with S$20 million of advanced systems capable of forecasting and optimizing the allocation of airline seats more effectively. Dubbed Krismax II, as part of the company's strategic knowledge networks, it was developed to cater to SIA's sales planning and marketing needs and helped to overcome the complexities of matching seat capacity with ever-changing customer demand. By employing PROS dynamic modeling and operations research techniques, the forecast demand for airline seats based on historical travel patterns and current booking trends are estimated more accurately, resulting in lower wastage, higher revenue, and more cost-effective deployment.

Human Technology KM Solution

Some organizations believe that the best practices of knowledge management (KM) involve huge investments in technological tools and machine-based solutions. On the contrary, managing knowledge is fundamentally about managing people. One vital lesson learned from successful knowledge enterprises is that people-centered priorities are most crucial. In order to implement effective KM solutions for innovation pursuits, one must recognize that machines are more adept at information tasks, such as collecting, categorizing, storing, processing, updating, and computing large amounts of data, and are less adept at knowledge tasks involving subjective interpretation or human judgment. People, not machines, are the real intelligent agents in KM solutions, regardless of how powerful they may be. The former are able to identify, assess, and act upon opportunities offered by new knowledge in order to bring organizations up the performance ladder, while the latter are only passive knowledge receptacles at best. Organizations that rely heavily on machine-based KM solutions but fail to complement technologies with human inputs would find themselves losing the very essence of what knowledge capital could offer. Like other forms of capital, knowledge capital depends on people to harness it.

SIA consistently has emphasized the importance of its people in its KM solutions. With a staff strength of 28,000, SIA understands that only people equipped with the right technological tools can intelligently assemble, interpret, and utilize knowledge for the purpose of advancing organizational innovations. Machine-based KM solutions cannot replace humans in knowledge codification for a specific industry (e.g., airline) or a particular business domain (onboard entertainment), and soft skills relating to cultural, political, social, and psychological dimensions of knowledge are all too crucial to be sidelined or, even worse, ignored. Since SIA's founding, one of its key competitive advantages has been its total commitment to talent. Compared to industry norms, SIA goes to extraordinary lengths to attract, develop, and retain the best people. To cite one aspect of SIA's people-centered philosophy, during declining air traffic (e.g., economic downturn during the Asian financial crisis), it made full use of the period to implement worldwide training programs for all of its employees in order to train them to be better knowledge workers in areas such as cabin crew, security, and airport services. Although these training investments cost millions, SIA emerged as a stronger airline.

Bottom-Up Knowledge Processes

While many innovative organizations have well-established structural knowledge processes in place to source, organize, and access resources for innovation, the majority of knowledge processes stem largely from informal sources. The act of creating knowledge innovation is a haphazard process and does not follow a predefined path whereby specific knowledge inputs certainly would lead to predetermined innovation outcomes. Rather, it is often difficult to direct the knowledge processes involved in any innovation. This leads us to discern two contrasting approaches to knowledge processes: top-down and bottom-up. The former is dictated by management and is restrictive in nature, involving highly structured processes of information flows, while the latter is employee-led and involves organic actions belonging to the domain of creative knowledge workers, who are usually resistant to the proverbial command-and-control model of hierarchy. The best knowledge innovations are often the results of creative chaos and reflective instincts and should best be carried out autonomously—with minimum top-down intervention and maximum bottom-up spontaneity.

In order to foster bottom-up knowledge processes, SIA consistently has exploited the advances of information technology (IT) and the availability of applicable tools and techniques. As new innovations are derived mainly from bottom-up knowledge processes, organizations should provide employees with ample resources and incentives to widen the applicability of their knowledge for various innovation projects. Very often, the best innovation ideas originate from intuitive jumps that are bottom-up in nature and not premeditated instructions that are top-down in nature. Successful knowledge innovations are not characterized by instruction-centered production tasks, and corporate leaders should reduce top-down structures that hinder productive knowledge processes. In order for any knowledge enterprise to be truly effective, the parochial hierarchy-based management style is not only inefficient but also detrimental to encouraging a knowledge-generating culture. In SIA's management hierarchy, the managers always have valued employees' feedback at every level through regular dialogue sessions and informal communications. One example was SIA's Staff Ideas Action scheme, which ensures that feedback from frontline workers always is taken into consideration when improving the delivery of services and products. Internal staff communication and information dissemination with employees also is encouraged through a variety of regular departmental newsletters and a companywide magazine.

Focus on Customer Success

In the new knowledge age, clients are more informed and more discerning, making their knowledge needs more sophisticated than before. To ensure that client needs are met, knowledge enterprises should be customer-centric (Davis & Botkin, 1999; Skyrme & Amidon, 1997). By that, we mean that each client should be treated as a customer, not just a consumer. The former is accorded a long-term partnership with the company whose success falls under the purview of organizational interests, while the latter is merely a buyer or purchaser of services or products. Because knowledge is vital to customer success, companies must be responsive in addressing their demands. New ventures now are revolving around customer-centric business models, whereby buzzwords like *consumer satisfaction* or *delighting customers* are outdated and can no longer sustain a comparative advantage in today's competitive environment; instead, highly successful organizations tend to focus on ensuring customer success. Organizations, therefore, should combine skills in knowledge codification and knowledge utilization with the ability to form productive relationships with customers to ensure the latter's success. Organizations now are convinced with the economic benefits of customer relationship management (CRM) software solutions as a KM tool, and more organizations are hurrying to join the race in order to acquire and develop in-house CRM capabilities.

In the competitive air travel industry, customer success means that the provision of products and services before ticket purchase and after passenger arrival must be taken seriously. SIA's managers understand that the best way to improve its corporate competitiveness is to maintain constant feedback and to establish open communication channels with customers at all times. The company makes a concerted effort to stay in touch with customers by listening actively to latent needs or unmet wants, ensure rapid responses to every customer's complaint, and act on the inputs of customer focus groups. In order to address customers' demands, SIA developed a quarterly Service Performance Index that provided a consolidation of statistics relating to customer services. The Index is closely monitored globally and benchmarked against the service standards of leading airlines. Environmental scans on consumer trends of competitors and similar services in related industries like hotels, car rental companies, and restaurants also are undertaken. For instance, SIA introduced short message service (SMS) remote check-in, and future plans include satellite news service and a cyber-cabin to allow passengers to surf the Internet and engage in electronic shopping and cyber entertainment while flying. In addition, in order to meet customers' pent-up demand for lower domestic airfares, as reflected by budget-conscious travelers' feedback, SIA launched a budget carrier, Tiger Airways, based on a no-frills business model. In summary, SIA's ability to ensure customer success has managed to secure its position as a global leader in air travel.

Conclusion

This chapter has examined and emphasized the significance of integrating concepts of KM and IM in order to harvest knowledge effectively for innovation. To provide further insights into how successful knowledge enterprises innovate, a set of evolving principles for leveraging knowledge innovation is highlighted and discussed. The proposition of six knowledge-centered principles and the case analysis of Singapore Airlines (SIA) is an attempt in search of these principles. Nevertheless, new principles, as they emerge in different business domains or industry contexts, may be included to further strengthen the proposition.

In order to better exploit the potential of knowledge innovation, three challenges have been identified. First, as KI involves different dimensions of knowledge assets, such as social, economic, and other forms of tacit knowledge, it requires the assimilation of human imagination, intuition, and creativity at all levels. The challenge is to permeate knowledge-based initiatives to all layers of society—industrial, organizational, and humanistic structures—to enable individuals to fully leverage intellectual capital to participate in fostering KI. Second, although the objective of knowledge innovation is to improve organizational performance, it should not be viewed as the magic cure for ailing organizations. Paradoxical as it may sound, the fewer KM practices an organization requires in its pursuit of innovation is a reflection that it has effectively embraced knowledge-centered principles in its activities. The challenge is to create knowledge assets continuously and to make them readily available for use by knowledge workers. Third, knowledge innovation should be fostered within an enabling environment. The challenge thus lies in strengthening the role of all stakeholders within a knowledge enterprise, promoting a knowledge-friendly environment, cultivating a knowledge-oriented culture, and nurturing a knowledge-sharing ecosystem. After all, the successful creation of a knowledge innovation depends ultimately on individuals who utilized the knowledge and not on the knowledge itself—the very trait that makes knowledge useful, beneficial, and valuable to society and mankind. In conclusion, like any form of capital, knowledge, too, often is managed under imperfect conditions. The overriding concern is to identify and deal with these conditions one at a time, and hence, the ultimate goal of effective management of knowledge innovation is perhaps one for which organizations should strive but one that is never to be completely accomplished.

References

Amidon, D. (1997). *Innovation strategy for the knowledge economy: The ken awakening.* Boston: Butterworth-Heinemann.

Barth, S. (2000). Defining knowledge management. *CRM Magazine.*

Blumentritt, R., & Johnston, R. (1999). Towards a strategy for knowledge management. *Technology. Analysis and Strategic Management, 11*(3), 287-300.

Davenport, T. (1993). *Process innovation: re-engineering work through information technology.* Boston: Harvard Business School Press.

Davenport, T. (1996). *What is a knowledge management project? Research note CBI311.* London: Ernst & Young LLP Centre for Business Innovation.

Davenport, T., & Prusak, L. (1998). *Working knowledge: How organizations manage what they know.* Boston: Harvard Business School Press.

Davis, S., & Botkin, J. (1999, September-October). The coming of knowledge-based business. *Harvard Business Review*, 165-170.

Drucker, P. (1988, January-February). The coming of the new organisation. *Harvard Business Review*, 45-53.

Fahey, L., & Prusak, L. (1998). The eleven deadliest sins of knowledge management. *California Management Review, 40*(3), 231-246.

Giget, M. (1997). Technology, innovation and strategy: Recent developments. *International Journal of Technology Management, 14*(6), 613-634.

Goh, A. (2002). Industrial policy focus of South East Asian nations: Technology development or innovation? *Journal for Institutional Innovation, Development & Transition, 6*(1), 89-91.

Goh, A. (2004). Enhancing competitiveness through innovation: Issues and implications for industrial policy-making. *International Journal of Applied Management & Technology, 2*(1), 88-113.

Gold, A., Malhotra, A., & Segars, A. (2001). Knowledge management: An organizational capabilities perspective. *Journal of Management Information Systems, 18*(1), 185-214.

Grossman, G., & Helpman, E. (1992). *Innovation and growth in the global economy.* Cambridge, MA: MIT Press.

Gupta, A., & MacDaniel, J. (2002). Creating competitive advantage by effectively managing knowledge: A framework for knowledge management. *Journal of Knowledge Management Practice, 3*(2), 40-49.

Harkema, S., & Browaeys, M. (2002). Managing innovation successfully: A complex process. In *Proceedings of the European Academy of Management Conference Proceedings, EURAM 2002.*

Liao. Z., & Chew, I. (2000). The development of innovation manpower for a knowledge-based economy: The Singapore approach. *International Journal of Innovation Management, 4*(1), 123-134.

Lindgren, R., & Henfridsson, O. (2002). Using competence systems: Adoption barriers and design suggestions. *Journal of Information & Knowledge Management, 1*(1), 65-78.

Malhotra, Y. (1997). Knowledge management in inquiring organisations. *Proceedings of the 3rd America Conference on Information Systems (Philosophy of Information Systems Mini-track),* Indianapolis, Indiana.

Malhotra, Y. (2000). Knowledge management and new organisation forms: A framework for business model innovation. In Y. Malhotra (Ed.), *Knowledge management and virtual organisations*. Hershey, PA: Idea Group Publishing.

Malhotra, Y. (2001). Knowledge management for e-business performance. *Information Strategy: The Executives Journal, 16*(4), 5-16.

Miller, W., & Morris, L. (1999). *Fourth generation R&D: Managing knowledge, technology and innovation.* Canada: John Wiley & Sons.

Mullin, R. (1996). Knowledge management: A cultural revolution. *Journal of Business Strategy, 17*(5), 56-60.

Nonaka, I. (1991, November-December). The knowledge-creating company. *Harvard Business Review*, 96-104.

Nonaka, I., & Takeuchi, H. (1995). *The knowledge creating company: How Japanese companies create the dynamics of innovation.* New York: Oxford University Press.

O'Dell, C. (1996). A current review of knowledge management best practice. *Proceedings of the Knowledge Management 96 Conference,* London.

OECD. (1996). *The knowledge economy.* Paris: OECD.

Skyrme, D. (1991). Knowledge networking: The intelligent enterprise. *Aslib, 1*(9/10), 9-15.

Skyrme, D., & Amidon, D. (1997). *Creating the knowledge-based business.* London: Business Intelligence.

Storey, J., & Barnett, E. (2000). Knowledge management initiatives: Learning from failure. *Journal of Knowledge Management, 4*(2), 145-156.

Strassmann, P. (1997). *The squandered computer: Evaluating the business alignment of information technologies*. New Canaan, CT: Information Economics Press.

World Economic Forum Study. (2003). *Global competitiveness report 2002-2003*.

Endnotes

a The Organisation for Economic Cooperation and Development (OECD) defines a knowledge economy as one in which the production, distribution, and use of knowledge are the main drivers of growth, wealth creation, and employment for all industries (OECD, 1996).

b Generally, organizations determine what information or data qualify as knowledge assets and what do not, depending on its industry context and business domain (Amidon, 1997; Drucker, 1988). Two significant differences separating knowledge assets from other assets are (1) the rate of accumulation and (2) the conditions under which they are accumulated.

c The higher returns are attributed to the externalities of knowledge as the utility increases significantly with bigger membership. The economic impact arising from knowledge usually far surpasses that resulting from physical resources both in terms of scale and scope in most instances.

d Singapore hopes in 30 to 40 years' time to be a first-league developed country. Based on current economic projection, the country's per capita GNP would match the Netherlands by 2020 and the United States by 2030.

e Singapore is a member of the Paris Convention, the Patent Co-operation Treaty, and the Budapest Treaty. Since 1995, the Intellectual Property Office of Singapore has granted patents and trademarks under the Patents Act and Patents rules and has administered the Geographical Indications Act and the Layout-Designs of Integrated Circuits Act.

f Knowledge management (KM) is defined as the systematic leveraging of data, information, skills, expertise, and various forms of assets and capital to improve organizational innovation, responsiveness, productivity, and competence (Barth, 2000; Davenport, 1996).

g IT economists have argued that there is no strong relationship between IT expenditures and company performance. Despite more than US$1 trillion spent on technology investments over two decades, U.S. industries have realized little improvement in the efficiency and effectiveness of its knowledge workers (Strassmann, 1997).

h These MNCs, through their branches and subsidiaries located in Singapore, have expanded the scope of business beyond mere manufacturing to areas like customer support services, merchandising, logistics management, and regional procurement.

i The critical areas of concern are that (1) organizations should adopt a mindset of knowledge-centered principles to maximize the value of human capital for innovation, (2) organizations should implement knowledge-sharing infrastructures through appropriate information and communication technologies (ICT) to enhance structural capital required in the pursuit of innovation, and (3) organizations should promote knowledge-based initiatives to better facilitate intellectual capital in order to create new knowledge innovation.

j In the 1980s, there were great expectations that computer-based information systems (e.g., decision support systems) may be exploited as KM tools in order to solve business problems. For two decades, the search for KM tools was centered on stand-alone solutions such as CASE tools or expert system shells. Presently, the Internet offers an efficient means for implementing enterprisewide knowledge-based initiatives through groupware systems (e.g., Lotus Notes and intranets).

k Among other service initiatives that have become industry norms, SIA was the pioneer of in-flight services in the 1970s, such as providing complimentary drinks and free headsets in the economy class. More advanced services and amenities include SIA's upgraded economy class with stereo soundtrack headphones, leading-edge gaming and entertainment systems, and a personal cellular phone.

l SIA has been a regular recipient of awards along the following themes, among others: preferred airline, best frequent flyer program, best cabin outfit, best catering, and best in-flight entertainment.

Chapter XIX

Knowledge Management Gap: Determined Initiatives, Unsuccessful Results

Ivy Chan, The Chinese University of Hong Kong, Hong Kong

Partrick Y. K. Chau, The University of Hong Kong, Hong Kong

Abstract

Knowledge increasingly is recognized to provide a foundation for creating core competences and competitive advantages for organizations, making effective knowledge management (KM) crucial and significant. Despite evolving perspectives and rigorous endeavors to embrace KM intentions in business agendas, it is found that organizations always cannot realize expected benefits and improve their performances. This study reports a case study of an organization in Hong Kong that shares the typical characteristics of other organizations with strong awareness and expectation of KM yet experienced failure of its program in two years. Our findings showed that KM activities carried out in the organization were fragmented and not supported by members. Based on this failure case, four lessons learned are identified for improving KM performance.

Introduction

Knowledge increasingly has been recognized as an important asset for improving organizational performance. The capability to manage knowledge is deemed crucial to advocating effective knowledge management (KM) programs/systems in large, small, and medium-sized organizations (Alavi & Leidner, 2001; KPMG, 2002; McAdam & Reid, 2001). While many KM success stories have been reported, there are also failure stories. As reported in many management research studies, the challenges of KM implementation are dependent not only on a company's technological abilities but also on its managerial and organizational capabilities (Akbar, 2003; King, Marks, & McCoy, 2002). In this chapter, we report on a case study of how an organization in Hong Kong initiated a promising KM project but failed in two years. We know the case, because we were asked by the company's top management to uncover the reasons why the KM initiative turned out to be a failure. Findings are discussed that reveal a gap between the KM initiatives and unmet practices. Derived from the failure results, we present what we can learn from it and finally conclude with implications for future KM theory and management actions.

Related Studies about Knowledge Management

KM researchers have suggested various key elements that contribute to KM success. The mainstream thoughts can be classified as follows:

Knowledge Classification

According to resource-based theory, knowledge is regarded as an object that can be identified and traded like other organizational resources and captured and documented in information systems (Fischer & Ostwald, 2001; Shin & Holden, 2000). Therefore, it is presumed that the more knowledge objects that organizations possess, the more likely they are to improve performance and productivity. According to the cognitive perspective, knowledge is viewed as a fluid mixture of experience, ideas, and capabilities that reside in the minds of individuals (Kim, 1993; Nonaka, 1994; Tuomi, 2000). Therefore, it is asserted that procedural design in enhancing individual learning and understanding to leverage knowledge to direct decision and action will improve performance. The social view asserts that knowledge is a social asset and is embedded in social context as a dynamic state of knowing leveraged from individuals to groups through collective interaction and learning by doing (Nonaka & Konno, 1998; Swan & Newell, 2000). Therefore, the effectiveness of KM primarily is encouraged by knowledge sharing among and between groups and individuals who are committed for common interests or trust.

Knowledge Management Frameworks

KM frameworks are categorized into two main groups: descriptive and prescriptive (Holsapple & Joshi, 1999). The descriptive framework characterizes the nature of KM phenomena: the fundamental capabilities that organizations manipulate in their KM activities. For example, APQC (2000) conceptualizes organizational members as engaging seven main KM processes, including creating, identifying, collecting, adapting, organizing, sharing, and using knowledge. It is stated that each process is designed and managed to support one another in order to ensure that the right knowledge gets to the right people at the right time to improve organizational performance. Other studies depict the core work of KM as relying upon the development of organizational memory (Appleyard & Kalsow, 1999) or fostering networked communities (Bowonder, 2000) in order to enable individuals to share and acquire knowledge in various aspects.

The prescriptive framework characterizes how organizations should structure effective KM implementation guidelines. For example, Allee (1997) suggests that traditional ways in managing physical resources (e.g., as raw materials) do not fit in the context of KM. She advances 12 principles for capitalizing the value of knowledge in regard to its fluid and diverse nature: knowledge is embedded with individuals and social networks; knowledge is not accountable to a single party, which means that it should be a responsibility of each employee. Lee and Kim (2001) propose four KM stages in which organizations nurture and grow their capabilities. The four stages include initiation, propagating, integration, and networking. The first stage—initiation—is regarded as the preparation for enterprisewide knowledge management efforts. The second stage is focused on the intraorganizational activation of knowledge activities (e.g., reward systems, KMS development). The third stage emphasizes integration of KM efforts to organizational outcomes, and the final stage expands knowledge activities with connection to external parties. They suggest variations and coordination in management plans and organizational members, and procedures are necessary for KM effectiveness.

Knowledge Management Enablers

Enabling factors facilitate KM activities, such as codifying and sharing knowledge assets among individuals. One enabler is organizational culture, which is critical to facilitating knowledge sharing norms and learning motivation among individuals (Amabile, 1997; Standing & Benson, 2000; Wong & Aspinwall, 2005). For example, Roberts (2000) explains that KM effectiveness is an integration of people relationship and technology. He furthers states that employees' enthusiasm and trust in others has direct influence on the ability of information and communication technology (ICT) to transfer knowledge across various departments. In addition, leadership and management initiatives are considered central in order to direct and evaluate knowledge management practices effectively (April, 2002; Brown & Woodland, 1999; Earl & Fenny, 1994). In a study of chief information officers and senior IS managers, Law and Lee-Partridge (2001) identified that the CEO could be an effective champion and key figure in breaking through these longstanding practices in daily work: encouraging employees to pay more attention to identifying knowledge, sharing best practices, and creating new thoughts for innovative products or services. Other studies emphasize the role of technology and information systems as essential to enabling knowledge acquisition and dissemination (Armbrecht et al., 2001; O'Leary & Selfridge, 2000; Sher & Lee, 2004). Marwick (2001) proposes that a number of IT tools be applied in regard to the various knowledge creation processes. For instance, he states that e-meetings can be an effective means for people to be able to chat and discuss in order to identify tacit knowledge, while document categorization is useful for employees to retrieve and access explicit knowledge.

Knowledge Management Strategies

KM strategies encapsulate the strategic directions in managing knowledge and its related processes. In general, there are two main orientations of KM strategy. First, technology-driven KM strategy is characterized by application of information systems such as knowledge directories and chat forums to maximize codification, connectivity, dissemination, and reusability of knowledge resources (Hansen, Nohria, & Tierney, 1999; May & Taylor, 2003; Swan & Scarbrough, 2001). In an empirical investigation of KM, Choi and Lee (2001) found that some organizations that emphasize the capability to store and use explicit and documented knowledge are more likely to put much attention on technology infrastructure and deployment. Management focuses on a specified set of rules and procedures to determine what knowledge and how knowledge should be manipulated. Different system emphasis, the human-driven KM strategy, is characterized with provision of channels (not necessarily technology enabled) such that people-to-people interactions, direct conversations, and social contact networks are fostered (Choi & Lee 2001; Connell, Klein, & Meyer, 2004; Oshri, Pan, & Newell, 2005). It is presumed that knowledge originates from social networks, storytelling, or experience sharing through dialogue. Other empirical studies are identified that support similar views of the significant role of humans as knowledge agents (e.g., knowledge providers, seekers, reusers) (Markus, 2001), and the trust and care among individuals to create knowledge communities in order to enable individuals to share, exchange, and explore knowledge through personal and unstructured ways (Bhatt, 2000; Von Krogh, Ichijo, & Nonaka, 2000).

These studies reveal that the field of KM proliferates with diverse approaches in research and practice. It is deemed that each study provides an explanation of a slice of the KM phenomena but not in a comprehensive manner (Alvesson, Kärreman, & Swan, 2002; Argote, McEvily, & Reagans, 2003). A concern is that KM practices within an organization may reflect several or a blend of those elements addressed in past studies. Thus, our case study aims to present and reveal such complexity. The case study illustrates a KM experience that starts with a sound initiative but is not sustained throughout its implementation.

Research Methods

The main focus of this research is to explore the underlying reasons why a sound KM initiative did not lead to its expected results. In regard to the complexity of KM issues, this study used case study methods to collect evidence from organizational records and in-depth interviews with employees at various organizational levels (Pettigrew, 1990; Weber, 1990; Yin, 1994). The analysis of organizational records, including employee logbooks, departmental minutes, productivity charts, and frontline supervisors' reports was conducted to diagnose the causes of ineffective organizational performance from which to imply the possible directions of KM programs. Discussion and clarification were made with the management in iterative rounds in order to develop a common discourse on KM issues. The discussion results in respect to knowledge categorization, KM enablers, and strategies are presented in Table 1.

Interviews were used to investigate the underlying reasons for ineffective KM practices in 2003. Taking into account the complexity of the issues, we sought insights from the key informants in various departments. From September to November 2003, there were 12 in-depth interviews conducted. Based on the guides in conducting case study method and qualitative research methods (Boyatzis, 1998; Denzin & Lincoln, 1994; Yin, 1994), the data were transcribed and scrutinized to identify eight flaws in the KM program.

Organization Background

Founded in 1983, BAGS.COM (the actual name of the company is disguised for confidentiality reasons) is a Hong Kong-based enterprise with a production plant in mainland China and engages primarily in the manufacture and export of handbags and leather premium products for the U.S. and European markets. Like many companies in Hong Kong, BAGS.COM centralizes all its strategic planning and decisions as well as sales and marketing functions at its head office in Hong Kong while doing production and assembly work across the border. The head office has 10 staff, including a CEO, a general manager, a sales manager, an operations manager, and six other administrative staff. The production plant in China has 450 staff, including 40 managerial, supervisory, or administrative staff, and 410 skilled workers. Over the years, BAGS.COM has expanded its range of products and production capacities and resources in order to seize market opportunities and has enjoyed healthy growth in terms of sales turnover and profits.

Business, however, began to decline with a double-digit revenue loss in 1998, primarily attributed to the fierce competition in the markets and soaring production costs. Because of this, the CEO and his senior management team began to plan the future of the company and to look into ways to improve the efficiency and productivity of its employees. To find out what had gone wrong, in 2001, the CEO formed a strategic task force, which consisted of all managers in Hong Kong, several key managers responsible for the production plant in China, and himself to look into the matter. After two weeks of exploration (including observation and communicating with other staff in the company), the strategic task force concluded that the ineffective performance could be attributed to the practice in managing the knowledge assets within the organization, with low knowledge diffusion and high knowledge loss as two key issues. Therefore, it was decided to do a more detailed and in-depth investigation.

The strategic task force was responsible for carrying out the analysis. After three months of investigation and observation, they asserted that knowledge should be the strategic assets utilized and developed in their business agenda, despite their lack of experience in managing knowledge. In order to seek more opinions and perspectives, the strategic task force determined that open communication and discussion were necessary and effective in order to examine further the KM problems and, therefore, called for a meeting with managers and supervisors.

The results of the meeting were encouraging, as many participants expressed their opinions and comments eagerly. In particular, staff in the meeting agreed that KM was neither an extension of information management nor solely a technology application to capture, organize, and retrieve information in order to evoke databases and data mining (Earl & Scott, 1999; Thomas, Kellogg, & Erickson, 2001). Instead, knowledge was embedded in people (e.g., skills and actions), tasks (e.g., production process), and the associated social context (e.g., organizational culture) that involved communication and learning among loosely structured networks and communities of people. Therefore, individuals/employees were crucial to drive KM initiatives by utilizing their knowledge and skills to learn, share, combine, and internalize with other sources of knowledge in order to generate new thoughts or new perspectives.

Knowledge Management in 2001

In spite of the determination to leverage knowledge assets, the analysis of organizational documents showed that there was little systematic mechanism to collate and assimilate various feedbacks and findings from the employees. For example, the organizational annual plan in 2002 had implicit emphasis and objectives to devise a KM program and institutionalize knowledge diffusion among employees and knowledge creation for quality products. The long-term goal remains broad and conventional with the aim to provide quality products at effective cost, in which the role of knowledge is not considerably stated or embraced. In essence, the KM program at BAGS.COM can be characterized with sound plan but fragmented and flamboyant process.

Fragmented Plan

Table 1 highlights the ineffective organizational performance and relates it to previously discussed KM elements. Taking into account the categorization of knowledge, it is found that there is neither a working definition of knowledge nor a clear categorization mechanism available to identify knowledge. There is an extensive pool of knowledge existing in BAGS.COM, as employees in different departments are required to record their tasks, procedures, and suggestions in their logbooks. However, there is scant appropriate policy to unify the presentation (handwritten notes, electronic document) and content. Management stated that they were reluctant to review the departmental logbooks, as it was a time-consuming exercise to reconcile the patchy ideas. In regard to the critical success factors of a particular product or sales project, there is little effort made to discuss and reconcile the diverse perceptions across departments. For example, the design team recorded the correct choice of color and pattern of leather that comply with those promoted in the latest fashion design, while the sales department stated the acute analysis of customers' preferences and effective control of cost that contributes to competitive selling price as the leading edge.

In regard to KM enablers, management did not put much effort into creating enablers for its KM activities. It was found that effective human resource plan was not made to enhance knowledge sharing. For example, there was a number of supervisors that stated that getting new employees to learn the skills is a painstaking process. The production department manager claims, "[W]ith inadequate time and people, and tremendous time pressure as I spend 14 hours a day on my primary tasks, I may not be a trainer ... subordinates are too passive and not willing to think out of box, they may not be effective learners or serious about improving their knowledge." Therefore, there are usual complaints that over-reliance on supervisors for advice and expertise affect their productivity in their managerial roles and tasks. Employees from the production and accounting departments unanimously reflected no strong culture in promoting personal knowledge within organization. The employee (production department) stated that "if you discuss with your colleagues about knowledge gained from our established sewing procedures, or break-through practices on improving productivity, most of them will see you are self-conceited or extraordinary. ... I do have the heart, and mind to embrace KM, but do not dare to implement it under a strong conformity environment."

In terms of KM strategies, it appears that BAGS.COM has no clear direction and inclination toward technology or human aspects for coordinating KM processes. For example, a merchandizing supervisor revealed that most skilled workers are competent in their tasks at hand but have little computer literacy. He states that "only a few of people, including one of my subordinates can manipulate the computers for recording our past experiences such as details during negotiation with suppliers, or search the web for prospective suppliers. ... Sometimes, I do worry about what will happen if he leaves our company." This implies that the knowledge plan also should account for the turnover of employees causing knowledge loss. Another merchandizing manager recalled that ex-employees have lots of good networks with suppliers or subcontractors, and "now they use their knowledge to defeat us for their new employers."

Trial Run

Based upon their understanding and investigation of KM, BAGS.COM intended to enhance employee acceptance and lessen resistance to change. Therefore, BAGS.COM chose to pilot the KM initiative on a new product series. As mentioned before, there is scant documentations detailing the KM programs. This section uses various departmental minutes and supervisor reports, and presents the results in the following four main aspects: strategic, organizational, instrumental, and output (Uit Beijerse, 1999).

In the strategic aspect, it was considered that knowledge available and possessed at BAGS.COM would fall short of the core competence necessary for business success (e.g., chic product design). Therefore, effort was needed to close this gap by acquiring knowledge from both external and internal sources. From the organizational side, it was thought that knowledge was valuable when it was shared and exchanged. Thus, a knowledge-friendly culture needed to be promoted by encouraging employees to socialize and converse their ideas and thoughts in order that new knowledge could be stimulated to broaden their knowledge

Table 1. Diagnosis of KM problems in BAGS.COM

Issues	Implications to KM initiatives
Knowledge Categorization ❖ Supervisors did not have unified standards to extract best practices from experiences. ❖ Employees encountered difficulties in identifying success stories or effective production techniques for respective clients.	❖ Knowledge was not appropriately defined, captured and retained. ❖ Knowledge is diverse and not consistently stated.
KM Enablers ❖ Supervisors complained about the heavy workload that kept them from training their team members. ❖ Supervisors had little interest in what other supervisors were doing and practicing, as they considered their tasks the most important agenda. ❖ Employees demonstrated passivity and taken-for-granted passion while they were learning new skills (e.g., they implemented instructions without asking).	❖ Knowledge was not shared across the company but kept by a small group of people. ❖ Learning initiatives among employees was low due to the silo effect of organizational structure.
KM Strategies ❖ When skilled workers left BAGS.COM, specific production techniques were acquired swiftly by other competitors who employed them. ❖ Employees did not have a strong willingness to learn new techniques and practices. ❖ Employees took a long time to acquire techniques and had a hard time retaining them.	❖ Knowledge was lost to competitors. ❖ Knowledge creation and development was not systematically encouraged, motivated, and nurtured.

repositories. At the instrumental level, it was thought that knowledge had to be acquired, stored, and disseminated in a systematic way to enable employees to access and reuse it easily. In so doing, essential knowledge such as key experiences in production skills and innovative ideas in product design could be captured and recorded. Individual employees or teams who contributed knowledge that was useful and relevant to BAGS.COM were to be rewarded. Last, but not least, from an output perspective, it was realized that periodic reviews were crucial for evaluating the KM effectiveness and for devising subsequent corrective action, if necessary. Performance indicators, such as production efficiency, adoption rate of good practices identified, and clients' satisfaction, were required.

An implementation plan was devised based on the previous analysis, which then was agreed upon and approved by the top management of BAGS.COM. The KM program was officially launched in April 2002.

Knowledge Management in 2003: A Failure

After 15 months of implementation, BAGS.COM found that the KM initiative did not have the expected positive impact on organizational performance. Organizational performance remained stagnant, revenue continued to be shrunk, and staff turnover rate stayed high. Our involvement with BAGS.COM as an external consultant began after the CEO decided to find out why and/or what happened. Our assistance to BAGS.COM was clear—to investigate the

Table 2. KM results in 2003

KM Focus	Initiatives in 2001	Results in 2003
Strategic ❖ To determine knowledge gaps	❖ Identified core knowledge that led to business success	❖ Unrealistic aims → created fallacies "All the best in BAGS.COM" to direct KM development ❖ Volatile support → undermined the KM climate
Organizational ❖ To establish knowledge-friendly culture	❖ Shared knowledge in various social and informal gatherings	❖ Unframed socialization → created more confusion or negative perceptions ❖ Ineffective human resources policy to retain knowledge workers → swifter loss of knowledge
Instrumental ❖ To acquire and stimulate knowledge creation	❖ Acquired knowledge in departmental handbook and rewarded knowledge-sharing behaviors	❖ Unlimited definitions or views of sources of knowledge → left individual knowledge untapped ❖ Emphasized monetary rewards to stimulate contributions → created self-defeating mechanism and unfriendly team culture ❖ Perceived IT as cutting-edge solution → led to undue investment on technology
Output ❖ To evaluate and audit KM development	❖ Conducted periodic review and measured organizational performance	❖ Reviewed infrequently → created pitfalls to learning from mistakes, then moved ahead ❖ Predisposed on efficiency and profitability → overwhelmed short-term benefits to exploit existing knowledge

situation, to uncover the mistakes, and to recommend remedies. Therefore, a series of 12 semi-structured interviews with key informants at managerial, supervisory, and operational levels was conducted. Table 2 summarizes our findings with respect to four KM focuses (Uit Beijerse, 1999). In essence, it is indicated that the initiatives designated in 2001 cannot be realized in 2003.

As indicated in previous research, a good start for a project does not guarantee its continuity and success (Davenport, Long, & Beers, 1998; De Vreede, Davison, & Briggs, 2003). First, two crucial reasons were identified as to why BAGS.COM was unable to bridge the knowledge gap. Most middle managers found KM too difficult to implement, as "the top management was too ambitious or unrealistic to grasp and incorporate the 'best' knowledge in industry into the company while we were starting as a small bush and couldn't grow into a forest within a short period of time". In addition, a number of operational staff stated that "there is insufficient role modeling to exhibit the desired behavior from our supervisors … we found KM is too vague." Similar to many other KM misconceptions, top management wrongly aimed to incorporate other enterprises' best practices (e.g., product design of the fad) or success stories (e.g., cost cutting and streamlining operational processes) into its repositories without considering the relevance, suitability, and congruence to its capabilities. Therefore, this chasing-for-the-best strategy soon became problematic and departed from its KM goals. BAGS.COM did not gain business advantages, such as unique product design and value-added services to customers, and was unable to respond swiftly to the marketplace.

Second, the mere presence of KM vision is not sufficient to guarantee KM success. Most employees commented that top management involvement in the KM implementation was volatile and appeared to be a one-shot exercise (Gold, Malhotra, & Segars, 2001). For example, the KM program started well with a noticeable initiative to identify untapped knowledge from various sources, yet it fell behind the expected goals, as top management involvement was remote (e.g., leaving the KM effectiveness to departmental responsibility), and support was minimal (e.g., time resources available for knowledge sharing and creation). One supervisor recalled that "the inauguration day for incorporating KM into our business agenda was great and impressive, yet we are not given explicit guides to assess and evaluate knowledge work." Another operational staff member (from the same department) stated that "at present, I am not sure how KM benefits me; also, I do not find management reports showing how KM helped organizational performance over the past two years." Therefore, it directly hampered the employees' dedication and belief in KM as a significant organizational move.

Third, from the organizational aspect, even though various social activities such as tea gatherings were used to foster a friendly and open organizational culture, we found that most of these knowledge-sharing activities were futile because no specific and/or appropriate guidelines for such sharing had been devised (Nattermann, 2000). As a result, instead of having discussions that were related directly to tasks or at least contributed to idea generation, frequent chats (e.g., gossiping) among employees and wandering around were found. Most respondents claimed that a sharing session is a time-killing exercise with superficial issues. One supervisor stated with disappointment, "I can hardly get a piece of useful ideas from my colleagues through those sharing sessions … their best practices are locked up in the ivory towers and cannot be reached." Some employees even perceived KM negatively as interfering activities in their daily tasks and resisted participating in such a temporary fad.

Fourth, the instruments used to help to acquire and stimulate knowledge creation and sharing encountered problems during implementation. The fallacy of knowledge acquisition with reliance on external sources (e.g., existing practices addressed by competitors) undermined employees' motivations to explore their own available but untapped knowledge (Bhatt, 2001; Nonaka, 1994). The use of information technology to drive knowledge storage and sharing was, in principal, acceptable to employees. Yet, the silo organizational structure of BAGS.COM with unintegrated databases for knowledge capture caused more harm than good. Some employees asserted that they did not have the incentive to access or utilize the departmental knowledge handbook and procedural guidance (available from databases), as it was a time-consuming endeavor to dig from the pile of information. Some employees found knowledge incomprehensible as stored using nonstandardized formats, jargons, and symbols.

Fifth, although a reward system was established for knowledge creation and/or sharing, the emphasis on extrinsic rewards such as monetary bonuses turned out to have an opposite and negative effect on cultivating the knowledge-sharing culture and trust among employees. Some employees commented that the no-free-lunch concept should be applied to the organizational KM program. Therefore, they stated that knowledge should be kept as a personal possession (i.e., not to be shared) until they felt that they would get the monetary reward when shared or recognized by management. Other employees found that harmony and cohesiveness within a team or among colleagues were destabilized as everyone maximized individual benefits at the expense of teamwork and cooperation.

Sixth, there was a misleading notion that IT could be *the* cutting-edge solution to inspire KM in the organization. Despite the introduction of IT tools to facilitate knowledge capture, codification, and distribution, it was found that IT adoption and acceptance remained low due to employee preference for person-to-person conversation and knowledge transfer instead of technology-based communication. In addition, the widespread low computer literacy caused employee hesitation in using new technology. Finally, given the insufficient support from management for IT training and practices, employees, particularly those who had served BAGS.COM for a long time, had a strong resistance to new working practices for facilitating KM.

Seventh, it was noted that the KM initiatives were left unattended once implemented. It was difficult to find existing accomplishments or to overcome pitfalls of the KM initiatives, as there was no precise assessment available. For instance, the last survey that evaluated the adoption of good practices from departmental knowledge was conducted a year ago, without a follow-up program or review session. A manager recalled that the survey "is the only form I completed about the KM progress … in fact, I do not see how effective KM can be promoted if you do not receive suggestions or comments from the third party." Another example was that the currency and efficacy of the knowledge recorded in the departmental handbook appeared obsolete, as there were no update or revision procedures for the handbook.

Last, but not least, there was undue emphasis and concern with how the what-best knowledge at BAGS.COM could be leveraged for short-term benefits (e.g., to exploit existing knowledge in order to achieve production efficiency) at the expense of long-term goals (e.g., to revisit and rethink existing knowledge, taken-for-granted practice in order to explore innovation and creativity opportunities). Some employees pointed out that they are inclined to modify the existing practices rather than create new approaches in doing the same or similar task, as recognition and positive impacts can be obtained promptly. One manager mused that "we are usually forced to imitate others' work particularly those management believe as quality practices, however, the real and innovative ideas would not be reinventing the wheel."

Implications of the Study

Many organizations try to instill KM into their business agendas and expect to improve organizational performance and profit. Our investigation of this failure has increased our understanding of the challenges and complexity of KM implementations (Choi & Lee, 2003; Gartner, 1999).

Research Implications

This study investigates an unsuccessful experience in implementing KM within a particular organization. Through document analysis and in-depth interviews, the study provides an understanding of KM complexity—a blend of the KM elements and focuses (knowledge categorization, KM enablers, KM framework, and KM strategies) that have been addressed separately by multidisciplinary researchers. The results indicate that each element contributes to KM success or failure and, therefore, should be harnessed in an integrative manner. In this regard, future researchers should expand this study to more organizations, industries, and KM initiative maturity in order to identify variations in the KM elements and their interrelationships in influencing organizational performance.

The results also show that a good and sound KM plan is only the beginning of a KM program, while the vital task for management is that of coordinating people and processes for effective implementation. It indicates that the human factor, with employees' perceptions, motivation, and participation toward KM work, is crucial to driving the KM process. In view of the advancement of IT and the expansion of business across different geographical territories, it is deemed appropriate to accommodate various research disciplines to foster KM discourse. Possible lines of inquiry (based upon Information Systems field) can be directed to an investigation of knowledge workers in the adoption and deployment of KM systems (e.g., how should interdepartmental systems balance between customization and standardization to maintain accuracy and consistency of knowledge and expertise?).

Management Implications

In the case of BAGS.COM, we argue that planning permeated with unrealistic expectation would undermine its efficacy to direct future actions. Therefore, it is suggested that a feasibility assessment of organizational infrastructures (e.g., financial resources, technology level) and organizational climate (e.g., employees' readiness to KM, resistance to change) first should be conducted to define the KM principles and goals. Inspirational aims, which can be accomplished reasonably and feasibly, encourage employees to assess their personal repositories and infuse others' practices to improve existing practices and to overcome new challenges.

In addition, employees from BAGS.COM revealed that vision without management support is in vain and temporary. Therefore, beyond the visionary leadership, management should not downplay its willingness to invest time, energy, and resources to promote KM. At its core, management could show its enthusiasm in a boundless and persistent way, including vocal support, speech, inaugural memo, and wandering around different business units to invite impulsive idea generation and knowledge creation from all staff levels. Also, management should champion the KM process and lead by example those employees who engage high receptive attitudes toward KM. Furthermore, in order to stimulate KM behaviors, specifically sharing and creation, it is important to assure a balanced reward system integrating monetary (e.g., bonus) and nonmonetary (e.g., acknowledge excellent performer in creating new work and thoughts through organizational newsletter) items that fit in various kinds of motivation.

Finally, it is deemed that KM requires continual, collective, and cooperative efforts to put various resources together in deployment. It is suggested that management direct an attitudinal change—remove or alleviate employees' negative perceptions toward KM. For example, the fear and misconception that KM is a means to downsize organizations, is a heavy workload that requires lots of IT expertise, or requires behavioral change, requires a supportive working environment in which employees can have ample time to engage in KM endeavors, sharing, and creation; a fair and positive culture that everyone is valued and possible to contribute to KM effectiveness is needed. We also advise, in the case of unexpected failure or unintended results, that management should address problems positively, such as calling for a break to identify the causes and to remedy solutions. Do not quit or look for someone to blame; otherwise, mutual trust and commitment for the KM processes will end.

Conclusion

To date, KM is considered an integral part of the business agenda. The dynamics of KM as human-oriented (Brazelton & Gorry, 2003; Hansen et al., 1999) and socially constructed processes (Brown & Duguid, 2001) require an appropriate deployment of people, processes, and organizational infrastructure. This failure case reflects the challenges that could be encountered and overcome in order to accomplish effective KM implementation. The people factor is recognized as the key to driving KM from initiation to full implementation. KM is a collective and cooperative effort that requires most, if not all, employees in the organization to support it. KM strategy and planning should be organized, relevant, and feasible within the organizational context. One's best practice may not be well-fitted to others unless evaluation and modifications are made. A balanced hybrid of hard and soft infrastructures (e.g., team harmony and organizational culture) is needed for success.

This study has the following limitations. First, the current study is based upon a single organization from which results may not be generalized to all other situations. Therefore, more organizations with KM initiatives need to be researched in order to identify the extent and significance of various KM elements. Second, though analysis checklists and iterative rounds of discussions for analyzing the organizational documents are used, there is a possibility of the investigators' subjective judgments being involved during the evaluation. In this sense, future work may involve additional assistants to validate data interpretation.

References

Akbar, H. (2003). Knowledge levels and their transformation: Towards the integration of knowledge creation and individual learning. *Journal of Management Studies, 40*(8), 1997-2021.

Alavi, M., & Leidner, D. E. (2001). Review: Knowledge management and knowledge management systems: Conceptual foundations and research issues. *MIS Quarterly, 25*(1), 107-136.

Allee, V. (1997). 12 principles of knowledge management. *Training & Development, 51*(11), 71-74.

Alvesson, M., Kärreman, D., & Swan, J. (2002). Departures from knowledge and/or management in knowledge management. *Management Communication Quarterly, 16*(2), 282-291.

Amabile, T. (1997). Motivating creativity in organizations: On doing what you love and loving what you do. *California Management Review, 40*(1), 39-58.

Appleyard, M. M., & Kalsow, G. A. (1999). Knowledge diffusion in the semiconductor industry. *Journal of Knowledge Management, 3*(4), 288-295.

APQC. (2000). *What is knowledge management?* Retrieved on December 11, 2000, from http://www.apqc.org/best/km/whatiskm.htm

April, K. A. (2002). Guidelines for developing a k-strategy. *Journal of Knowledge Management, 6*(5), 445-456.

Argote, L., McEvily, B., & Reagans, R. (2003). Managing knowledge in organizations: An integrative framework and review of emerging themes. *Management Science, 49*(4), 571-582.

Armbrecht, R. F. M. Jr., Chapas, R. B., Chappelow, C. C., Garris, G. F., Friga, P. N., Hartz, C. A., et al. (2001). Knowledge management in research and development. *Research Technology Management, 44*(4), 28-48.

Bhatt, G. D. (2000). Organization knowledge in the knowledge development cycle. *Journal of Knowledge Management, 4*(1), 15-26.

Bhatt, G. D. (2001). Knowledge management in organizations: Examining the interaction between technologies, techniques, and people. *Journal of Knowledge Management, 5*(1), 68-75.

Bowonder, B. (2000). Technology management: A knowledge ecology perspective. *International Journal of Technology Management, 19*(7/8), 662-685.

Boyatzis, R. E. (1998). *Transforming qualitative information: Thematic analysis and code development*. Thousand Oaks, CA: Sage.

Brazelton, J., & Gorry, G. A. (2003). Creating a knowledge-sharing community: If you build it, will they come? *Communications of the ACM, 46*(2), 23-25.

Brown, J. S., & Duguid, P. (2001). Knowledge and organization: A social-practice perspective. *Organization Science, 12*(2), 198-213.

Brown, R. B., & Woodland, M. J. (1999). Managing knowledge wisely: A case study in organizational behavior. *Journal of Applied Management Studies, 8*(2), 175-198.

Choi, B., & Lee, H. (2001). Justification on knowledge management strategies: A new perspective on knowledge creating process. Proceedings of the Pacific Asia Conference on Information Systems, Seoul, Korea, 108-122.

Choi, B., & Lee, H. (2003). An empirical investigation of KM styles and their effect on corporate performance. *Information & Management, 40*(5), 403-417.

Connell, N. A. D., Klein, J. H., & Meyer, E. (2004). Narrative approached to the transfer of organizational knowledge. *Knowledge Management Research & Practice, 2*(3), 184-193.

Davenport, T. H., Long, D. W. D., & Beers, M. C. (1998). Successful knowledge management projects. *Sloan Management Review, 39*(2), 43-57.

Denzin, N. K., & Lincoln, Y. S. (1994). *Handbook of qualitative research.* London: Sage.

De Vreede, G. J., Davison, R. M., & Briggs, R. O. (2003). How a silver bullet may lose its shine. *Communications of the ACM, 46*(8), 96-101.

Earl, M. J., & Fenny D. F. (1994). Is your CIO adding value? *Sloan Management Review, 35*(3), 11-20.

Earl, M. J., & Scott, I. A. (1999). What is a chief knowledge officer? *Sloan Management Review, 40*(2), 29-38.

Fischer, G., & Ostwald, J. (2001). Knowledge management: Problems, promises, realities, and challenges. *IEEE Intelligent Systems, 16*(1), 60-72.

Gartner Group Inc. (1999). *Knowledge management from academic concepts to fundamental business practices.* Stamford: InfoEdge Inc.

Gold, A. H., Malhotra, A., & Segars, A. H. (2001). Knowledge management: An organizational capabilities perspective. *Journal of Management Information Systems, 18*(1), 185-214.

Hansen, M. T., Nohria, N., & Tierney, T. (1999). What's your strategy for managing knowledge? *Harvard Business Review, 77*(2), 106-116.

Holsapple, C. W., & Joshi, K. D. (1999). Description and analysis of existing knowledge management frameworks.In *Proceedings of the Thirty-Second Hawaii International Conference on System Sciences.* Maui, Hawaii.

Kim, D. H. (1993). The link between individual and organizational learning. *Sloan Management Review, 35*(1), 37-50.

King, W. R., Marks P. V. Jr., & McCoy, S. (2002). The most important issues in knowledge management. *Communications of the ACM, 45*(9), 93-97.

KPMG Consulting. (2002). *Knowledge management research report 2000,* 1-24.

Law, D.Y.F. and Lee-Partridge, J.E. (2001). Exploring knowledge management perceptions among information systems managers- Empirical sense-making through focus group research. *Australian Journal of Information Systems,* Special Edition (Knowledge Management), 42-55.

Lee, J. H., & Kim, Y. G. (2001). A stage model of organizational knowledge management: a latent content analysis. *Expert Systems with Applications, 20*(4), 299-311.

Markus, M. L. (2001). Toward a theory of knowledge reuse: Types of knowledge reuse situations and factors in reuse success. *Journal of Management Information Systems, 18*(1), 53-97.

Marwick, A. D. (2001). Knowledge management technology. *IBM Systems Journal, 40*(4), 814-830.

May, D., & Taylor, P. (2003). Knowledge management with patterns. *Communications of the ACM, 46*(7), 94-98.

McAdam, R., & Reid, R. (2001). SME and large organization perceptions of knowledge management: Comparisons and contrasts. *Journal of Knowledge Management, 5*(3), 231-241.

Nattermann, P. M. (2000). Best practice does not equal to best strategy. *The McKinsey Quarterly, 2*, 22-31.

Nonaka, I. (1994). A dynamic theory of organizational knowledge creation. *Organization Science, 5*(1), 14-37.

Nonaka, I., & Konno, N. (1998). The concept of "ba". Building a foundation of knowledge creation. *California Management Review, 40*(3), 40-54.

O'Leary, D. E., & Selfridge, P. (2000). Knowledge management for best practices. *Communications of the ACM, 44*(11), 281-292.

Oshri, L., Pan, S. L., & Newell, S. (2005). Trade-offs between knowledge exploitation and exploration activities. *Knowledge Management Research & Practice, 3*(1), 10-23.

Pettigrew, A. M. (1990). Longitudinal field research on change: Theory and practice. *Organizational Science, 1*(3), 267-291.

Roberts, J. (2000). From know-how to show-how? Questioning the role of information and communication technologies in knowledge transfer. *Technology Analysis & Strategic Management, 12*(4), 429-773.

Sher, P. J., & Lee, V. C. (2004). Information technology as a facilitator for enhancing dynamic capabilities through knowledge management. *Information & Management, 41*(8), 933-945.

Shin, M., & Holden, T. (2000). Exploring the dimensions of knowledge flow: A preliminary process model. In *Proceedings of the Americas Conference on Information Systems*, Long Beach, California, 1343-1348.

Standing, C., & Benson, S. (2000). Organizational culture and knowledge management. In *Proceedings of the Fourth Pacific Asia Conference on Information Systems*, Hong Kong, 1103-1113.

Swan, J., & Newell, S. (2000). Linking knowledge management and innovation. In *Proceedings of the Eighth European Conference, on Information Systems 2000*, Vienna, Austria.

Swan, J., & Scarbrough, H. (2001). Knowledge, purpose and process: Linking knowledge management and innovation. In *Proceedings of the Thirty-Fourth Hawaii International Conference on System Science*, Maui, Hawaii.

Thomas, J. C., Kellogg, W. A., & Erickson, T. (2001). The knowledge management puzzle: Human and social factors in knowledge management. *IBM Systems Journal, 40*(4), 863-884.

Tuomi, I. (2000). Data is more than knowledge: Implications of the reversed knowledge hierarchy for knowledge management and organizational memory. *Journal of Management Information Systems, 16*(3), 103-117.

Uit Beijerse, R. P. (1999). Questions in knowledge management: Defining and conceptualizing a phenomenon. *Journal of Knowledge Management, 3*(2), 94-109.

Von Krogh, G., Ichijo, K., & Nonaka, I. (2000). *From managing to enabling knowledge. Enabling knowledge creation: How to unlock the mystery of tacit knowledge and release the power of innovation.* New York: Oxford University Press.

Weber, R. P. (1990). Basic content analysis. California: Sage.

Wong, K. Y., & Aspinwall, E. (2005). An empirical study of the important factors for knowledge-management adoption in the SME sector. *Journal of Knowledge Management 9*(3), 64-83.

Yin, R. K. (1994). *Case study research design and methods.* Thousand Oaks, CA: Sage.

Chapter XX

The Lifecycle of a Knowledge Management System for Organizational Learning: A Case Study

Lynne Cooper, California Institute of Technology, USA

Teresa Bailey, California Institute of Technology, USA

Rebecca Nash, California Institute of Technology, USA

Tu-Anh Phan, California Institute of Technology, USA

Abstract

This chapter describes the development and operation of a knowledge system to support learning of organizational knowledge at the Jet Propulsion Laboratory (JPL), a US national research laboratory whose mission is planetary exploration and to do what no one has done before. JPL 101 is a Web-accessible database of general organizational knowledge captured in a series of quizzes. The heart of JPL 101 is the content that is encoded as questions and annotated answers with connections to related information and resources. This chapter describes the requirements generation process, implementation, and rollout of the JPL 101 system. Data collected over 19 weeks of operation were used to assess system performance with respect to design considerations, participation, effectiveness of communication mechanisms, and individual-based learning. Analysis of content three years after primary operations assessed the degree of knowledge obsolescence in the system. These results are discussed in the context of organizational learning research and implications for practice.

Background

The Jet Propulsion Laboratory (JPL) is a United States Federally Funded Research and Development Center (FFRDC) managed by the California Institute of Technology (Caltech) under contract with the National Aeronautics and Space Administration (NASA). JPL's primary mission is to explore our own and neighboring planetary systems. In pursuit of this mission, JPL has a rich program of technology development, science, and mission development (the three value-adding processes of the laboratory) as well as an extensive infrastructure to support Research and Development.

Setting the Stage

The JPL 101 system described in this chapter is a Web-accessible database of general organizational knowledge that is encoded as questions and annotated answers with connections to related information and resources and captured in a series of quizzes. JPL 101 was conceived as both a learning resource and a knowledge repository. The motivation for the system was twofold: to improve the connection between different communities at the laboratory spanning value-adding and enabling processes; and to share valuable insights on stakeholder issues and basic operations gained through previous knowledge capture activities. To perform the planetary exploration mission and to do what no one has done before, large numbers of technical and professional disciplines must be integrated to support innovation (the value-adding processes). In addition, infrastructure and support services are required to perform routine organizational functions (the enabling processes). While cross-functional project teams have become a common approach to integrating multidisciplinary knowledge in support of product development (Brown & Eisenhardt, 1995), less attention has been paid to bridging gaps between value-adding and enabling processes.

In established firms, emergent knowledge processes (EKPs) (Markus, Majchrzak, & Gasser, 2002) such as product development take place within the context of the organization's bureaucracy. The clash between those tasked with operating the bureaucracy and those who must work within it can be viewed as another flavor of "thought world." Dougherty (1992) describes thought world differences among members from the marketing, engineering, and manufacturing functions in new product development teams. Areas such as human resources, contracting, accounting, and information technology also draw from different professional disciplines, focus on different critical issues, and use different approaches to define and solve problems. While cross-functional teams serve to bridge thought worlds by creating a shared vision of a successful, marketable product, there are few resources (e.g., mission statements) that are effective at providing the same sort of actionable focus for the organization as a whole.

Thought-world related problems, such as conflict and miscommunication, can be mitigated by helping people to learn about other domains and to recognize and exploit differences (Dougherty, 1992). Knowledge management systems (KMS) have the potential to support this type of learning. Knowledge-based approaches have been used to support transfer of best practices (Markus, 2001), knowledge reuse for innovation (Majchrzak, Cooper, & Neece, 2004), identifying experts, and a variety of business processes (Davenport, Jarvenpaa, & Beers, 1996).

Therefore, JPL 101 was envisioned as an educational resource for laboratory personnel and a way to assist them in exploring the abundance of electronic and other resources available to them. The orienting question that guided development was "How do you help people to make sense of the 'big picture,' given that direct work-related exposure may be minimal (or nonexistent)?"

Case Description

This case covers the life cycle to date of a knowledge management system (KMS) developed to support organizational learning at the Jet Propulsion Laboratory. JPL 101 evolved over an 11-month period through a series of paper prototypes, requirements generation, and implementation. Primary operations lasted 12 weeks, with an additional seven weeks of monitoring. While still available to users more two years later, the system stopped being maintained after the initial 19 weeks of operation. The following case describes the concept definition, requirements generation, implementation, and rollout of the JPL 101 system. Data collected over 19 weeks of operation were used to assess system performance with respect to design considerations, participation, effectiveness of communication mechanisms, and individual-based learning. Analysis of content three years after primary operations assessed the degree of knowledge obsolescence in the system.

Concept Definition

From the beginning, JPL 101 was conceived as a quiz. The name was chosen as a tongue-in-cheek reference to beginners' classes in college to emphasize the educational nature of the resource and to convey that much of the content is basic material that employees should know. The quiz metaphor seemed like a natural approach in an organization that values education as highly as JPL does.

A beta test version consisting of a paper prototype was implemented to demonstrate the concept. Over the course of one week, the team brainstormed questions; experimented with different formats, difficulty, and wording of questions; and had a lot of fun creating wrong answers. The resulting list of 81 questions was divided into three roughly equal groups. Beta-test participants were given the three sets of questions in different orders to make sure that all the test questions would have at least a subset of the group looking at them. Timed tests then were conducted in which people worked their way through the quizzes. As expected, there were the occasional chuckles as people viewed the more humorous entries.

Beta testing of content provided insight into the types of questions that had the potential to be controversial—primarily those that asked about absolutes such as firsts, only, or bests. This led to standards for structuring a good question and guidelines for a reasonable amount of material to include in the answer.

Following the internal beta test, organizations within JPL that were perceived as potential stakeholders of the eventual system were contacted: internal communications, human resources, and the ethics office. Additionally, a shortened, improved set of questions was tested as a demonstration model on actual work groups from the team's home organizations. The response was overwhelmingly enthusiastic. People were anxious to share the information with their colleagues and to contribute questions and answers, and they considered it both valuable and fun. Everyone, including people who had been with the organization for multiple decades, indicated that they learned something either through the questions or the supporting information given in the answers. In addition to encouraging proceeding with development, people also began suggesting questions that they thought would be good to include.

JPL 101 users were defined as employees and affiliates of JPL that possessed access to our internal laboratory intranet. Given this definition, the only major constraint that needed to be addressed was timekeeping: How much time could people commit to participating before we needed to provide them with an account code? This was resolved through the ethics office and resulted in a requirement that each individual quiz take 15 minutes or less. Also, our Ethics Office confirmed that JPL personnel could participate but that the Human Resources Department would have to make a determination on whether contractors could participate. Based on this definition of users, the system needed to be capable of supporting up to 8,000 participants.

In addition to the requirement to keep participation under 15 minutes, we set specific goals for the system. First, we wanted to make the quizzes challenging but not burdensome. To be safe, we aimed for a quiz that would take five to 10 minutes. Beta-test results indicated that the upper limit for user comfort would be 15 questions in 10 minutes. Second, we needed to keep the number of questions that needed to be developed to a reasonable amount, so we settled on five questions per quiz. Finally, we wanted to have a broad mix of questions that included some aspect of all the different work areas of the laboratory in order to appeal to the broad user base.

By the end of the concept definition phase, the following was achieved:

- Confirmation that the basic concept was sound and likely to be received positively by the laboratory population.
- A cadre of stakeholders interested in seeing the system implemented.
- A clear understanding of what constituted a well-formulated question: clear, concise, simple structure, cautious use of absolutes, and humorous wording.
- A practical approach to ensure correctness of the question either by triangulating an answer (two-sources to confirm) or by verification through an unimpeachable source.

- A requirement from the Knowledge Management Project that the system encourage employees to explore the JPL intranet.
- A clear definition of the potential users of the system.
- Guidelines on how to structure the quizzes.

Implementation

JPL 101 is a Web-accessible database of general organizational knowledge. Knowledge is encoded as questions, answers, and connections to related information and resources (see Cooper, 2003a for a detailed discussion of the use of the quiz interface). The system is organized into quizzes containing five to 10 multiple choice and matching questions each. The deployment of the system took place over 12 weeks, after which it entered steady-state operation. During each of the first 12 weeks, a new quiz was added. Following the 12-week initial deployment of the content, the system provided access to the full set of past quizzes.

The implementation of JPL 101 was relatively simple with a minimal amount of user functions. Due to rapidly dwindling support from the KM project, low maintenance costs were essential, and the questions and answers needed to be robust with regard to obsolescence. In addition to question and answer fields, the JPL 101 database also included administrative fields for identifying the category, originator, quiz, and validation date for each question.

During the initial 12-week deployment, the entry page for JPL 101 featured a direct link to the current week's quiz. Access to previous quizzes, background information, and feedback mechanisms were provided through pull-down menus. Following the 12-week deployment period and continuing on, the entry page provided a direct link to the list of previous quizzes as well as to the menu-accessible items.

Design Considerations

JPL 101 was designed based on the assumptions that the general JPL population had access to a computer, was able to effectively use a Web interface, and would find the use of a quiz-based model for the knowledge acceptable. The first two are reasonable, given the proliferation of Web-based institutional applications for general exchange of information, support of business and administrative functions, and organizational communications. The third assumption was validated during preliminary beta testing of the concept.

Based on the assessment of the organization and with guidance from ethics, human resources, and internal communications offices, several constraints were incorporated into the design process. First, the overall set of quizzes was made representative of concerns across the wide range of disciplines on Lab so that no group would feel ignored in the process and to ensure that the thought-world issues were addressed. Second, in order to avoid potential problems with timekeeping rules, the quizzes were kept short. Third, people were ensured that they could participate at their convenience and that pragmatics, such as individuals being on travel, would not limit participation. Fourth, since participation would be voluntary; there had to be motivations to use the system. Fifth, the goal of the system was learning; therefore, it was critical that there were mechanisms for assessing whether people actually

Table 1. JPL 101 question categories

Area	Description	Rationale	Example
Basics (n=22)	General knowledge about how JPL operates at and below the level of published procedures	Make it easier for employees to learn about things that make it easier to get their job done (and correct misconceptions)	What is the number to call if you're having computer hardware- or software-related problems? (A: x4-HELP)
History (n=6)	Knowledge of key accomplishments and of individuals who contributed greatly to the lab	Establish a connection to the past and share accomplishments that contribute to a sense of pride. Share the excitement of space exploration, which is the reason for existence for the lab	Who was the director of GALCIT, and co-founder of JPL? (A: Theodore von Karman)
Missions (n=10)	Knowledge about missions, which are the primary product of the laboratory and the focus of our work		What is the name of the rover that explored the surface of Mars in 1997? (A: Sojourner)
Product Development (n=9)	Knowledge about how the laboratory builds and operates space missions and instruments	The three JPL core processes represent the reason the lab exists: our mission of space exploration. All work at the laboratory contributes either directly to one of these three areas or is responsible for supporting these processes.	Where could you go at JPL to evaluate your spacecraft under environmental conditions that are similar to those found in space? (A: 25-foot Space Simulator)
Science (n=5)	Knowledge about key scientific principles of importance in space exploration		What is the most active volcanic body currently known in the solar system? (A: Jupiter's moon, Io)
Technology (n=4)	Knowledge about the development of technology of importance in space exploration		What is the name of the substance nicknamed "frozen smoke"? (A: Aerogel)
Stakeholders (n=10)	Knowledge about external entities that impact or are impacted by JPL	JPL is answerable to multiple constituencies and is often highly constrained in the way it can operate. It is critical for JPL personnel to understand these factors and how they impact their work.	Who is the President of Caltech? (A: Dr. David Baltimore)

benefited from the system. Finally, it was important that people not feel that they were being graded or assessed in any way. Therefore, it was necessary to ensure that participants could take the quizzes without fear of violating their privacy. This limited the type of performance and participation data that could be collected.

Content

The heart of JPL 101 is the content. The content categories were chosen carefully to emphasize areas that were important to the laboratory, essentially representing the different thought worlds. Table 1 provides a description of the different categories, the rationale for including them, and an example of each.

Over the course of the 12 weeks, a total of 66 questions were presented. Each question went through a rigorous quality check to ensure accuracy and that it met the standards for a well-formulated question. The distribution of questions across categories is also provided in Table 1.

Two areas received special attention in developing the questions: JPL Basics and Stakeholders. The 21 questions in the Basics category covered material ranging from how to get help with computer problems to knowledge on new institutional resources and local restaurants available after hours. This is the type of knowledge that generally doesn't get high visibility but contributes to the overall work environment. The Stakeholder category consisted of 10 questions that covered the multiple constituencies to which JPL is responsible. Because JPL is a National Laboratory operated for NASA by the California Institute of Technology, there is a wide spectrum of stakeholders that influence the operations of the Laboratory. Understanding the nature of these stakeholder relationships and the various legal, contractual, and public trust concerns of the laboratory is important for efficient operation.

Data Collection

Data were collected during two timeframes: during the primary operations of the quiz (through week 19 of operations) and three years after initial deployment. The initial collection provided data on performance, participation, and users. The second collection provided data on the status of the content.

Two primary methods were used for collecting performance, participation, and user data: background collection of usage statistics and quiz answers, and user participation in the form of e-mail feedback, an online survey, and an online form to submit comments. The background data collection was performed using a commercial monitoring package associated with the Web server. It provided information such as hit rates, IP addresses, number of unique visitors, amount of time spent onsite, and time distributions of users. In addition, the quiz database recorded the answers submitted each time someone took a quiz.

The online survey was used to collect basic organizational demographics (tenure, organizational unit, job category, and whether a manager or not) and responses to two questions: Did you learn anything from the questions? and Did you learn anything from the answers? Taking the survey was voluntary, as was responding to the demographic questions. The second anonymous response method was an online feedback form. Users could submit comments, problems, feedback, and candidate questions for the system. While most users decided to remain anonymous, some made the effort to include their names and contact information. Finally, the e-mail-based feedback form was available to contact the development team directly. This was not anonymous and was the least used form of feedback.

The content status data were generated by reviewing the content to assess obsolescence. We reviewed each question to determine if the question and answer were still valid. In addition, we evaluated the extended information provided in the answer to assess the state of the information provided and the status of the associated links.

Results

JPL 101 premiered on January 13, 2003, and ran for 12 weeks, ending its initial deployment on April 6. It remains in operation, although new content currently is not being developed. Results are presented based on analysis of the data collected during the initial 12 weeks and extending through week 19 of operations relative to the following: design considerations, usage, motivation for use, learning results, and general reaction. Results also are presented for the three-year post evaluation, which addresses issues of content obsolescence.

Design Considerations

Background usage and database data were analyzed to assess how well the design considerations were met. Background usage data indicated success in meeting the participation time goals of the system. The average time spent in the system each workday ranged from 2:01 minutes to 8:21 minutes, with the mean being 3:53, which are within the limits recommended by JPL Ethics and Human Resources offices.

A second consideration was that the quizzes needed to be challenging but not too hard. Figure 1 shows the average quiz scores for the 12 quizzes based on data from the entire operational period. With the exceptions of weeks five and eight, the average quiz scores stayed between 70% and 90%, meeting the goal.

Figure 1. Average quiz score per quiz

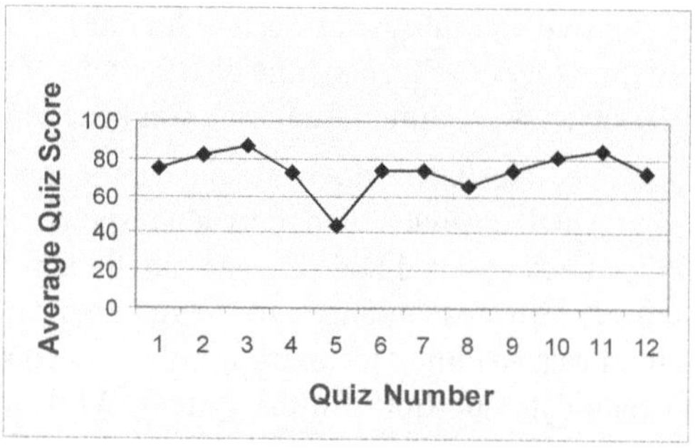

Additionally, there was a concern with question quality. Because the JPL culture is such that participants would readily point out any errors in the questions, evaluation of question quality was based on the number of corrections required. Two inputs regarding the accuracy of questions were received, one of which resulted in a minor change (attributing an additional source for information in an answer). Given the volume of material in 66 questions plus all the associated ancillary information, two minor comments were well within the range for acceptable performance.

Participation

Ultimately, a measure of success for a system is the number of people who use it (DeLone & McLean, 1992). Given that this is a voluntary-use resource and not required for anyone's job, participation statistics are critical for gauging overall success. Background usage statistics were collected, including hit rates and unique visitors based on IP addresses, modified to filter out members of the development team and automated Web crawlers. During the 19 weeks of operation covered in this study, a total of 2,144 employees participated, roughly 40% of the laboratory population. Figure 2 shows the usage statistics over time for the 19 weeks.

In addition to reaching a large audience, the goal was to reach a broad audience. Although privacy and user-burden concerns prevented automatic collection of organizational demographics on general participants, a voluntary survey instrument was used to collect some data. Five hundred and fifty surveys have been received to date (the response rate during operations was 25%). The organizational tenure for participants ranged from brand new (0 years) to a maximum of 47 years, with an average of 15.1 years and a standard deviation of 10.5 years. Users spanned the entire laboratory, with participation concentrated most heavily in the Technical and Administrative divisions, where the majority of laboratory personnel are assigned. Participants were distributed across technical, administrative, and science disciplines, and included both managers and nonmanagers. Taken in total, the data collected via the online survey indicate a a broad and substantial audience.

Figure 2. Participation by week, annotated to show key communication activities

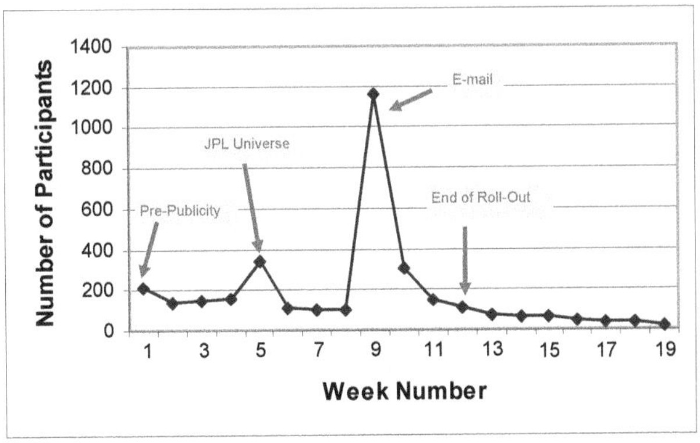

Impact of Communication Mechanisms

Because JPL 101 is a voluntary-use system, providing general rather than job-specific knowledge, several institutional communication mechanisms were employed to let people know this resource existed. These mechanisms were the following:

- **JPL Universe:** A traditional, bi-weekly organizational newspaper distributed to personnel through interoffice mail. There was a multicolumn story about JPL 101 plus a sample quiz the week before rollout.
- **Cafeteria monitors:** Closed-circuit television screens in the cafeterias that broadcast announcements. Consisted of teaser questions—shorter versions of quiz questions—plus the URL for the site for three days prior to rollout.
- **Daily Planet:** Electronic daily newspaper for JPL personnel. Accessible via intranet. Publicity was via a small graphic posted on the sidebar of the page that linked to JPL 101 and started the first day of rollout and continuing through the 12-week rollout period. In addition, a short informational article was placed in the center column news-item area during week five of rollout.
- **Inside JPL portal:** Web portal that provides central access to JPL Web space for internal users. A link to JPL 101 was included in sections for new employees and institutional knowledge management during the first week.
- **This Week:** Electronically distributed (e-mail announcement with link to Web page) weekly newsletter that highlights personnel announcements, organizational changes, upcoming talks, and events. A one-paragraph blurb about JPL 101 plus access information was included several times throughout the 12-week rollout.
- **All.Personnel e-mail:** A tightly controlled list that sends e-mail to entire laboratory population. A single all.personnel e-mail was sent during week nine.

Publicity for JPL 101 began one week prior to its rollout. Pre-release publicity included an article in the JPL Universe and announcements on the JPL monitors. In partnership with the internal communications office, the primary entry point for JPL 101 was the Daily Planet. Unfortunately, higher priority events limited entry to a single sidebar icon during the initial weeks. This icon remained until the end of the initial 12-week run. Later during the first week, access was added via the Inside JPL portal. These links continued throughout the entire period.

The impact of each of these devices can be seen in the usage statistics shown in Figure 6. The first spike in the graph occured during week five and corresponds to the publication of the Daily Planet article. Additionally, a smaller increase, not visible in the weekly statistics but present in the daily statistics, occurred when links were added to the Inside JPL portal. The most prominent feature of the graph, however, is the gigantic spike that occured during week nine. This corresponds to the sending of the all.personnel e-ail publicizing JPL 101. This spike is due almost entirely to the day the e-mail was sent.

Learning Results

The primary goal of the system was individual learning. Success was assessed in attaining this goal in two ways. The first and most direct way was to use the survey to simply ask participants if they learned anything. More than 90% of the survey respondents indicated that they had learned something from either the questions, the answers, or both. No significant correlations were found between learning, tenure, or job category, as shown in Table 2.

The second approach to evaluating learning was to look at the quiz response data. Figure 1 shows the average scores for each of the 12 quizzes. These data indicate that, on average, people missed one to two questions per quiz, indicating that a learning opportunity existed. Detailed analysis of individual questions shows that the number of respondents getting a specific question right varied from a low of 33% to one question for which everyone who answered got it right.

There was also interest in how well people performed across the different categories of questions and in what questions were skipped. Table 3 provides a summary of the performance in each of the categories. Inspection of Table 3 data indicates that JPLers performed well on questions relating to the three value-adding processes, slightly below average on Basics, History, and Missions, and significantly below average on Stakeholder questions. While JPL 101 is not intended as a diagnostic system for organizational knowledge, these results suggest a gap in knowledge about stakeholders that should be remedied. Inspection of the data on questions that were skipped clearly showed that matching-type questions were skipped more often than multiple-choice questions, with all five matching questions placing within the bottom six response rates. We believe this was due to the extra effort required to answer these questions.

Feedback via e-mail and through the online form was overwhelmingly positive. (The sole negative comment received via any of the feedback mechanisms was a complaint about the use of the all.personnel e-mail.) For example, one respondent wrote, "This is great and I love it! I learned more about JPL in the past few weeks just by taking these quizzes then the 3

Table 2. Correlation data

	N	(1)	(2)	(3)	(4)	(5)	(6)
(1) Years	550	1.0					
(2) Learned	550	-.083	1.0				
(3) Technical	285	.035	-.054				
(4) Administrative	143	.002	.068				
(5) Scientist	26	.059	-.047				
(6) Other	51	-.104*	.041				
(7) Manager	77	.373**	-.056	.251**	-.229**	.040	-.105
(8) Nonmanager	104	-.373**	.056	-.251**	.229**	-.040	.105

*Note: * $p < .05$, ** $p < .01$, Note Items (3) – (6) and (7)-(8) represent categorical data; therefore, intracorrelations are not shown*

years I have been here. Thank you." Several constructive comments were made about how to improve the system. Respondents were pleased with the quiz-type presentation, and one suggested that "JPL 101 is the paradigm that should be used for all training and knowledge dissemination at JPL."

One area of disappointment was the lack of suggestions for questions. During beta testing for JPL 101, one of the most surprising results was the level of excitement individuals had over the idea of the quiz and their desire to contribute questions and make suggestions for material. Because of this response, the feedback form in the system included a field specifically for submitting potential questions. Only three suggestions were received, resulting in two new questions.

Content Obsolescence

The content consisted of questions, answers, and ancillary information. The questions were assessed to determine if they were still valid. Of the 65 questions presented in the quizzes, 52 were judged still valid. Two additional questions were deemed valid but out-of-date. Specifically, these questions referred to events in the Cassini mission as in the future (they have since occurred successfully) and referred to the TOPEX mission as in progress (the mission ended in January 2006). A total of 11 questions were considered obsolete. Personnel changes (and impending changes) accounted for six of the 11. Reorganizations led to two questions becoming obsolete. Finally, three questions were deemed obsolete due to cancellation of a project, changes in IRS mileage allowances, and the closing of a popular local restaurant. In total, this led to 17% obsolescence after three years of nonmaintenance on the content, as summarized in Table 4.

As a hedge against eventual obsolescence, ancillary information was included for both the right and wrong answers to a question. This information included links to other Web-based resources both internal to JPL or NASA and external. We considered a link broken if (a) it resulted in a could-not-be-found error, (b) the specific content referenced by JPL 101 was no longer available on the site, or (c) if the URL now pointed to a completely different site. We considered a link still valid if (a) it pointed to the referenced information, even if it now required a password for access, or (b) the site moved, but the URL provided a pointer to the new site. Of the 158 referenced links, 121 (77%) remained valid, while 37 (23%) were considered broken. This rate of URL obsolescence is consistent with other findings, such

Table 3. Summary of performance across question categories

	Basics	History	Missions	Prod Dev	Science	Stake-holders	Technology	Total/ Avg
Number of Qs	22	6	10	9	5	10	4	66
Avg % Skipped	2.1	1.7	1.4	0.8	0.8	1.5	0.6	1.3
Avg % Right	73.2	70.9	75.6	83.5	85.2	66.0	85.1	77.1

Table 4. Summary of question obsolescence

Still Valid	Valid but out of date	Impending Obsolete		Obsolete		
		Per. Change	Other	Per. Change	Reorganization	Other
52	2	1	1	5	2	2
54		2		9		
54 = 83%		11 = 17%				

as the 29% of inactive URLs within two years found in the Kitchens and Mosley (2000) study cited by Wales (2005). External links were broken with roughly the same frequency as internal links (24% vs. 23%), while links associated with obsolete questions were broken with roughly the same frequency as those associated with valid questions. These results are summarized in Table 5.

Summary

In summary, the variety of data collected during the 19 weeks of operation for JPL 101 and at the three-year point following operations provided valuable information that hopefully can be applied to future efforts. Although unable to collect all the data, as originally planned, sufficient data were collected for a pragmatic approach that is reasonable for practitioner analysis. The following section discusses these results and the potential learning to be gained from them.

Current Challenges Facing the Organization

JPL 101 was a small effort created to share special information and promote intraorganizational appreciation for the different areas that need to work together to accomplish the JPL mission. When JPL controls spacecraft en route to other planets, small forces applied in

Table 5. Summary of link obsolescence

	Counts				Percentage			
	Internal		External		Internal		External	
	Valid	Broken	Valid	Broken	Valid	Broken	Valid	Broken
Obsolete	16	3	6	4	10.1	2	3.8	2.5
Valid	80	26	19	4	50.6	16	12	2.5
Subtotal	96	29	25	8	60.8	18	15.8	5.1
Grand Total	158				100			

the right direction at the right time are the difference between reaching the destination and missing by hundreds of kilometers. These efforts are viewed in a similar light.

As with many KM systems, the effects of the knowledge conveyed through JPL 101 cannot be measured directly (Cooper, 2003b). Conditions before and after remain virtually indistinguishable. The differences, if any, were small and below the surface; for example, less frustration when following a policy, a little more respect for others doing their jobs, and a greater sense of community. By having a positive individual impact, we expect to have a positive organizational impact, as suggested by Jennex and Olfman (2002). While we cannot measure it, the net effect of JPL 101 was that nearly half of the employees learned something new that is relevant to the organization. And that should be a good thing.

As noted by Kuchinke (1995), "organizations have in fact little control over *whether* learning takes place, but they do have potentially substantial amounts of control over the *kind* of learning that occurs within their bounds" (p. 309). In this respect, JPL 101 provides a learning opportunity in which the content, by its mere presence, indicates a degree of organizational importance, and the system serves as an intervention aimed at reducing thought world differences between personnel.

Insights

There is a number of valuable lessons for the organization to be gained from JPL 101. First, fun works. The use of humor and clever construction of questions and answers did not diminish the fundamental value of the content but, instead, contributed to user satisfaction.

Second, there were remarkable differences in the effectiveness of different institutional communications channels, as evidenced by the usage data. While one must be cautious about extrapolating from a small number of experiences, the data for JPL 101 imply that specific channels are more effective in motivating participation than others. In this case, the all.personnel e-mail (which was short and clearly indicated that participation would take a small time investment with high potential for payoff) resulted in orders of magnitude increases in participation. The e-mail message differed from the other mechanisms because it was initiated by the team and sent directly to the users rather than requiring the users to initiate contact, for example, by visiting the Inside JPL Web portal. It essentially caught their attention without requiring any effort on their part at a time when they were logged on to the intranet and reading their e-mail, and thus had easy access to the system.

Third, the differences in successful response rates for different question categories provide a level of diagnostic information regarding gaps in individual knowledge about the organization. The particularly low scores in the stakeholder category reinforced the concern about general awareness of stakeholder issues. This information could be used to modify communication and training activities to place special emphasis on areas with subpar performance.

Fourth, the feedback responses were overwhelmingly positive, particularly with respect to the quiz interface. Given the JPL culture, it was felt that this was a good approach (Cooper, 2003a), but there was surprise at the level of enthusiasm and with the degree of frustration expressed regarding other online training interfaces. This result indicates that modifications to existing training approaches may be warranted.

Finally, the future value of a KMS is dependent upon continued support. Management support (e.g., funding) for JPL 101 stopped immediately after the initial 12-week deployment. While JPL 101 continues to remain available to users, no new content has been developed nor has the existing content been updated to correct for obsolescence following the original operations period.. This was anticipated, and the questions were designed to minimize obsolescence; the system incorporated mechanisms to make content maintenance easy (e.g., on the order of minutes to update questions or answer content). Despite these efforts, over the course of three years, 17% of the question content and 23% of the associated links became obsolete. The content was affected primarily by personnel changes at the executive level and reorganizations at both JPL and NASA. The data indicated that both internal and external links broke with similar frequencies.

Individual Learning and Organizational Learning

JPL 101 is first and foremost a system for individual learning. If one adopts the perspective of Huber (1991) that an organization learns if "any of its units acquires knowledge that it recognizes as potentially useful to the organization" (p. 89), then JPL 101 also can be seen as supporting organizational learning. Using the framework of Senge, Kleiner, Roberts, Ross, and Smith (1994), JPL 101 supports organizational learning by the following:

1. **Mental models:** Contributing to the development and maintenance of mental models of how the organization operates and why it operates that way. For JPL, the natures of the work and of the institution both drive and constrain the work environment in many different ways. Mental models that accurately predict the behavior of this complex environment will contribute to improvements in peoples' ability to work more effectively. JPL 101 attempted to contribute to mental model development, for example, by providing information about how JPL's special status as an FFRDC affects operations. A number of policies that might not make sense under a for-profit business model appears much more logical with a fuller understanding of FFRDC status.
2. **Personal Mastery:** JPL 101 provides a mechanism both for validating the personal knowledge of individuals who are well-informed about how the lab operates and for guiding less experienced personnel to important material. The privacy afforded by the quiz interface allows individuals to assess their own knowledge in a nonthreatening environment. There are no penalties for getting answers wrong. By structuring JPL 101 for self-learning and by keeping the general tone light and fun, individuals were encouraged to test themselves strictly for their own knowledge.
3. **Shared Vision:** JPL 101 served to provide insights into the culture as shaped by past accomplishments and an understanding of important components of current projects. Simply asking a question in a given area sends the message that this area is important. JPL 101 was intended to help bridge boundaries between different groups at the laboratory. By highlighting critical issues associated with different disciplines, JPL 101 served to expose participants to areas outside of their normal working environments.

4. **Team Learning:** JPL 101 is an individual learning tool that also can be used in a shared mode. For example, there were instances in which versions of the quizzes were used as an activity during group meetings. One common approach was to print out a quiz and to complete it as an exercise at a group meeting. Members would share with each other their own insights to the answers presented in the quiz, debate answers, and describe their personal experiences relative to a topic covered in the quiz. Several requests were received to generate longer versions of the quizzes with special groupings of questions to support larger organizational meetings.
5. **Systems Thinking:** JPL 101 contributes to systems thinking by providing insights into the internal structure, processes, and players as well as external influences. All of the areas covered in the quiz contribute to JPL's overall mission. A better understanding of the competing constraints, differing perspectives, and the coupling between different functions leads to a better ability to make sense of the organization.

Questions for Future Research

The work reported in this chapter raises several questions regarding both knowledge management and organizational learning:

1. JPL 101 was designed to support boundary spanning between different technical and administrative disciplines and to promote sharing of cultural information. While the literature on cross-functional teams has looked at the benefits of integrating technical disciplines for new product development, the cross-organizational integration of knowledge attempted by JPL 101 represents an underexplored boundary.
2. The relationship between individual and organizational learning is the subject of debate in the literature (Argyris, 1999). How does learning about the organization, as supported by JPL 101, relate to organizational learning?
3. There are obvious connections between learning and knowledge management systems. JPL 101 is a KMS explicitly created to provide a learning opportunity. It collects knowledge and codifies it in a way to make it appealing to a broad audience. It also provides a starting point for deeper exploration of the topics presented in the quizzes. Based on the JPL 101 experience, the use of a quiz interface provides a mechanism to transform a KMS into a tool for learning. While this proved true at JPL, additional research is needed to identify general approaches to merging KMSs with learning support.
4. The JPL 101 experience clearly demonstrated that different communication media have different results with respect to increasing participation. The huge increase in participation following the all.personnel e-mail indicates that at JPL, this is a powerful tool for instigating initial attention. However, in this environment, broadcast e-mail could only be used once during the 12 weeks of primary operations due to internal communications policies. Questions remain regarding how effective e-mail would be if employed on subsequent occasions, how to increase long-term participation (i.e.,

as compared to the significant fall-off in participation in weeks after the e-mail blip), and theories for how to predict which communication mechanism would be most effective in general.

5. The categories of questions included in JPL 101 were chosen specifically for this organization and weighted toward topics (i.e., Basics and Stakeholders) in which it was felt there was both unique information to share and a special need to do so. The generalizability of these categories, however, remains an open question. From a process-based perspective, all organizations have a reason to exist that represents value to external customers and need internal processes to enable the organization to function. In addition, most organizations develop a culture that influences how work gets done. From this perspective, four general categories can be identified that are common across organizations: value-adding processes, enabling processes, external interfaces, and culture. From within these broader categories, organizations can focus on areas of particular importance to them.
6. The JPL 101 content was designed to minimize obsolescence, given that there was no support for maintenance of the system. Despite these efforts, 17% of the content and 23% of the associated links became obsolete. This is, however, only a single data point, and from such, it is hard to determine whether this represents good or bad performance. An area for future research is to understand the dynamics of content obsolescence and to develop performance standards against which to estimate the performance of individual resources.
7. Finally, the question of whether organizational learning actually occurred is unanswered. While there is clearly evidence that individuals learned—and considered what they learned to be valuable—organizational learning could not be assessed. For example, it is not known if or how knowledge conveyed by JPL 101 may have changed attitudes or work behavior. It also isn't known if there was a sufficient enough change (e.g., from many individuals) to have a measurable impact on the organization.

Conclusion

This case contributes to the ongoing discussion of knowledge management and organizational learning by providing a detailed description of the deployment and operation of an organizational knowledge-based resource specifically targeted to support general learning. A clear goal for knowledge management systems is to expand the knowledge base of the organization; in other words, learning. The work presented in this chapter describes one instance of the deployment of such a knowledge management system and provides some lessons learned that can be applied to future systems.

Acknowledgments

The work described in this chapter was carried out at the Jet Propulsion Laboratory, California Institute of Technology, under contract with the National Aeronautics and Space Administration. We would like to acknowledge the contributions of Eric Ramirez in the implementation and administration of JPL 101 and offer special thanks to Barbara Amago, Winston Gin, Cara Cheung, Sanjoy Moorthy, and Angela McGahan for their contributions. An earlier version of this chapter appeared in the *International Journal of Knowledge Management.*

References

Argyris, C. (1999). *On organizational learning* (2nd Ed.). Malden, MA: Blackwell Business.

Brown, S. L., & Eisenhardt, K. M. (1995). Product development: Past research, present findings, and future directions. *Academy of Management Review, 20*(2), 343-378.

Cooper, L. P. (2003a). The power of a question: A case study of two organizational knowledge capture systems. In *Proceedings of the 36th Annual Hawaii International Conference on System Sciences.*

Cooper, L. P. (2003b). A research agenda to reduce risk in new product development through knowledge management: A practitioner perspective. *Journal of Engineering and Technology Management, 20*, 117-140.

Davenport, T. H., Jarvenpaa, S. L., & Beers, M. C. (1996, Summer). Improving knowledge work processes. *Sloan Management Review 37*(4), 53-65.

DeLone, W. H., & McLean, E. R. (1992). Information systems success: The quest for the dependent variable. *Information Systems Research, 3*(1), 60-96.

Dougherty D. (1992). Interpretative barriers to successful product innovation in large firms. *Organization Science, 3*(2), 179-202.

Huber, G.P. (1991). Organizational learning: The contributing processes and the literatures. *Organization Science, 2*(1), 88-115.

Jennex, M. E., & Olfman, L. (2002). Organizational memory/knowledge effects on productivity, a longitudinal study. In *Proceedings of the 35th Annual Hawaii International Conference on System Sciences.*

Kitchens, J. D., & Mosley, P. A. (2000). Error 404: Or, what is the shelf-life of printed Internet guides? *Library Collections, Acquisitions & Technical Services, 24*(4), 467-478.

Kuchinke, K. P. (1995). Managing learning for performance. *Human Resource Development Quarterly, 6*, 307-316.

Majchrzak, A. Cooper, L., & Neece, O. (2004). Knowledge reuse for innovation. *Management Science, 50*(2), 174-188.

Markus, M. L. (2001). Toward a theory of knowledge reuse: Types of knowledge reuse situations and factors in reuse success. *Journal of Management Information Systems, 18*(1), 57-93.

Markus, M. L., Majchrzak, A., & Gasser, L. A. (2002). Design theory for systems that support emergent knowledge processes. *MIS Quarterly, 26*(3), 179-212.

Senge, P. Kleiner, A., Roberts, C., Ross, R., & Smith, B. (1994). *The fifth discipline fieldbook: Strategies and tools for building a learning organization.* New York: Currency Doubleday.

Wales, T. (2005). Library subject guides: A content management case study at the Open University, UK. *Program, 39*(2), 112-121.

About the Authors

Murray E. Jennex is an associate professor at San Diego State University, USA; editor-in-chief of the *International Journal of Knowledge Management*; editor-in-chief of Idea Group Publishing's "Advances in Knowledge Management" book series; and president of the Foundation for Knowledge Management (LLC). Dr. Jennex specializes in knowledge management, system analysis and design, IS security, e-commerce, and organizational effectiveness. Dr. Jennex serves as the KM systems track co-chair at the Hawaii International Conference on System Sciences. He is the author of over 80 journal articles, book chapters, and conference proceedings on knowledge management, end user computing, international information systems, organizational memory systems, ecommerce, security, and software outsourcing. He holds a BA in chemistry and physics from William Jewell College, an MBA and an MS in software engineering from National University, an MS in telecommunications management and a PhD in information systems from the Claremont Graduate University. Dr. Jennex is also a registered professional mechanical engineer in the state of California and a certified information systems security professional (CISSP).

* * *

Vittal S. Anantatmula, DSc PMP CCE, has worked in the petroleum and power industries for several years as an electrical engineer and project manager. As a consultant, he worked with the World Bank, Arthur Andersen, and other international consulting firms. Anantatmula is a certified project management professional and certified cost engineer. He is a member of PMI and AACE. His academic qualifications include BE (electrical engineering), MBA, MS in engineering management, and DSc in engineering management. Anantatmula is a faculty member of the College of Business, Western Carolina University, USA. Before joining Western Carolina University, he served as program director of the Project Management Graduate Degree Program, School of Business, The George Washington University, Washington, DC.

Ahmed Ashour is a professor of management at Alexandria University, Egypt. He has a PhD in organizational studies from the University of Minnesota. His research has been published in journals such as *Human Relations*, *Journal of Applied Psychology*, *Organizational Behaviors and Human Performance*, and *Journal of Personality and Social Psychology*. He was the director general of The Arab Administrative Development organization from 1991 to 1999.

Teresa Bailey has been a Jet Propulsion Laboratory (JPL) technical librarian for more than 22 years. In addition to performing typical library functions such as cataloging and reference work, she is the program development coordinator for the Library, Archives, and Records Section, which allows her to use her creative and leadership talents in activities that include networking, marketing, outreach, and community building. She has an MLS from the University of Southern California and is currently a doctoral student at the Fielding Institute, where she is researching the contribution of storytelling to organizational learning and knowledge sharing.

Casey Cegielski, PhD, is an associate professor of management information systems in the College of Business on the Faculty of Auburn University in Auburn, Alabama, USA. He earned a doctorate in business administration with a concentration in management information system from the University of Mississippi. Additionally, he earned a master of accountancy and a bachelor's degree from the University of Alabama. His current research interests are in the areas of innovation diffusion, emerging information technology, computer-facilitated speech recognition, and the strategic use of information technology. His research has appeared in several international information systems journals

Ivy Chan is an instructor at The Chinese University of Hong Kong. She received her PhD in business administration from the School of Business at the University of Hong Kong. Her research interests include knowledge management, information systems planning, and organizational learning.

Patrick Y. K. Chau is a professor of information systems at The University of Hong Kong. He received his PhD in business administration from the Richard Ivey School of Business at the University of Western Ontario, Canada. His research interests include IS/IT adoption and implementation, decision support systems, and information presentation and model visualization. He has published papers in journals such as *MIS Quarterly*, *Communications of the ACM*, *Journal of Management Information Systems*, *Decision Sciences*, *Information and Management*, and *Journal of Organizational Computing and Electronic Commerce*.

Lynne Cooper is a Senior Engineer at JPL, where she divides her time between developing Mars science instruments and knowledge management to support the JPL science and technology proposal process. She received her BS in electrical and computer Engineering from Lehigh University and an MS in computer engineering from USC. Her work has been published in *Management Science* and the *Journal of Engineering and Technology Management*. She is currently a PhD candidate in industrial and systems engineering at the University of Southern California, investigating how risk operates within project teams. Her awards include the NASA Exceptional Service Medal for her work in automation and the Best Paper, Academy of Management Organizational Communication and Information Systems Division (2001).

David Croasdell is an assistant professor of management information systems in the Accounting and Computer Information Systems Department at the University of Nevada, Reno, USA. Croasdell's research interests are in distributed knowledge systems, knowledge networks, knowledge management, organizational memory, and inquiring organizations. Croasdell has published more than 25 papers in outlets such as the *Information Systems Management Journal*, the *Communications of the Association of Information Systems*, the *Australian Journal of Information Systems*, and *Annals of Cases on Information Technology*. He is currently co-chair of the research cluster on Knowledge Management, Organizational Memory and Organizational Learning for the Hawaii International Conference on System Sciences. Before embarking on his academic career, Croasdell worked at Los Alamos National Laboratory, where he managed a computer-based training laboratory and supervised computer-assisted software engineering efforts across multiple local area networks.

Andrew Goh earned his PhD and MSc degrees from the University of London. He has written more than 50 professional publications, including internationally-refereed and peer-reviewed academic articles in reputable American, Asian, Australian, and European journals. He is the founding editor of the *International Journal of Applied Knowledge Management* (IJoAKM); the book editor of the *International Journal of Knowledge and Learning* (IJKL), and a consulting editor of the *Australian Journal of Information Systems* (AJIS). He also serves on the international editorial boards of the *Journal of Knowledge Management Practice* (JKMP) and the *Electronic Journal of Knowledge Management* (EJKM). Currently, he is an adjunct head at the Management Development Institute of Singapore and lectures in Human Capital Management for the Master of Science (Knowledge Management) Program at Nanyang Technological University, Singapore. His current research interests lie in knowledge management and innovation management.

Dianne Hall is an assistant professor of management information systems at Auburn University, USA. She received her doctorate at Texas A&M University. She has served as an instructor of MIS, computer science, and economics at Texas A&M University in College Station, Corpus Christi, and Kingsville and has served as a consultant. Her work has appeared in academic and practitioner journals and books. Her current research interests include applications of information technologies in support of multiple-perspective and value-based decision making.

Clyde Holsapple holds the Rosenthal endowed chair in MIS and is a professor of decision science and information systems at the University of Kentucky, USA. His research focuses on supporting knowledge work, particularly in decision-making contexts. He has authored more than 100 research articles in journals such as *Decision Support Systems, Decision Sciences, Operations Research, Journal of Management Information Systems, Group Decision and Negotiation, Journal of Operations Management, Organization Science, Communications of the ACM, Journal of American Society for Information Science and Technology, Knowledge and Process Management, International Journal of Knowledge Management, Journal of Knowledge Management,* and *IEEE*. His many books include *Foundations of Decision Support Systems, Decision Support Systems: A Knowledge-Based Approach,* and the two-volume *Handbook on Knowledge Management,* a basic reference work. He is editor-in-chief of the *Journal of Organizational Computing and Electronic Commerce,* area editor of *Decision Support Systems*, and has served as an inaugural area editor for the *INFORMS Journal on Computing* and associate editor for *Management Science.*

Kiku Jones is an assistant professor in the School of Accounting and MIS at the College of Business Administration of the University of Tulsa, USA. She has a BS in computer information systems, an MBA from Western Kentucky University, and a PhD in information systems from the University of Kentucky. Her research interests include knowledge management, human computer interaction, electronic commerce, decision support systems, and information systems strategy. Her publications have appeared in *Knowledge and Process Management, International Journal of Information Management*, and the *Journal of Computer Information Systems*.

Atreyi Kankanhalli (http://www.comp.nus.edu.sg/is/bio/atreyi.html) is assistant professor in the Department of Information Systems, School of Computing at the National University of Singapore (NUS). She received her PhD in information systems from NUS. She has been a visiting scholar at the Haas Business School, University of California Berkeley and the Indian Institute of Science, Bangalore. Prior to joining NUS, she has considerable experience in industrial R&D. She has consulted for a number of organizations, including World Bank. Kankanhalli's work has been published in journals such as the *MIS Quarterly, Journal of the American Society for Information Science and Technology, Communications of the ACM, Decision Support Systems,* and *International Journal of Information Management*. Her research has been presented at conferences including the ICIS, HICSS, and WITS. She has served or is serving on several information systems conference committees such as PACIS, ICKM, and IRMA and serves on the editorial boards of the *International Journal of Knowledge Management, Journal of Global Information Management,* and *Journal of Information Privacy and Security*. Her research interests include knowledge management, e-government, virtual teams, and information systems security. She was awarded the President's Graduate Fellowship, the Dean's Graduate Award, and the Infocomm Development Authority Gold Medal at NUS. She is also the winner of the ACM-SIGMIS ICIS 2003 Best Doctoral Dissertation award.

Omar Khalil is a professor of information systems and associate dean at the University of Massachusetts-Dartmouth. Has a PhD in information systems from the University of North Texas. His publications have appeared in journals such as the *Journal of Global Information Management*, *Journal of Organizational and End-User Computing*, *Information Resources Management Journal*, *International Journal of Production and Economics*, *International Journal of Man-Machine Studies*, *Journal of Business Ethics*, and *Journal of Informing Science*. His research interests include information systems effectiveness, global information systems, information quality, and knowledge management.

Lutz Kolbe has headed the Competence Center Customer Management (CC CM) since July 2002 and teaches at AACSB-accredited University of St. Gallen, Switzerland. His research interests are customer relationship management and security management, as well as advanced technologies in the residential environment. After having worked as financial consultant, Kolbe studied information management at Brunswick Technical University, Germany, where he received a master's degree. He worked on his dissertation at Freiberg Technical University, Germany, and the University of Rhode Island, USA. He received his PhD in 1997. After that, he worked at Deutsche Bank in Frankfurt and New York, where he became managing director in 2001.

Stefan Kremer works as a consultant with the Information Management Group (IMG AG), Switzerland. He received a PhD from the University of St.Gallen, Switzerland, in information management in 2004, and a joint master's degree in electrical engineering and economics from the University of Paderborn, Germany, in 2000. He worked as a research assistant and lecturer at the Department of Information Management of the University of St.Gallen, Switzerland, in the fields of knowledge management, process analysis and design, information architecture, portals and information retrieval. Kremer has several years of experience in numerous industrial knowledge management, portal and search engine projects and has published several articles on these topics.

Shih-Chen Liu is an associate professor in the Department of International Business at Chihlee Institute of Technology, Taiwan. Her primary areas of research include the assessment of information systems effectiveness and value, with an emphasis on the support of knowledge management and effect of learning. She has published articles in The Americas Conference on Information Systems and The Chinese Conference on Human Resources Development. Liu earned an MBA from Katz Graduate School of Business at the University of Pittsburgh and a PhD from the School of Information Science at Claremont Graduate University.

Anne Massey is the dean's research professor and Lilly faculty fellow of information systems in the Kelley School of Business at Indiana University. She received her PhD from Rensselaer Polytechnic Institute. Her research interests include knowledge management, computer-mediated communication and virtual teams, technology implementation, and related topics. Her research has been published in *MIS Quarterly*, *Academy of Management Journal*, *Journal of Management Information Systems*, *Decision Sciences*, and *IEEE*

Transactions on Engineering Management, among others. Professor Massey is a member of the Association for Information Systems, IEEE, the Academy of Management, and the Decision Sciences Institute.

William Money, PhD, PMP, is an associate professor with the Department of Information Systems and Technology Management, School of Business, at The George Washington University, USA (1992-present). His publications and recent research interests focus on business process analysis and engineering, information system development tools and methodologies, including the WWW, Web workflow and expert systems; and developing collaborative data sharing, distribution, and decision-making solutions to complex management problems. His work analyzes the use of process documentation tools and process engineering for collecting and documenting organizational knowledge and memory, and develops teaching and training techniques that prepare students to use GSS tools in complex organizations and dynamic work environments that are experiencing significant change. He also is developing an initiative to research and design information systems to effectively implement Edge organizations that perform highly integrated tasks without step-by-step guidance from hierarchical decision-making structures. Money has more than 12 years of management experience in the design, development, testing, and implementation of management information systems (1980-1992) and has been engaged as a consultant to the government and to industry on a number of software development programs.

Mitzi Montoya-Weiss is a professor of marketing and innovation in the Business Management Department at North Carolina State University, USA. She received her PhD from Michigan State University. Her research focuses on innovation processes and strategies and the role of technology as an enabler of decision making. Her publications have appeared in *MIS Quarterly, Management Science, Marketing Science, Academy of Management Journal,* and *Journal of the Academy of Marketing Science,* among others. Montoya-Weiss is a member of the American Marketing Association, Academy of Marketing Science, and Product Development and Management Association.

Rebecca Nash is a senior software engineer devoted to technical communications and institutional computing at the Jet Propulsion Laboratory. She received her BS in biological sciences from California State University at Los Angeles and her MS in interactive telecommunications from the University of Redlands. Nash designs interfaces from Web sites to applications and helps organizations improve the usability of their products.

Mark Nissen is an associate professor of information systems and management at the Naval Postgraduate School, USA. His research focuses on knowledge dynamics. He views work, technology, and organizations as an integrated design problem and has concentrated recently on the phenomenology of knowledge flows, culminating in a new book titled *Harnessing Knowledge Dynamics: Principled Organizational Knowing & Learning* (IRM Press, 2006). Nissen's publications span information systems, project management, organization studies, knowledge management, and related fields. In 2000, he received the Menneken Faculty

Award for Excellence in Scientific Research, the top research award available to faculty at the Naval Postgraduate School. In 2001, he received a prestigious Young Investigator Grant Award from the Office of Naval Research for work on knowledge-flow theory. From 2002 to 2003, he was visiting professor at Stanford, integrating knowledge-flow theory into agent-based tools for computational modeling. In 2004, he established the Center for Edge Power for multi-university, multidisciplinary research on what the military terms *command & control.* Before his information systems doctoral work at the University of Southern California, he acquired more than a dozen years' management experience in the aerospace and electronics industries.

Lorne Olfman is dean of the School of Information Systems and Technology at Claremont Graduate University, USA, Fletcher Jones chair in technology management, and co-director (with Terry Ryan) of the Social Learning Software Lab (SL2). His research interests are in designing effective collaboration, learning, and knowledge management technologies. To this end, Olfman and his SL2 colleagues are conducting research on a variety of topics, including the design of an intelligent online discussion board, the development of an integrated set of tools to facilitate The Claremont Conversation for the 21st Century, and the design of a virtual dialogue system. Olfman has been integrating the use of wiki technology into his research and teaching for the past couple of years.

Todd Peachey is a doctoral candidate in the Management of Information Technology and Innovation program at Auburn University, USA. He has published one article in the *International Journal of Knowledge Management* and presented papers at HICSS and AMCIS. His primary research interest is knowledge management.

Loo Geok Pee (http://www.comp.nus.edu.sg/~peelooge) is a second-year graduate student in the Department of Information Systems, School of Computing, National University of Singapore (NUS). She received a bachelor's degree in information systems from NUS. She has participated in workshops such as HCI Research in MIS. Her research interests include knowledge management in public and private sectors and information system security.

Tu-Anh Phan is a senior software engineer at the Jet Propulsion Laboratory, where she is responsible for the implementation of many Web enabled database systems, such as those that support JPL flight projects and proposal activities. Currently, she is leading the development of the NASA Program and Project Management Support System. She received her Bachelor of Science in mathematics from the University of California, Los Angeles and has been with JPL since 1994.

V. Ramesh is an associate professor in the Information Systems Department and Ford Motor Company teaching fellow at Indiana University's Kelley School of Business, USA. He is the director of the MS in information systems program. His research interests are in data modeling, heterogeneous databases, virtual teams and groupware, usability in mobile systems and software engineering. His research has appeared in leading journals, such as *MIS Quarterly, ACM Transactions on Information Systems, Communications of the ACM, Journal of Management Information Systems, Information Systems, IEEE Expert*, among others.

Vincent Ribière received his Doctorate of Science in knowledge management from The George Washington University, and a PhD in management sciences from the Paul Cezanne University, Aix en Provence, France. He is an assistant professor of information systems at the Management School of the New York Institute of Technology in New York, USA. He teaches and conducts research in the area of knowledge management and information systems. He is the program director at The George Washington University Institute for Knowledge and Innovation. Over the past years, he has presented various research papers at different international conferences on knowledge management, organizational culture, information systems and quality, as well as published in various refereed journals.

Terry Ryan is an associate professor in the School of Information Systems and Technology and co-director (with Lorne Olfman) of the Social Learning Software Lab (SL^2) at Claremont Graduate University, USA. His teaching and research interests are in the design, development, and evaluation of information systems to support teaching and learning, online discussions and dialogues, and preparing for and responding to emergencies. He has published articles in *Communications of the AIS, Data Base, Information & Management, International Journal of Human-Computer Studies, International Journal of Knowledge Management, Journal of Computer Information Systems, Journal of Database Management, Journal of Information Systems Education,* and other outlets.

David G. Schwartz is a senior lecturer and head of the Information Systems Division of the Graduate School of Business Administration at Bar-Ilan University, Israel. Since 1998 he has served as editor-in-chief of the internationally acclaimed *Journal of Internet Research.* Schwartz's research has appeared in publications such as *IEEE Intelligent Systems, International Journal of Human-Computer Studies, IEEE Transactions on Professional Communications, Kybernetes*, and the *Journal of Organizational Behavior*. His books include *Encyclopedia of Knowledge Management* (IGP, 2006), *Cooperating Heterogeneous Systems* (Kluwer, 1995), and *Internet-Based Organizational Memory and Knowledge Management* (IGP, 2000). Schwartz received his PhD from Case Western Reserve University, his MBA from McMaster University, and his BSc from the University of Toronto, Canada.

Ahmed Seleim is a PhD candidate and a lecturer at the Management Department at Alexandria University, Egypt. He is a visiting scholar at the Wharton School at the University of Pennsylvania. He holds an MBA and BCom degrees from Alexandria University, Egypt. His research has been published in journals such as *Journal of Global Information Management, The Learning Organization*, and *Arab Journal of Administration Sciences*. His research interest areas include management information systems, knowledge management, and intellectual capital.

Meenu Singh is an assistant professor in information systems at the Murray State University, USA. He holds an MS in computer science from Western Illinois University and a PhD in management information systems from the University of Kentucky. He has many years of teaching experience and has been selected for inclusion in *Who's Who Among America's Teachers*. He has published in *Information and Management, Journal of Organizational Computing and Electronic Commerce, Knowledge and Process Management, Expert Systems with Applications, Journal of Computer Information Systems*, and others. His current research interests include application of knowledge management for competitive advantage, electronic commerce, and supply chain management.

Stefan Smolnik is a post-doctoral research fellow, senior lecturer, and project manager at the European Business School (ebs), Oestrich-Winkel, Germany. Smolnik received a master's degree in computer science as well as a PhD in information systems from the University of Paderborn, Germany. He has been working in the research fields of knowledge management, semantic technologies, and collaborative computing for a couple of years. Smolnik is an internationally recognized specialist in the research domain of topic maps and has published several articles on the topics of knowledge and process management and information visualization.

Eric W. Stein is an associate professor of management science and information systems at the Penn State Great Valley School of Graduate and Professional Studies (Malvern, Pennsylvania, USA). His areas of research and consulting are organizational memory and learning, human expertise, knowledge management, artificial intelligence business applications, and high-tech venture development. He is director of the New Ventures and Entrepreneurial Studies option in the MBA program. Stein has published in the *Journal of Management Information Systems, The International Journal of Expert Systems, The International Journal of Information Management, Information Systems Research*, and *Expert Systems with Applications*, among others. His works include commentaries and book chapters, including a chapter in *Expertise in Context* (AAAI/MIT Press). He reviews for several journals and conferences and is a member of AoM and DSI. Stein received his doctorate in managerial science from the Wharton School of the University of Pennsylvania and his undergraduate degree in physics from Amhert College.

Bernard Cheng Yian Tan (http://www.comp.nus.edu.sg/~btan) is a professor and head of the Department of Information Systems at the National University of Singapore (NUS). He received his PhD in information systems from NUS. He has won teaching and research awards at NUS. He has been a visiting scholar in the Graduate School of Business at Stanford University and the Terry College of Business at the University of Georgia. Tan is on the editorial boards of *MIS Quarterly* (senior editor), *e-Service Journal* (senior editor), *Management Science, Journal of the AIS, IEEE Transactions on Engineering Management, Information and Management, Journal of Global Information Management*, and *International Journal of Distance Education Technologies*. His research has been published in journals such as *ACM Transactions on Computer-Human Interaction, ACM Transactions on Information Systems*,

ACM Transactions on Internet Technology, Communications of the ACM, Decision Support Systems, European Journal of Information Systems, IEEE Transactions on Engineering Management, IEEE Transactions on Professional Communication, IEEE Transactions on Systems, Man, and Cybernetics, Information and Management, Information Systems Research, International Journal of Human-Computer Studies, Journal of Management Information Systems, Journal of the AIS, Journal of the American Society for Information Science and Technology, Management Science, and *MIS Quarterly*. He is an Asia-Pacific councillor for the Association for Information Systems. His research interests are cross-cultural issues, knowledge management, virtual communities, and software project management.

Hazel Taylor is an assistant professor at the Information School, University of Washington, Seattle, USA. She holds a PhD from Queensland University of Technology, Brisbane, Australia. Prior to joining the Information School, Taylor taught at the University of Waikato in New Zealand and at the Hong Kong University of Science and Technology, and conducted research in Hong Kong on risk management and tacit knowledge in IT projects. Her teaching and research focuses on IT project management and risk management, and information systems analysis and development, with an emphasis on tacit knowledge and decision making in these areas. A secondary teaching focus is in the area of research methods. Prior to her academic career, Taylor worked in industry with manufacturing, construction, and government organizations, both as a systems manager and an IT project manager.

Francis Tuggle holds a BS degree from MIT and an MS and PhD from Carnegie Mellon University. Presently, he is a professor in the George L. Argyros School of Business and Economics at Chapman University, USA. Previously, he was the Robert J. and Carolyn A. Waltos Jr. Dean at Chapman University, and before that, he was dean of the Jones Graduate School of Management at Rice University (and Jesse H. Jones Professor of Management) as well as dean of the Kogod School of Business at American University. He also has held faculty appointments at the University of Kansas. He has held tenured professorships in departments of business, psychology, and computer science. He has written two books and more than 50 refereed journal articles. He consults broadly and sits on several corporate boards, including the audit committee of a NYSE listed firm. A longtime student of processes, he is presently most interested in knowledge management and organizational cultures.

Arch Turner is a retired naval aviator. Turner presently works as a senior consultant with the Office of Naval Research in Arlington, Virginia. He is responsible for assessing the applicability of advanced technologies to the Department of the Navy's future military requirements. Turner's work exposes him to a broad range of modern technologies. He has worked recently in the areas of unmanned vehicles; advanced capability electric powered systems and weapons, information technology, and distributed sensing. Turner has published and spoken on topics in information technology and advance high resolution sensing. Turner has a B.S. in aerospace engineering, an MSs in both operations research and information systems, and is presently a doctoral candidate at the Department of Information Systems and Technology Management, School of Business, The George Washington University, USA.

Index

J

K

Q

R

S

T

U

W

Y

www.ingramcontent.com/pod-product-compliance
Lightning Source LLC
LaVergne TN
LVHW062110200926
848484LV00007B/1629